# Advances in Intelligent and Soft Computing 138

Editor-in-Chief: J. Kacprzyk

# Advances in Intelligent and Soft Computing

**Editor-in-Chief**

Prof. Janusz Kacprzyk
Systems Research Institute
Polish Academy of Sciences
ul. Newelska 6
01-447 Warsaw
Poland
E-mail: kacprzyk@ibspan.waw.pl

Further volumes of this series can be found on our homepage: springer.com

Vol. 123. Yinglin Wang and Tianrui Li (Eds.)
*Knowledge Engineering and Management, 2011*
ISBN 978-3-642-25660-8

Vol. 124. Yinglin Wang and Tianrui Li (Eds.)
*Practical Applications of Intelligent
Systems, 2011*
ISBN 978-3-642-25657-8

Vol. 125. Tianbiao Zhang (Ed.)
*Mechanical Engineering and
Technology, 2012*
ISBN 978-3-642-27328-5

Vol. 126. Khine Soe Thaung (Ed.)
*Advanced Information Technology
in Education, 2012*
ISBN 978-3-642-25907-4

Vol. 127. Tianbiao Zhang (Ed.)
*Instrumentation, Measurement, Circuits
and Systems, 2012*
ISBN 978-3-642-27333-9

Vol. 128. David Jin and Sally Lin (Eds.)
*Advances in Multimedia, Software Engineering
and Computing Vol.1, 2011*
ISBN 978-3-642-25988-3

Vol. 129. David Jin and Sally Lin (Eds.)
*Advances in Multimedia, Software Engineering
and Computing Vol.2, 2011*
ISBN 978-3-642-25985-2

Vol. 130. Kusum Deep, Atulya Nagar,
Millie Pant, and Jagdish Chand Bansal (Eds.)
*Proceedings of the International Conference
on Soft Computing for Problem Solving
(SOCPROS 2011) December 20–22, 2011, 2012*
ISBN 978-81-322-0486-2

Vol. 131. Kusum Deep, Atulya Nagar,
Millie Pant, and Jagdish Chand Bansal (Eds.)
*Proceedings of the International Conference
on Soft Computing for Problem Solving
(SocProS 2011) December 20–22, 2011, 2012*
ISBN 978-81-322-0490-9

Vol. 132. Suresh Chandra Satapathy, P.S. Avadhani,
and Ajith Abraham (Eds.)
*Proceedings of the International Conference on
Information Systems Design and Intelligent
Applications 2012 (India 2012) held in
Visakhapatnam, India, January 2012, 2012*
ISBN 978-3-642-27442-8

Vol. 133. Sabo Sambath and Egui Zhu (Eds.)
*Frontiers in Computer Education, 2012*
ISBN 978-3-642-27551-7

Vol. 134. Egui Zhu and Sabo Sambath (Eds.)
*Information Technology and Agricultural
Engineering, 2012*
ISBN 978-3-642-27536-4

Vol. 135. Honghua Tan (Ed.)
*Knowledge Discovery and Data Mining, 2012*
ISBN 978-3-642-27707-8

Vol. 136. Honghua Tan (Ed.)
*Technology for Education and Learning, 2012*
ISBN 978-3-642-27710-8

Vol. 137. Jia Luo (Ed.)
*Affective Computing and Intelligent Interaction,
2012*
ISBN 978-3-642-27865-5

Vol. 138. Gary Lee (Ed.)
*Advances in Intelligent Systems, 2012*
ISBN 978-3-642-27868-6

Gary Lee (Ed.)

# Advances in Intelligent Systems

Selected Papers from 2012 International
Conference on Control Systems (ICCS 2012),
March 1-2, Hong Kong

 Springer

*Editor*
Gary Lee
Information Engineering Research Institute
Newark, DE
USA

ISSN 1867-5662      e-ISSN 1867-5670
ISBN 978-3-642-27868-6      e-ISBN 978-3-642-27869-3
DOI 10.1007/978-3-642-27869-3
Springer Heidelberg New York Dordrecht London

Library of Congress Control Number: 2011945318

Printed on acid-free paper

Springer is part of Springer Science+Business Media (www.springer.com)

# Preface

2012 International Conference on Environment Science and 2012 International Conference on Computer Science (ICES 2012/ICCS 2012) will be held in Australia, Melbourne, 15–16 March, 2012.

The topics of the conference are focused on environment science and computer science. Relevant materials are also included into the conference programs. From the 386 pieces of submissions, 98 pieces of abstracts are selected for presentations in the conference. We have divided the papers into two volumes:

Volume 1 Advances in Computational Environment Science contains the 2012 International Conference on Environment Science (ICES 2012) papers. Environmental science is an interdisciplinary academic field that integrates physical and biological sciences, (including but not limited to Ecology, Physics, Chemistry, Biology, Soil Science, Geology, Atmospheric Science and Geography) to the study of the environment, and the solution of environmental problems. Environmental science provides an integrated, quantitative, and interdisciplinary approach to the study of environmental systems. Related areas of study include environmental studies and environmental engineering. Environmental studies incorporate more of the social sciences for understanding human relationships, perceptions and policies towards the environment. Environmental engineering focuses on design and technology for improving environmental quality. Environmental scientists work on subjects like the understanding of earth processes, evaluating alternative energy systems, pollution control and mitigation, natural resource management, and the effects of global climate change. Environmental issues almost always include an interaction of physical, chemical, and biological processes. Environmental scientists bring a systems approach to the analysis of environmental problems. Key elements of an effective environmental scientist include the ability to relate space, and time relationships as well as quantitative analysis.

Volume 2 Advances in Computational Environment Science contains the 2012 International Conference on Computer Science (ICCS 2012) papers. Volume 2 contains some topics in intelligent system. There are 51 papers were selected as the regular paper in this volume. It contains the latest developments and reflects the experience of many researchers working in different environments (universities, research centers or even industries), publishing new theories and solving new technological problems. The purpose of volume 2 is interconnection of diverse scientific fields, the cultivation of every possible scientific collaboration, the exchange of views and the promotion of new research targets as well as the further dissemination, the diffusion of intelligent system, including but not limited to Intelligent System, Neural networks, Machine Learning, Multimedia System and Applications, Speech Processing, Image & video Signal Processing and Computer - Aided Network Design the dispersion.

The proceedings contain a selection of peer-reviewed papers that will be presented at the conference. We are thankful to all authors and reviewers who make the present volume possible. The rigorous review of the submissions by the invited reviewers has been highly acknowledged. Their comments and suggestions are important to improve the quality of the publications. We hope the book will be a useful guide environment science, computer science and the relevant fields.

On behalf of the organizing committee, I would like to extend my thanks to all the members who serve in the IERI committee. Their suggestions, impressive lectures and enlightening discussions have been critically important for the success of the conference. The sponsorship of Springer is gratefully acknowledged.

Garry Lee

# Organization Committee

## Honorary Chair

Wei Lee — Melbourne ACM Chapter Chair, Australia

## General Chairs

Jun Zhang — Huazhong University of Science and Technology, China
Minli Dai — Suzhou University, China

## Publication Chair

Garry Lee — Information Engineering Research Institute, USA

## Organizing Chairs

Jun Zhang — Huazhong University of Science and Technology, China

## International Committee

Minli Dai — Suzhou University, China
Ying Zhang — Wuhan University, China
Zhenghong Wu — East China Normal University
Tatsuya Akutsu — ACM NUS Singapore Chapter, Singapore
Aijun An — National University of Singapore, Singapore
Yuanzhi Wang — Anqing Teachers' University, China
Yiyi Zhouzhou — Azerbaijan State Oil Academy, Azerbaijan
Gerald Schaefer — Loughborough University, UK
Biswanath Vokkarane — Society on Social Implications of Technology and Engineering
Jessica Zhang — Information Engineering Research Institute, USA
David Meng — Information Engineering Research Institute, USA
Kath David — Information Engineering Research Institute, USA

# Contents

## Advances in Intelligent Systems

# Performance Analysis of Virtual Force Models in Node Deployment Algorithm of WSN

Ren Xiaoping[1], Cai Zixing[1], Li Zhao[1], and Wang Wuyi[2]

[1] School of Infor. Science & Engineering,
Central South University, Changsha, China
[2] Shanxi Institute of Metrology, Supervision & Verification,
Taiyuan, China
xiaopingren@gmail.com, zxcai@mail.csu.edu.cn,
357970907@qq.com, wangwuyi1964@sohu.com

**Abstract.** Node deployment is an essential and important issue in WSN for it not only determines the energy cost and communication delays for sensors network, but also affects how efficient and maximum a region is covered and monitored by sensors. Virtual force models receive more interests in coverage algorithms of node deployment of WSN. In this paper we analyze the parameters of four virtual force models, and then choose four evaluation factors, which are coverage increment scale, iterative number, coverage efficiency and average movement distance of nodes, to evaluate these virtual force models. We proved that virtual force model of three parameters was a better one, as well as force model which don't considering its force value but direction.

**Keywords:** WIreless sensor network, virtual force, force model, network coverage, node deployment.

## 1   Introduction

WSNs have the advantage of strong synergies[1], its most important task is to monitor the surrounding environment[2], and transmit the information to the console. According to the environmental monitoring, deploying the sensor nodes rationally will help improving coverage and target detection efficiency of WSNs and reduce energy consumption. Therefore, the coverage method is an important issue in WSNs, which reflects the quality of the networks' perceived service [2]. In order to ensure coverage capabilities and control costs, the researchers proposed a hybrid network by adding mobile robot or sensor nodes into WSN. The hybrid-perception networks, which consist of static nodes and mobile nodes, adjusting the locations of nodes dynamically and renovate networks' blind coverage areas through re-deploying of mobile nodes, and then improving the network performance effectively [3].

In recent years, virtual force algorithm for deployment of sensor nodes has attracted extensive attentions [1][4][5]. This idea is inspired by combination of virtual repelling forces in order to estimate the directions and locations of nodes' movement. It can maximize the coverage of sensor nodes. Akkaya proposed a distributed actuator configuration algorithm, ensuring coverage as well as connectivity between nodes.

G. Lee (Ed.): Advances in Intelligent Systems, AISC 138, pp. 1–9.

However, this algorithm still has non-covered blind area. This paper, which was based on [7], analyzed the effects of parameters on several commonly used virtual forces models, and then compared the performance of these four models, hoping to provide a reference for virtual force model selection of the coverage algorithm of WSN.

This paper is organized as follows: the next section discusses the system model, assumptions, and four evaluation standards. Details of four virtual forces models are discussed in section 3 and the experiment of these four models is in section 4 Section 5 concludes the paper with a summary.

## 2 Problem Description and System Modeling

Principles of coverage algorithm based on virtual force model can be described as follows: the repulsion and gravitation exist among sensor nodes as well as barriers to nodes. Nodes spread outwards under repulsion reducing the coverage redundant area, on the other hand, they gather together under the force of gravity, reducing the blind coverage area [3]. Each node moves according to the resultant force it suffers, and regulates the deployment strategy adaptively through the balance of forces [1].

### 2.1 Question Model and Assumption

Assume that WSN network, which consists of $m$ fixed nodes and $n$ mobile nodes, randomly deployed in a two-dimensional square area $A$. $V$ is the set of mobile nodes, and each node $i$ has a global coordinate $(x_i, y_i)$. Note that two mobile nodes do not exist at the same location. Additionally, all these nodes have the same communication distance $r_c$. Node $i$ and $j$ are neighbors if $j$ is in the transmission scope of $i$, and their edge is $e(i,j)$, $d_{ij} = d(i,j)$ is their Euclidean distance. Node $i$'s coverage scope is a circle, that is $S_i = \{k \in A \mid d(k,i) \le r, i = 1,2,..,n\}$. Mixed network's topology is denoted by $G(r_c) = (V, E(r_c))$, while $E(r_c) = \{e(i,j) \mid d_{ij} < r_c \text{ and } i, j \in V\}$. $G(r_c)$ is the union of all nodes' coverage scope, denoted by $S_G$, thus, $S_G = \bigcup S_i$. Let $\lambda_i$ be the degree of node $i$, denoting the number of neighbors around, and $L_i$ denotes neighbor list of node $i$.

### 2.2 Performance Evaluation Standards

- Coverage Increase Scale

Total area coverage of all nodes is $S_G$ before movement and $S_G'$ after reconfiguration, then coverage increase scale as $\varphi$ is shown in (1):

$$\varphi = (S_G' - S_G) / S_G \tag{1}$$

We can see that $\varphi$ is an important indicator to evaluate the nodes' coverage performance of coverage algorithm in WSN.

- Coverage Efficiency

The coverage efficiency $\eta$, which is used to evaluate the utilizing rate of node's perception scope, can be defined as: ratio of the union $S_G{}'$ to the sum of all nodes' coverage scope in area.

$$\eta = S_G{}' / (n \cdot \pi r_c^2) \qquad (2)$$

- Nodes Average Movement Distance

$s_1^i$ is start of node $i$ and $s_n^i$ is destination .The node location sequence during move can be denoted by $s_1^i, s_1^i, \ldots, s_{n-1}^i, s_n^i$, and then the average movement distance $\zeta_i$ of node $i$ is expressed as[2]:

$$\zeta_i = \frac{1}{\lambda_i} \sum_{j=1}^{\lambda_i - 1} d(s_j^i, s_{j+1}^i) \qquad (3)$$

Thus, nodes average moving distance $\delta$ is denoted by $\dfrac{1}{n} \sum_{i=1}^{n} \zeta_i$ ,and the smaller $\delta$ is, the less energy the network consumes.

- Iterative Number

Besides, the iterative number $\chi$ is also an important standard to measure an algorithm.

## 3    Selection of Virtual Force Models

Assuming that there are only three nodes $N_1$ , $N_2$ , $N_3$ which are deployed in $A$ randomly. $\overrightarrow{f_{12}}$ is the repelling force on $N_2$ from $N_1$ ,and $\overrightarrow{f_{32}}$ is repelling force on $N_2$ from $N_3$. While the direction angle $\theta_{ij}$ is $\arctan(k_{ij})$, $k_{ij}$ is line slope from node $i$ to $j$. Since there may be several forces effecting on an actor node, the node should get the combined forces based on vector addition method.

Akkaya defined $f(d_{ij}) = (r_c - d_{ij})/2$ in [6], they didn't analyze the impact of selection of virtual force models in coverage algorithm. Generally speaking, the selection of force model in coverage algorithm is inspired by forces models in physics, such as gravity, intermolecular forces, as well as elastic forces defined by Hooke's Law, etc[5]. We choose some forms of force models of representation to analyze their performance.

Let:

$$vf_1 = \begin{cases} 0, & d_{ij} \geq r_c \\ \dfrac{r_c{}' - d_{ij}}{2}, & d_{ij} < r_c \end{cases} \qquad (4)$$

where $d_{ij}$ denotes the distance between $i$ and $j$ [7].

Coulomb force between charges is also a force model frequently used in node deployment algorithm, it can be expressed as [5]:

$$vf_2 = \begin{cases} 0, & d_{ij} \geq r_c \\ \dfrac{\kappa Q_i Q_j}{d_{ij}^2}, & d_{ij} < r_c \end{cases} \tag{5}$$

where $Q_i$ and $Q_j$ are the quantities of electric charge; $\kappa$ is the proportional coefficient. Here $Q_i$ and $Q_j$ is degree of nodes, equal $\lambda_i$ and $\lambda_j$.

Another kind of virtual force model is to adjust the attributes of interaction force between sensor nodes by using $d_{th}$, threshold of distance. Expression of $vf_3$ is shown in (6), where $w_A$, $w_R$ and $d_{th}$ are three parameters.

$$vf_3 = \begin{cases} 0, & d_{ij} \geq r_c, d_{ij} = d_{th} \\ w_A(d_{ij} - d_{th}), & r_c > d_{ij} > d_{th} \\ w_R(\dfrac{1}{d_{ij}} - \dfrac{1}{d_{th}}), & d_{ij} < d_{th} \end{cases} \tag{6}$$

Assume that the coordinates of nodes $N_1, N_{2old}, N_{2new}, N_3$ are $(x_1, y_1)$, $(x_2, y_2)$, $(x_2', y_2')$ and $(x_3, y_3)$ respectively. $k_{12}$ and $k_{32}$ can be computed as follows[7].

$$\begin{cases} k_{12} = (y_2 - y_1)/(x_2 - x_1) \\ k_{32} = (y_3 - y_2)/(x_3 - x_2) \end{cases} \tag{7}$$

From (7) we can compute $\sin\theta_{12}$, $\cos\theta_{12}$, and the same to $\sin\theta_{32}$ and $\cos\theta_{32}$. Then we can compute the final addition of two vectors $\overrightarrow{f_{12}}$ and $\overrightarrow{f_{12}}$ on $O_2$ through $\overrightarrow{F_2} = \overrightarrow{f_{12}} + \overrightarrow{f_{12}}$. That means:

$$|\overrightarrow{F_2}| = \sqrt{f_{12}^2 + f_{32}^2 + 2f_{12}f_{32}(\sin\theta_{12}\sin\theta_{32} + \cos\theta_{12}\cos\theta_{32})} \tag{8}$$

The direction of $\overrightarrow{F_2}$ is:

$$k_2 = \frac{\sqrt{1+k_{32}^2}f(d_{12}) + \sqrt{1+k_{12}^2}f(d_{32})}{k_{12}\sqrt{1+k_{32}^2}f(d_{12}) + k_{32}\sqrt{1+k_{12}^2}f(d_{32})} \tag{9}$$

From all above forces models and computing method, we can see that the distance and movement direction of nodes depend on virtual forces, including its value and direction. Along the direction of composition forces $\overrightarrow{F_2}$ ,node $N_2$ move from $(x_2, y_2)$ to a new location $(x_2', y_2')$ .However, for all the virtual forces models are stimulate the forces among nodes, however it doesn't exist. Here we consider another kind of force model $vf_4$ .While we have computed $\sin\theta_{12}$ , $\cos\theta_{12}$ , $\sin\theta_{32}$ and $\cos\theta_{32}$ . When computing the forces, we assume all the forces are unit vectors.

Then we can compute the final slope of $\overrightarrow{F_2}$ is:

$$k_2 = \frac{\sin\theta_{12} + \sin\theta_{32}}{\sqrt{(\sin\theta_{12} + \sin\theta_{32})^2 + (\cos\theta_{12} + \cos\theta_{32})^2}} \tag{10}$$

Then in the program, we use *Step* value to control nodes how far to move while in one iterative computing. That means every node moves *Step* far away from current location to the stop location along the direction decided by $k_2$ .In the next time, every nodes will compute new $k_2$ and move again. Expression of $vf_4$ can be described as follows:

$$vf_4 = \begin{cases} 0, & d_{ij} \geq r_c \\ 1, & d_{ij} < r_c \end{cases} \tag{11}$$

## 4    Simulaton Experiment

We realized the simulation of algorithm based on MATLAB7.9 and NS2 simulator. $A$ is a square area of $300 \times 300\ m^2$ ,and mobile nodes which are generated by NS2 setdest tool, are randomly deployed in $A$ .The number of sensors, respectively, are 100,200,…,600. $r_c$ is $20\ m$ ,and algorithm procedure refers to [7].

In the simulation, *Step* value of program is associated with computing time. We should adopt a proper *Step* value with less loss of simulation accuracy and at the same time reduce the computing time. For the max nodes number is 600, we set *Step* equals 1.

### 4.1    7B Influences of $\kappa$ on $vf_2$

As shown in Figure 1, $n = 160$, $\kappa$ ranges from $10^0$ to $10^6$ ,we can see there's no significant change in iterative number, which reflects that $\kappa$ value has no effect on .

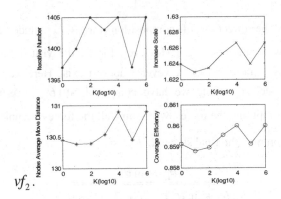

$vf_2$ .

**Fig. 1.** Influences of $\kappa$ on $vf_2$

## 4.2   8BInfluences of $d_{th}$, $w_A$, $w_R$ on $vf_3$

Parameters influences can be discussed in three situation: when $w_A = 1$, $w_R = 3$, $d_{th}$ influences on $vf_3$ ;   when $w_R = 3$ , $d_{th} = 0.8 r_c$ , $w_A$ influences on $vf_3$ ; when $w_A = 1$ , $d_{th} = 0.8 r_c$ , $w_R$ influences on $vf_3$ , $n = 300$ .The simulation result was shown in a figure with the same evaluation standard , note that we only show the scale of $d_{th}$ ,which is $0.1, 0.2, ..., 15$ .

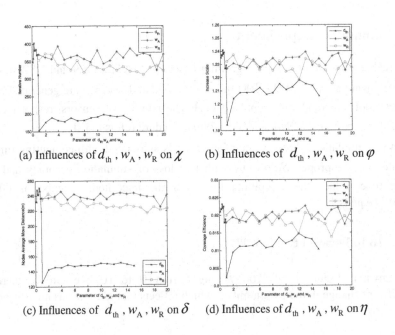

(a) Influences of $d_{th}$ , $w_A$ , $w_R$ on $\chi$     (b) Influences of $d_{th}$ , $w_A$ , $w_R$ on $\varphi$

(c) Influences of $d_{th}$ , $w_A$ , $w_R$ on $\delta$    (d) Influences of $d_{th}$ , $w_A$ , $w_R$ on $\eta$

**Fig. 2.** Influences of $d_{th}$ , $w_A$ , $w_R$ on $vf_3$

From the results of simulation in figure 2, we can find that when $d_{th} > r_c$, $\chi$, $\varphi$, $\eta$ and $\delta$ gradually increase with increasing of $d_{th}$; when $d_{th} < r_c$, $\chi$, $\varphi$, $\eta$, $\delta$ gradually reduce with increasing of $d_{th}$; $\chi$ and $\delta$ gradually increase with increasing of $w_A$, although the rate is not obvious; $\chi$ and $\delta$ gradually reduce with increasing of $w_R$. Both $\eta$ and $\varphi$ haven't changed along with $w_A$ and $w_R$, so we consider that $w_A$ and $w_R$ have no relationship with $\eta$ and $\varphi$. Thus, we draw the conclusion that $d_{th}$ can be used to adjust the network's coverage performance. $w_A$ and $w_R$ can control the algorithm's execution time and network's total energy consumption[9].

### 4.3    9BPerformance Analysis of Four Virtual Forces Models

In this section, we analyze the performance of four virtual force models under varied number of nodes. Considering that nodes number are not very large, in this section we set $Step = 0.1$ to ensure the accuracy of algorithm. Parameters are set as follows: $\kappa = 1$, $w_A = 1$, $w_R = 5$, $d_{th} = 0.8r_c$.

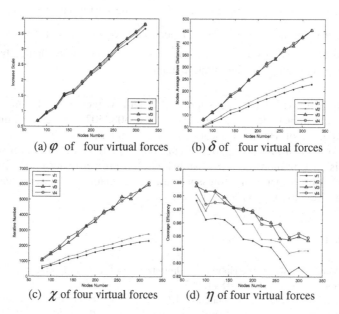

(a) $\varphi$ of four virtual forces          (b) $\delta$ of four virtual forces

(c) $\chi$ of four virtual forces          (d) $\eta$ of four virtual forces

**Fig. 3.** Comparison of performance of four virtual force models

From figure3 we can obviously find that $vf_3$ and $vf_4$ are better than $vf_1$ and $vf_2$, $vf_2 > vf_1$. $vf_3$ and $vf_4$ have similar performance, as is shown in table 1. It declares the average value in this stimulation experiment.

**Table 1.** Comparision of $vf_1, vf_2, vf_3$ and $vf_4$

| Virtual Model | Iterative Number $\chi$ | Coverage Increase Scale $\varphi$ | Average Moving Distance $\delta$ | Coverage Efficiency $\eta$ |
|---|---|---|---|---|
| $vf_1$ | 1502.84 | 2.1588 | 146.8 | 0.8464 |
| $vf_2$ | 1760.46 | 2.2158 | 164.9 | 0.8579 |
| $vf_3$ | 3560.15 | 2.2522 | 270.9 | 0.8661 |
| $vf_4$ | 3656.61 | 2.2575 | 271.5 | 0.8653 |

## 5    Conclusion

In this paper, we analyze four kinds of virtual force models used in node deployment algorithm of WSN. $vf_1$ has no parameters and it can not adjust performance of coverage; $\kappa$ value of $vf_2$ has no relation with deployment result; $d_{th}$, $w_A$ and $w_R$ are three important factors of $vf_3$; then we proposed $vf_4$ without considering value of force but only using its directions is of the same performance to $vf_3$, and it is convenient for its no need to select parameters. Although the virtual force model $vf_4$ has better results in algorithm, it is also necessary to evaluate synthetically from the network energy consumption and iterative number.

**Acknowledgment.** This work was supported in part by the National Natural Science Foundation of China under Grant 90820302 and 60805027, Research Fund for the Doctoral Program of Higher Education under Grant 200805330005, Academician Foundation of Hunan Province under Grant 2009FJ4030, and also in part by Quality and Supervision Commonweal Profession Research Project under Grant 200810002. Finally, thanks my supervisor Professor Cai Zixing, thanks my parents and also my wife Zhichao Xue, loving you forever.

## References

1. Alaiwy, M.H., Alaiwy, F.H., Habib, S.: Optimization of actors placement within wireless sensor-actor networks, pp. 179–184 (2007)
2. Akkaya, K., Thimmapuram, A., Senel, F., Uludag, S.: Distributed recovery of actor failures in wireless sensor and actor networks. In: Proceedings of the 32nd IEEE Conference on Local Computer Networks, pp. 496–503 (2007)
3. Akkaya, K., Younis, M.: C2AP: Coverage-aware and connectivity-constrained actor positioning in wireless sensor and actor networks. In: IEEE International Performance, Computing, and Communications Conference, 2007, pp. 281–288 (2007)

4. Selvaradjou, K., Dhanaraj, M., Siva Ram Murthy, C.: Energy efficient assignment of events in wireless sensor and mobile actor networks. In: 14th IEEE International Conference, vol. 2, pp. 1–6 (September 2006)
5. Xiong, J.-F., Tan, G.-Z., Dou, H.-Q.: Formation control of swarm robots based on virtual force. Computer Engineering and Applications 43(5), 185–188 (2007)
6. Akkaya, K., Janapala, S.: Maximizing connected coverage via controlled actor relocation in wireless sensor and actor networks. Computer Networks 52(14), 2779–2796 (2008)
7. Ren, X., Cai, Z.: A distributed actor deployment algorithm for maximum connected coverage in WSAN. In: 2009 Fifth International Conference on Natural Computation, pp. 283–287 (2009)
8. Fu, Z.-X., Xu, Z.-L., Huang, C., Wu, X.-B.: Survey on sensor deployment problem in wireless sensor networks. Transducer and Microsystem Technologies 27(3), 116–120 (2008)
9. Liu, W., Cui, L., Huang, C.: EasiFCCT:A fractional coverage algorithm for wireless sensor networks. Journal of Computer Research and Development 45(1), 196–204 (2008)

# Design of WEB Video Monitoring Center Server

Jiang Xuehua

School of Engineering,
Linyi Normal University,
Linyi, Shandong,
276000, China
jxhyx@163.com

**Abstract.** According to the extensive application of the video monitoring system, the application of applying the Web technology to the video monitoring system is studied. Based on the analysis of the concept, development situation and the necessary software used for system development, the total structure of the video monitoring system based on Web technology is given. According to the function, the system is divided into four function modules which are communication, video controlling services, data storage and client control. The design method of center server of the video monitoring system based on Web is discussed mainly, its composition structure and working process are analyzed, and the main functions of the center server are realized by programming. The software running shows that the users can control the remote working fields by browser.

**Keywords:** video monitor, center server, total structure, ASP.NET.

## 1   Introduction

Along with the social progress and rapid development of technology of computer, image progressing, communications and network, monitoring technology have updated continuously. New kinds of security equipment emerged, and monitoring systems have changed dramatically. The so-called video surveillance system is the combinative system of monitoring and controlling relative region through images. Monitoring is to exam the site that is interested in or in need of security, and controlling is the necessary operation on certain equipment. Motoring system has played an important role on ensuring the safety of people's life, property and the automotive production processing from the date of birth.

As the rapid development of network technology, a number of monitoring system has developed into the system with multimedia network function mainly based on Web, which is not only can transfer images, audio, text, etc., but also can interact information from multiple monitors, and the center controlling room can check the end of each monitoring information[1].

This paper discusses the video surveillance system components based on Web, and analyzes the function and design of the server.

G. Lee (Ed.): Advances in Intelligent Systems, AISC 138, pp. 11–17.
springerlink.com      © Springer-Verlag Berlin Heidelberg 2012

## 2    Web Video Controlling System

### A.    Structure of the Remote Controlling System

Web-based remote video monitoring system is composed of center server, video server and client. The three parts are connected as the organic whole through Internet, which all play an important role. The structure of the system is shown as Fig. 1.

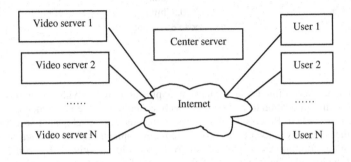

**Fig. 1.** Total structure of web video controlling system

In this system, the role of the only one center server is to play a center node, and the controlling points are more scattered in various controlling points through the center server together, which form a unified video surveillance group to monitor multiple controlling points.

Video server uses an industrial controlling computer including video capture cards which connect with video output of the camera to realize video capture. Industrial computer connects to a dedicated decoder through a serial port, and the decoder connects to controlling input of the camera to control a variety of head and camera movement, the alarm signal processing. Similar to the video server, the center server operating system is installed on Microsoft Windows 2000 server and network information server is Microsoft IIS. The relative software installed on the server includes Microsoft.NET Framework, server software of video surveillance, video monitoring web pages and software communicating with the center server[2].

Client is the operating interface of users which mainly refers to the web users browsing. The client includes website monitoring group home page, controlling point management page, multi-site monitoring screen preview page, single site video monitoring page. In the above pages, single site monitoring video page is placed on the monitoring server, and the other pages are placed on the center server.

### B.    System Function Module

From the functional module, the system can be divided into communication module, video monitoring service module, data storage module and the client control module.

In the system communication module is composed of the client in the video server and server in the center server of web communication software, and its function is to link the distributed video servers together into a video surveillance system.

Data storage module is the database systems SQL in the center server, mainly used to store video server information and user information. Video monitoring service module refers to the video surveillance server, and its function includes video services, video capture, motion detection, video, head and camera control[3].

Management module of the client is composed of group home page in the center server, video surveillance web in various video servers and the browser on the client. Its main function includes user login management, video surveillance point control, picture preview of all controlling points, video surveillance of a single controlling point, control of manipulating console and camera, parameter setting of motion detection and video parameter setting.

## 3    Design and Implementation of Center Server of Video Controlling System

### A.    Function of Center Server

Center server is responsible for the registration of each user and rights management, recording all operation of the video server, checking and recording the video server working state including alarm and work abnormality. It is equivalent to headquarters which can make decisions, monitor and record all actions for inquiries.

The function of center server is as follows:

1) work as  video server in management systems.
2) manage the users and their permissions of the whole system.
3) monitor and manage system operations and video streaming activities.
4) focus on achieving the user's authentication and authorization.
5) release system resources news.
6) store video files in video servers and support document retrieval and playback.
7) manage system log.

It shows that the management server function is cumbersome and important, so in the actual implementation, we can simplify the unnecessary function to make the design as simple as possible considering stability and reliability of the system. Additionally, we can decompose the function, and use the distributed management to increase system reliability.

### B.    Structure of Center Server

Center server is a center WEB server of a video monitoring group, which is used to provide relative information of video monitoring web and video link services. It is a main site of the remote video monitoring system and the site where the WEB is. Before the provision of services, all controlling points must register to the center server and add to the video surveillance base[4].

The operating system installed on the center server is Microsoft Windows 2000 Server and the network information server is Microsoft IIS. The relevant software installed on the server is Microsoft NET Framework. As the page in the WEB server uses NET framework, so the installation of Microsoft .NET Framework is main to provide a good environment for the normal operation and serve of the object code. Microsoft SQL Server 2000 is mainly used to store relevant user information, the center server information, information of video surveillance group related to WEB. HTML is used to provide user access interface. Software communicating to video server is used to receive information from video surveillance point to the center server, and it mainly includes IP of controlling points and storage location of capture picture.

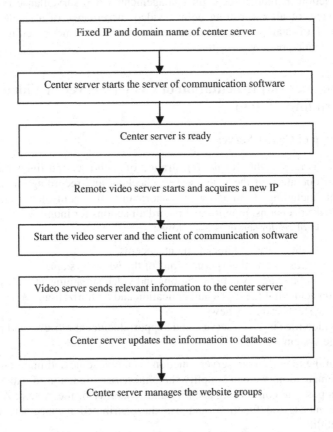

**Fig. 2.** Formative progress of website video controlling group

## C.   Working Process of Center Server

System workflow can be divided into two parts: formation process of video monitoring group and the process of user access controlling points. The following introduces the former. Fig. 2 is the formation process of the video monitoring group.

After the formation of video surveillance group, the user can achieve the visit to the control group through accessing the home page of a center server. Fig. 3 is the process of the user accessing web video monitoring group.

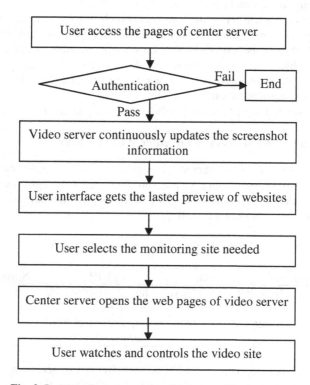

**Fig. 3.** Process of user accessing video controlling group

## 4    Function Implementation of Center Server

Function of the center server is simpler than video server. It mainly plays as the primary site. It has two main functions. Firstly, it is the preview function of all controlling points. Through this function, users visiting the home page can see screenshots of all sites regularly, so you can simultaneously monitor multiple controlling points. Secondly, all of the monitoring sub-site addresses achieve redirection through a center server, so that users can access the needed controlling points through the main site.

### A.    Site Setup of Center Server

The normal running of ASP.NET site needs web pages, relevant programs, and site configuration file. We open Visual Basic.Net 2003, construct the new site, the system will automatically creates the configuration file of this site. We can simply modify the file[5].

In this paper, the configuration is as follows.

1) Set DengLu.aspx as the starting page of the site. Before the user successfully logs in, no matter the user browses any page, he will be redirected to the log file DengLu.aspx and required to log on.

2) Set the authentication strategy *Forms*, which is form patterns. Authentication strategy is more useful, in this document we set authentication strategy as *Forms*. It must be explained that form authentication model is tested by the cookie object, so it can normally work only when the client starts cookie. In addition, the document also set Cookie object named CheckPwd which is used to achieve authentication form mode.

3) Set the site to refuse to authorize the person without entering account.

## B.  Implementation of Preview Function of Center Server

1) All the video server is numbered and registered, and the registration number will write to the database table in the sub Website. The registration information includes video server number, name, IP address, the link address of a screenshot, status and notes, as shown in Tab. 1.

**Table 1.** Register information of video service

| ID | Nam | URL | ImgURL | State | Remark |
|----|-----|-----|--------|-------|--------|
| 001 | Site 1 | http://192.168.0.55/x/ | http://192.168.0.55/pic/img.gif | 1 | -- |
| 002 | Site 2 | http://192.168.0.66/x/ | http://192.168.0.66/pic/img.gif | 1 | -- |

2) Center server receives IP address sent by video server and link address information of screenshot and updates the database through communication software of video server. The function of the video server communication module is to send number, name, IP address, the link screenshot, current status, the related information to the center server. The specific implementation of the communication module is to use NMHTTP of DELPHI to complete HTTP communication.

3) When the user opens the center server home page, the center server can monitor the home page according to information dynamic website of database. In the home page, it can display more than one screenshot images of controlling point. Client generates the relevant preview image of multiple controlling points according to the number of the controlling point, and direct the address of images to URL of monitoring screenshot file, and set the hyperlink of image as URL of monitoring page. When a screenshot of controlling point is refreshed, you can achieve the preview screen of multiple monitoring points as long as the client refreshes the image regularly. When the user wants to monitor the designated site, he can log on to response site as long as he clicks the specified image with the mouse. Preview page of multiple controlling points is shown in Fig. 4.

**Fig. 4.** Preview of multi-site

## 5   Conclusion

Video monitoring system based on Web combined with technology of ASP.NET, database, multimedia video, computer network achieve networking and intelligent network management, enabling users to connect to remote video surveillance site through browser on the Internet, and realize video transmission.

This paper discusses the basic structure and work flow of video monitoring system based on WEB, designs center server based on the database SQL SERVER and interface ASP.NET, in which the table creation and data is the core of the center server. In the existing video surveillance we introduce Web serve according to the characteristics of Web serve, to standardize the data and simplify the management process. We design a new architecture for the management server to make the system maintain more secure, reliable and easier.

## References

1. Wilson, C.: Web-based video offers more choice. Telephony 247, 18–19 (2006)
2. McKinion, J.M., Turner, S.B., Willers, J.L.: Wireless technology and satellite Intem connectivity in precision agriculture. Agricultural Systems 81, 201–212 (2004)
3. Guthery, S.B., Cronin, M.J.: Mobile Application Development with SMS and the SIM Toolkit. Post & Telecom Press, Beijing (2003)
4. Cao, W., Wang, G.-H.: Design of Transformer Terminal Unit Based on GSM. Journal of Harbin University of Science and Technology 12, 1–4 (2007)
5. Louderback, J.: Simulcasting the World with Web Video. PC Magazine 25, 65–66 (2006)

# An Artificial Intelligence Central Air-Conditioning Controller

Chunhe Yu and Danping Zhang

Department of Electronic and Information Engineering,
Shenyang Aerospace University,
Shenyang, 110034 P.R. China
chunhe_yu@tom.com

**Abstract.** In order to obtain a comfortable indoor environment, a fuzzy central air-conditioning controller is designed by using fuzzy control theory, and a digital sensor SHT11 is used to measure temperature and relative humidity. For the purpose of reducing electric wires in the room, the radio frequency technology is applied in the circuits of the sensor and the controller, through which data and commands are transferred or received. In order to obtain a finer performance, a fuzzy algorithm is proposed in the controller, which controls the three-speed fan coil motor of the central air-conditioning to make a suitable environment through an output parameter. The practical operation has tested the controller is reliable and stable.

**Keywords:** fuzzy control, air-conditioning controller, radio frequency technology.

## 1 Introduction

Compared with a general air-conditioning, a central air-conditioning has the characteristics of energy saving, comfortable, convenient adjustment, low noise and little vibration [1~3], and it controls the temperature and relative humidity through the speed of the fan coil motor. In order to obtain a comfortable environment and achieve auto-regulation function, an artificial intelligence controller is designed according to the room's temperature and relative humidity. Every room needs one sensor for achieving independent control. The temperature and the relative humidity are regarded as the input variables to control the speed of the fan coil motor, which is the control part of the central air-conditioner. In order to reduce the number of electric wires in a room, the wireless network technology is used based on the integrated chip CC2420 [4]. For getting an excellent control effect, a fuzzy algorithm is adopted in the controller.

In this paper, a single chip ATmega128L (8 bits) is used, and it is designed with the 2.4GHz wireless module CC2420 to build a communication knot. The rest of the paper is organized as follows: The hardware structure of the controller, the fuzzy algorithm, and the summary of the controller.

## 2 Work Principle and Hardware Structure

As shown in figure 1, the hardware of the controller system includes two parts: sensor modules and one controller of the fan coil motor (the central air-conditioning). The

G. Lee (Ed.): Advances in Intelligent Systems, AISC 138, pp. 19–25.

integrated chip SHT11 in the sensor module is selected to measure temperature and relative humidity in a room, and the parameters are sent to the controller of the fan coil motor module through the RF module. After receiving the parameters, the controller of the fan coil motor is selected a suitable speed to work by fuzzy algorithm. In order to obtain the good performance, the fan coil motor is select three-speed to control. ATmega128L is selected as the MCU in the system, which has enough capability and low energy in the controller. Once the ATmega128L is out of order, its watch dog circuit will be reset and ensure the sensor module and the controller operate steadily.

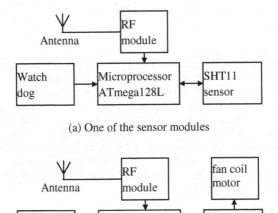

(a) One of the sensor modules

(b) The controller of fan coil motor

**Fig. 1.** Hardware structure of the controller system

## 2.1    Temperature and Relative Humidity Sensor

SHT11 is an integrated digital sensor with two wires bus, which is used to measure the temperature and relative humidity at same time. It can measure the temperature range from -40° to 123° with ±0.1° accuracy; and the relative humidity range from 0 to 100% RH. The ATmega128L of the sensor module picks up the temperature and relative humidity every 3 minutes and then sends the two parameters through the RF module by inquiring mode.

## 2.2    RF Module

In the RF module [4], we use the chip of CC2420 to accomplish the data transmission, which provides a spreading 9 dB gain and an effective data rate of 250 kbps by a digital direct-sequence spread-spectrum modem. The CC2420 is a low-cost, highly integrated solution for the robust wireless communication in the 2.4 GHz unlicensed ISM band. In our application it is used together with the ATmega128. According to the designation of the hardware circuit, the program idea is: using the SPI port of the MCU to initialize CC2420 firstly, and then transmitting or receiving the two parameters individually in the two RF modules. The program flow charts are shown in figure 2 and figure 3, in which the process of the sending massage and the receiving massage are described. The RF receiver module of the controller adopts inquiring mode.

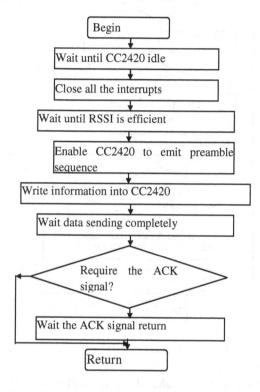

**Fig. 2.** The flow chart of sending massage program

## 2.3 The Control Circuti of the Fan Coil Motor

In our central air-conditioning system, a three-speed motor is applied, which is controlled by the command output of the controller. The motor of the fan coil is controlled by the control circuit, in which there are three set of control chips: two-way SCR TLC336A and two-way SCR drive circuit MOC3041.The output of the command is calculated from the fuzzy algorithm. .

## 3 Fuzzy Algorithm

The fuzzy algorithm [5~9] of the controller system has two input variables: temperature and relative humidity.

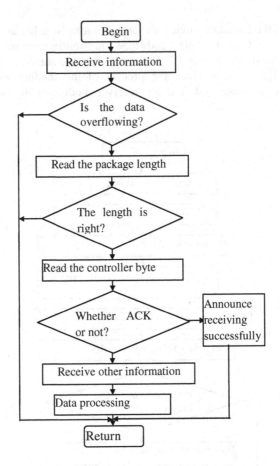

**Fig. 3.** The flow chart of receiving massage program

The output is the speed command of the fan coil motor. The algorithm can be separated into three parts: fuzzy input variable, fuzzy reason & judgment, and fuzzy output variable. According to the temperature and relative humidity, the fuzzy algorithm will make the input variable to change fuzzy quantity in order to carry out fuzzy reason and decision-making.

The controller algorithm is designed as follows.

### 3.1 Making Precise Quantity to Fuzzy Quantity

This system adopts the two-dimensional fuzzy controller algorithm which is widely used at present. Suppose the fuzzy variables are that: "$e$" stands for the temperature difference, "$el$" stands for the relative humidity difference, "$f$" stands for the output variable. The corresponding fuzzy variable values are selected as follows:

```
e= {NB, NS, ZO, PS, PB}
el= {NH, NL, ZO, PL, PH}
f = { L, M, H }
```

To blur the quantity, it needs to obtain subordination function of various words collection.

1) Blurring subset subordination function of the temperature

In the controller, the temperature difference is divided into 7 grades, {-3, -2 -1 0, 1, 2, 3}. The statistic table 1 is obtained through statistics as follows.

**Table 1.** Fuzzy Variable Assignment Table

| Quantizing grade | $\mu(x)$ ╲ e  Quantity domain | NB | NS | ZO | PS | PB |
|---|---|---|---|---|---|---|
| -3 | $-\infty < e \leq -3$ | 1 | 0 | 0 | 0 | 0 |
| -2 | $-3 < e \leq -1$ | 0.5 | 0.5 | 0 | 0 | 0 |
| -1 | $-1 < e \leq 0$ | 0 | 1 | 0.5 | 0 | 0 |
| 0 | $e = 0$ | 0 | 0 | 1 | 0 | 0 |
| +1 | $0 < e \leq 1$ | 0 | 0 | 0.5 | 1 | 0 |
| +2 | $1 < e \leq 3$ | 0 | 0 | 0 | 0.5 | 1 |
| +3 | $3 < e \leq \infty$ | 0 | 0 | 0 | 0 | 0.5 |

2) Fuzzy subset subordination function of the relative humidity

In the controller, the relative humidity is also divided into 7 grades, that is {-3,-2 -1 0, 1, 2, 3}, the statistic table 2 is obtained as follows to show the variable "e1".

**Table 2.** Fuzzy Variable "EL" Assignment Table

| Quantizing grade | $\mu(x)$ ╲ e1  Quantity domain | NB | NS | ZO | PS | PB |
|---|---|---|---|---|---|---|
| -3 | $-\infty < e1 \leq -3$ | 1 | 0 | 0 | 0 | 0 |
| -2 | $-3 < e1 \leq -1$ | 0 | 1 | 0 | 0 | 0 |
| -1 | $-1 < e1 \leq 0$ | 0 | 1 | 0 | 0 | 0 |
| 0 | $e1 = 0$ | 0 | 0 | 1 | 0 | 0 |
| +1 | $0 < e1 \leq 1$ | 0 | 0 | 1 | 1 | 0 |
| +2 | $1 < e1 \leq 3$ | 0 | 0 | 0 | 1 | 1 |
| +3 | $3 < e1 \leq \infty$ | 0 | 0 | 0 | 0 | 1 |

3) Fuzzy subset subordination function of output variable

The range of the output quantification is divided into 7 grades through statistics: {0, 1, 2, 3, 4, 5, 6 }, which is shown in table 3.

**Table 3.** Output Variable "F" Evaluation

| $\mu(x)$    $f$ <br> Quantity domain | L | M | H |
|---|---|---|---|
| 0 | 0 | 0 | 0 |
| 1 | 1 | 0 | 0 |
| 2 | 1 | 0 | 0 |
| 3 | 0 | 1 | 0 |
| 4 | 0 | 1 | 1 |
| 5 | 0 | 0 | 1 |
| 6 | 0 | 0 | 0 |

## 3.2    The Formation of Fuzzy Control Rule

Considering the fuzzy controller with double inputs and one output, its control rule can be written in the following form:

if $e=A_i$ and $e1=B_j$ then $f=C_{ij}$ $(i=1,2,...,n; j=1,2,..., m)$

Where, "$A_i$" is the fuzzy subset of temperature difference; "$B_j$" is the fuzzy subset of the relative humidity changing rate $e1$; "$C_{ij}$" is the fuzzy subset of the output variable. According to the practical running experience, we obtain the following control rule, which is shown in table 4.

**Table 4.** Fuzzy Reasoning Rule

| $e$   $f$   $e1$ | PH | PL | ZO | NL | NH |
|---|---|---|---|---|---|
| PB | H | H | M | L | L |
| PS | H | M | L | L | L |
| ZO | M | L | L | L | M |
| NS | L | L | L | M | H |
| NB | L | L | M | H | H |

## 3.3    Fuzzy Reasoning

The rule of table 4 can be summarized as a fuzzy relation $R$, that is:

$$R = \underset{ij}{Y}(A_i \times B_j \times C_{ij})$$

$R$'s grade of membership is:

$$\mu_R(x, y, z) = \overset{i=n, j=m}{\underset{i=1, j=1}{\vee}} (\mu_{Ai}(x)\mu_{Bj}(x)\mu_{Cij}(x))$$

When the input fuzzy variable selects "$A_i$", "$B_i$" individually, the output value "$C_{ij}$" can be obtained by the fuzzy reasoning.

$$C_{ij} = (A_i \times B_j)oR$$

Then the output value $C$'s grade of membership is:

$$\mu_C(z) = \bigvee_{x \in X, y \in Y} (\mu_R(x, y, z) \wedge \mu_A(x) \wedge \mu_B(y))$$

According to the last formula, and the grade of membership of output, we adopt the method of weighted mean, so the precise quantity is:

$$f = \sum (\mu_f(fi) \cdot fi) / \sum \mu_f(fi)$$

## 4 Conclusion

The practical experiment has testified the controller system is reliable and stable. Based on the wireless technology and the fuzzy control algorithm, the central air conditioning can save energy and get a comfortable environment for users.

## References

1. Ahot, G.P., et al.: Greenhouse climate control. Academic Press, San Diego (1996)
2. Xu, X.N., Ding, Y.F.: Intellectual controlling and energy-saving management for terminal equipment of air-conditioning system. Building Science 24(8), 43–46 (2008)
3. Fu, S.B., Chen, X.Z., Tao, S., et al.: The fuzzy controller in the application of the temperature control of centre air-conditioning system. Control & Automation 21(4), 36–37 (2005)
4. Chipeon, A.S.: Smart RF CC2420. Preliminary Datasheet (rev 1.2) vol. 6 (2004)
5. Passino, K.M., Yurkovich, S.: Fuzzy Control. Addison Wesley Longman, Menlo Park (1998)
6. Shi, L., Shao, L.H., Feng, D.Q., et al.: The prediction of grain state by using the soft measuring based on fuzzy inference system. Journal of Zheng Zhou Institute of Technology 25(2), 71–74 (2004)
7. Duan, Y.H.: Research of predictive fuzzy-PID controller in temperature of central air-conditioning system. Journal of System Simulation 20(3), 620–622+626 (2008)
8. Zhao, R.J., Wang, X.L.: Application of fuzzy-PID control in central air-conditioning system. Computer Simulation 33(11), 331–333+339 (2006)
9. Li, J., Gong, J.H., Huo, X.P.: Study on the method of fuzzy neural network adaptive control using wavelet neural network as identifier applying in the VAV air condition system. Refrigeration Air Conditioning & Electric Power Machinery 123(29), 12–15+38

# Application of GMDH to Short-Term Load Forecasting

Hongya Xu[1], Yao Dong[2], Jie Wu[2,*], and Weigang Zhao[2]

[1] School of Physical Science & Technology Lanzhou University,
Lanzhou, China
xuhy07@lzu.cn
[2] School of Mathematics & Statistics Lanzhou University,
Lanzhou, China
wuj19870903@126.com

**Abstract.** Daily power load forecasting plays a significant role in electrical power system operation and planning. Therefore, it is necessary to find automatic interrelations of data and select the optimal structure of model. However, obtaining high accuracy by using single model for short-term load forecasting (STLF) is not easy. In this paper, Group Method of Data Handling (GMDH) is applied to forecast electric load demand of New South Wales (NSW) in Australia from January 17, 2009 to January 18, 2009. Compared with outcomes obtained by ARIMA, we demonstrate that GMDH is a better method for STLF.

**Keywords:** Group Method of Data Handling (GMDH), short-term load forecasting (STLF), ARIMA.

## 1 Introduction

Electric load demand plays a significant role in the planning and secure operation of modern power systems, more and more researches have been dedicated to it. In general, load forecasting can be classified into three categories: short-term, mid-term and long-term load forecasting. Compared with mid-term and long-term load forecasting, short-term load forecasting (STLF) plays a more important role for its less inherent difficulties.

A great number of models and methods have been used for STLF in recent years. Artificial neural networks evolved by a genetic algorithm for short-term load forecasting was presented in [1], the best-evolved artificial neural network were found to accurately forecast one-day ahead hourly loads not only on weekdays, but also on weekends. A new class of nonlinear models which are known as smooth transition periodic autoregressive (STPAR) models were used in [2] for STLF, and the performance of a simple autoregressive process, a naïve benchmark and a feed forward, back propagation artificial neural network for STLF were listed in the paper to show the predictive ability of STPAR. Different parameter combination using a support vector regression (SVR) model will cause different accuracy for load forecasting, chaotic particle swarm optimization algorithm was applied by Wei-Chiang Hong in [3] to choose suitable

---

* Corresponding author.

G. Lee (Ed.): Advances in Intelligent Systems, AISC 138, pp. 27–32.
springerlink.com     © Springer-Verlag Berlin Heidelberg 2012

parameter combination for a SVR model. A newly developed site-independent technique using a parameterized rule base and a parameter database for STLF was analyzed in [4].

In this paper, group method of data handling (GMDH) is used to forecast electric load demand of New South Wales (NSW) in Australia from January 17, 2009 to January 18, 2009. At first, Load demand data from January 1, 2009 to January 11, 2009 are employed as the training set, the ones from January 6, 2009 to January 16, 2009 are applied as the checking set. Then we forecast short-term load by GMDH. Finally, compared with the outcomes by ARIMA, it is obvious to give a clear seeing that the performance of GMDH for STLF is better than ARIMA.

The structure of this paper is organized as follows. A brief description about GMDH is given in section 2, the outcomes and analysis of GMDH are represented in section 3. Conclusion remarks are shown in section 4.

## 2    GMDH Nerual Networks

GMDH is a learning machine based on the principle of heuristic self-organizing proposed by Ivakhnenko in the 1960s. It is an evolutionary computation technique, in which a series of operations of seeding, rearing, crossbreeding, selection and rejection of seeds correspond to the determination of the input variables, structure and parameters of model, and selection of model by principle of termination [5].

The classical GMDH algorithm can be represented as a set of neurons in which different pairs of them in each layer are connected through a quadratic polynomial and thus produce new neurons in the next layer [6].

General connection between inputs and output variables can be expressed by a complicated discrete form of the Volterra functional series in the form of [7]:

$$y = a_0 + \sum_{i=1}^{m} a_i x_i + \sum_{i=1}^{m}\sum_{j=1}^{m} a_{ij} x_i x_j + \sum_{i=1}^{m}\sum_{j=1}^{m}\sum_{k=1}^{m} a_{ijk} x_i x_j x_k + \cdots, \tag{1}$$

which is known as the Kolmogorov–Gabor polynomial, where $X = (x_1, x_2, \ldots, x_m)$ is the input vector, and $y$ is the output variable. This full form of mathematical description can be represented by a system of partial quadratic polynomials consisting of only two variables (neurons) in the form of:

$$\hat{y} = G(x_i, x_j) = a_0 + a_1 x_i + a_2 x_j + a_3 x_i x_j + a_4 x_i^2 + a_5 x_j^2. \tag{2}$$

The main purpose is to make $\hat{y}$ as much as possible close to actual output $y$.

An outline of the GMDH algorithm can be listed as follows [8]:

**Step 1:** Determine all neurons (estimate their parameter vectors with training data set $M$ ) whose inputs consist of all the possible couples of input variables, i.e. $m(m-1)/2$ couples (neurons).

**Step 2:** Using a validation data set, not employed during the parameter estimation phase, select several neurons which are best-fitted in terms of the chosen criterion.

**Step 3:** If the termination condition is fulfilled (the network fits the data with desired accuracy or the introduction of new neurons did not induce a significant increase in the approximation abilities of the neural network), then STOP, otherwise use the outputs of the best-fitted neurons (selected in Step 2) to form the input vector for the next layer, and then go to Step 1.

GMDH works by building successive layers with complex links (or connections) that are the individual terms of a polynomial. The initial layer is simply the input layer. The first layer created is made by computing regressions of the input variables and then choosing the best ones. The second layer is created by computing regressions of the values in the first layer along with the input variables. This means that the algorithm essentially builds polynomials of polynomials [9].

## 3    Discussion and Analysis

### 3.1    Evaluation of Prediction Performance

The loss function can be served as the criteria to evaluate the prediction performance relative to power load value including mean absolute percentage error (*MAPE*), the prediction performance is better when the loss function value is smaller. The loss function is expressed as follows:

$$MAPE = \frac{1}{T}\sum_{t=1}^{T}\left|\frac{\hat{y}_t - y_t}{y_t}\right| \times 100\%, \tag{3}$$

where T demonstrates the total time, $y_t$ and $\hat{y}_t$ represent the original value and prediction value when time is $t$, respectively.

### 3.2    The Analysis of Procedure

The proposed GMDH model is tested using a case study of forecasting power load. The power load data were collected on a half-hourly basis (48 data points per day) from January 1, 2009 to January 18, 2009 which are got from New South Wales in Australia.

When we build a network, the data from January 1 to 11 are training data, from 6 to 16 are testing data, load values before six days are used to predict the seventh day at the same time, thus, the input layer has 6 neurons, according to the algorithm mentioned in section II, GMDH neural network can be constructed through the connection of the given number of neurons, as shown in Fig. 1. The results of training and checking stage can be seen in Fig. 2. After the good network has been built, the data form January 11 to 17 can be drawn into the network to forecast the data from January 17 to 18.

Table 1. The Weights of Network Nodes

| Index (m,n,k) | (1,1,5) | (1,1,6) | (1,3,6) | (2,1,3) |
|---|---|---|---|---|
| $a_0$ | 9741.47 | 7326.66 | -8696.44 | 5615.07 |
| $a_1$ | -1.0729 | -2.0401 | 0.2578 | -1.1349 |
| $a_2$ | -0.3611 | 1.0255 | 3.0599 | 0.7434 |
| $a_3$ | 0.0004 | 0.0002 | -0.0002 | -0.0008 |
| $a_4$ | -0.0001 | 0 | 0.0001 | 0.0005 |
| $a_5$ | -0.0001 | -0.0001 | 0 | 0.0004 |

As you can see from Table 1, all coefficients in (2) have been calculated, so mathematical descriptions between two neighboring layer neuron can be obtained, where $m$ represents the network layer number, namely, layer $m$; $(n, k)$ represents the node belongs to the combined number of upper layer, such as (1,1,5) represent that the value of first living cell in the first network layer $y_1$ is determined by the values of input variables $x_1$ and $x_5$, the corresponding relation formula can be expressed as follows:

$$y_1 = 9741.47 - 1.0729x_1 - 0.3611x_5 + 0.0004x_1x_5 - 0.0001x_1^2 - 0.0001x_5^2.$$

## 3.2 The Discussion of Results

Simulation and prediction systems have been frequently adopted to forecast the power load. Nevertheless, the prediction accuracy is not satisfied. This work demonstrates the strong performance of *GMDH* model.

We consider the empirical analysis of the load data applying GMDH. Compared with ARIMA (1, 1, 1) model, it is clear that *MAPE* using GMDH is lower than ARIMA model in Table 2. Also, it can be seen from Fig. 3 that the results obtained from GMDH agree with the original electricity load exceptionally well. The prediction accuracy obtained from *GMDH* is higher than the traditional *ARIMA*. For STLF, the proposed forecasting procedures using GMDH leads to **19.472%** reductions in total *MAPE* in comparison with ARIMA model. These results demonstrate again that GMDH performs better in STLF.

Table 2. The Outcomes of GMDH and ARIMA

| Model | Stage Segment | MAPE(%) |
|---|---|---|
| GMDH | 2009.01.01-2009.01.11(training stage) | 4.42 |
| | 2009.01.06-2009.01.16(checking stage) | 5.29 |
| | 2009.01.17-2009.01.18(forecasting stage) | **5.82** |
| ARIMA | 2009.01.17-2009.01.18(forecasting) | **23.75** |

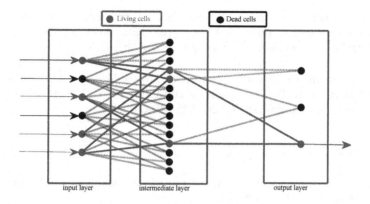

**Fig. 1.** The structure of the GMDH neural network

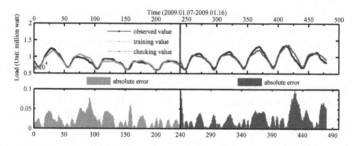

**Fig. 2.** The comparison between observed value and training, checking value

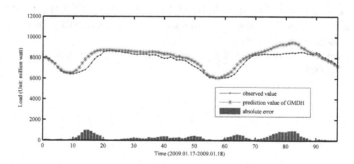

**Fig. 3.** The comparison between observed value and prediction value

## 4    Conclusions

Accurate daily load forecasts are significant for secure and profitable operation of modern power utilities. There are two advantages of the proposed GMDH. Firstly, the proposed GMDH is essentially automatic and does not need to make complicated decision about the explicit form of approach for each particular case. Secondly, the proposed methodology creates commendable improvements that are relatively satisfactorily for current research. With proper characteristic selection, it is obvious that GMDH is a more efficient and straightforward approach.

**Acknowledgment.** The research was supported by the Ministry of Education overseas cooperation "Chunhui Projects" under Grant (Z2007-1-62012) and the NSF of Gansu Province in China under Grant (ZS031-A25-010-G).

# References

1. Srinivasan, D.: Evolving artificial neural networks for short term load forecasting. Neuro Computing 23, 265–276 (1998)
2. Amaral, L.F., Souza, R.C., Stevenson, M.: A smooth transition periodic autoregressive (STPAR) model for short-term load forecasting. International Journal of Forecasting 24, 603–615 (2008)
3. Hong, W.C.: Chaotic particle swarm optimization algorithm in a support vector regression electric load forecasting model. Energy Conversion and Management 50, 105–117 (2009)
4. Rahman, S., Hazim, O.: Load forecasting for multiple sites: development of an expert system-based technique. Electric Power Systems Research 39, 161–169 (1996)
5. Xiao, J., He, C.Z., Jiang, X.Y.: Structure identification of Bayesian classifiers based on GMDH. Knowledge-Based Systems 22, 461–470 (2009)
6. Nariman-zadeh, N., Darvizeh, A., Darvizeh, M., Gharababaei, H.: Modelling of explosive cutting process of plates using GMDH-type neural network and singular value decomposition. Journal of Materials Processing Technology 128, 80–87 (2002)
7. Kalantary, F., Ardalan, H., Nariman-Zadeh, N.: An investigation on the $S_u$-$N_{SPT}$ correlation using GMDH type neural networks and genetic algorithms. Engineering Geology 104, 144–155 (2009)
8. Witczak, M., Korbicz, J., Mrugalskia, M., Pattonb, R.J.: A GMDH neural network-based approach to robust fault diagnosis: Application to the DAMADICS benchmark problem. Control Engineering Practice 14, 671–683 (2006)
9. Srinivasan, D.: Energy demand prediction using GMDH networks. Neuro Computing 72, 625–629 (2008)

# Recognition of Control Chart Patterns Using Decision Tree of Multi-class SVM

Xiaobing Shao

School of Information Science and Technology,
Jiujiang University, Jiujiang, Jiangxi province, 332005, China
wxh_jj_mail@163.com

**Abstract.** This paper aims to realize the automatic recognition of abnormal patterns of control charts in a statistical process control system. A novel multi-class SVM is proposed to recognize the control chart patterns, which include six basic patterns (i.e. normal, cyclic, up-trend, down-trend, up-shift, and down-shift pattern). Unlike the commonly used One-Against-All (OAA) implementation methods, the structure of proposed multi-class SVM is same as a special decision tree with each node as a binary SVM classifier, which is built via recursively dividing the training dataset of six classes into two subsets of classes. The proposed multi-class SVM can increase recognition accuracy and resolve the unclassifiable region problems caused by OAA methods. Based on this, Monte Carlo simulation is used to generate training and testing data samples. The results of simulated experiment show that the problem of false recognition has been addressed effectively, and the proposed decision tree of multi-class SVM is more effective in detecting unnatural patterns on control charts than the traditional OAA methods.

**Index Terms:** pattern recognition, multi-class SVM, OAA methods, control chart.

## 1 Introduction

Statistical process control (SPC) is usually used to improve the quality of products and reduce rework and scrap so that the quality expectation can be met [1]. Shewhart control charts are the most popular charts widely used in industry to detect abnormal process behavior. The most typical form of control charts consists of a central line and two control limits representing the specifications of the product and the variant range limits. However, these control charts do not provide pattern-related information because they focus only on the latest plotted data points [2].

Grant and Leavenworth divided control charts into six various basic patterns, namely, normal, upward trend, downward trend, upward shift, downward shift and cycle pattern [3]. Over these years, most of the studies in the pattern recognition of control charts emphasized the recognition of these basic patterns. The methods used in this field can be divided into two parts. One is statistical classification and expert system method [4,5]. The other is artificial intelligent method [6,7], which takes neural networks, expert system etc. as powerful tools for the application of pattern classification, recognition

G. Lee (Ed.): Advances in Intelligent Systems, AISC 138, pp. 33–41.

and prediction. Most of the researchers considered the neural network methods have better performance than that of statistical classification methods.

From the overview of the past literatures, one may find that most of the ANN used by the studies is back propagation (BP) network. Although BP network has some good performances in pattern recognition of control charts, it has many disadvantages by nature, such as slow training and weak model adaptation ability. Once a new pattern appears, it has to be trained again by both the old samples and new ones. Recently, support vector machine (SVM) is gaining applications in area of pattern recognition because of excellent generalization capability. In this paper, we have used the generalization capability of SVM and proposed a special decision tree of multi-class SVM for the control chart patterns data [8].

The paper is organized as follows. Section 2 explains the six control chart patterns. Section 3 describes the classifier. Section 4 describes the data generation. Section 5 shows some simulation results. Finally Section 6 concludes the paper.

## 2    Interpretation of Control Chart Patterns

It is known that a quality attribute of one machining process can be described as

$$y(t) = u + x(t) + d(t), \tag{1}$$

where $y(t)$, $u$, $x(t)$ and $d(t)$ represent the observed value of quality attribute, the mean value of quality attribute, random and systematic disturbance, respectively. If there is systematic disturbance at a machining process, the abnormal pattern of observed data will be presented in the control chart. Therefore, recognizing abnormal patterns of the control chart is an essential issue for detecting systematic disturbances related to the machining process and then eliminating the root causes of these disturbances. From the Equation (1), the key issue of recognizing the control chart patterns is to identify the fluctuation of $d(t)$. The types of patterns in this study are described as follows.

(1) Normal pattern (class A) is shown in Figure 1($a$). It can be seen that the points of control chart are arranged randomly, which can be described as follows:

$$d(t) = 0, \tag{2}$$

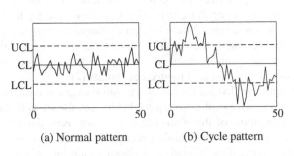

(a) Normal pattern          (b) Cycle pattern

**Fig. 1.** Six patterns in control charts

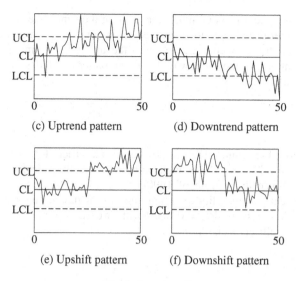

(c) Uptrend pattern          (d) Downtrend pattern

(e) Upshift pattern          (f) Downshift pattern

**Fig. 1.** (*continued*)

(2) Cycle pattern (class B) is shown in Figure 1(b). It is characterized by short trends in the data, which occurs in repeated patterns, described as follows:

$$d(t) = a \times sin(2\pi t/\Omega), \tag{3}$$

where $a$ and $\Omega$ represent amplitude and cycle value of the cycle pattern, respectively.

(3) Trend patterns include up-trend (class C) and downtrend (class D), as shown in Figure 1(c) and (d), respectively. Trend patterns are the patterns that fail to balance themselves about the centerline, which are defined as a long series of points with no change in direction, described as follows:

$$d(t) = \pm \rho \times d \times t, \tag{4}$$

where positive and negative symbols represent up-trend and downtrend, respectively; $\rho$ is a binary value, which represents whether there is a trend pattern or not (No trend $\rho = 0$, otherwise, trend $\rho = 1$); $d$ is the trend slope and $d \in [0.1\sigma, 0.26\sigma]$.

(4) Shift patterns include up-shift (class E) and downshift (class F), as shown in Figure 1(e) and (f), respectively. Shifts are the patterns that a series of points deviate from their previous points in control chart with large magnitude, described as follows:

$$d(t) = \pm \upsilon \times s, \tag{5}$$

where positive and negative symbols represent up-shift and downshift, respectively; v is a binary value, which represents whether there is a trend pattern or not (before shift $\upsilon = 0$, after shift $\upsilon = 1$); $s$ is the shift magnitude, and $s \in [1\sigma, 3\sigma]$.

In this study, the patterns lying in the same side of the centerline are regarded as the same kind of patterns in order to decrease the incorrect recognition to the similar graph, such as up-shift and up-trend etc, e.g., the up-shift and up-trend patterns are regarded as up pattern. In the same way, the down pattern includes the downshift and downtrend patterns.

# 3    Classifier

We have proposed a multi-class SVM based classifier for abnormal patterns recognition of control chart. SVMs were introduced on the foundation of statistical learning theory. Since the middle of 1990s, the algorithms used for SVMs started emerging with greater availability of computing power, paving the way for numerous practical applications. The basic SVM deals with two-class problems; however, it can be developed for multi-class classification.

## 3.1    Binary SVM (BSVM)s

BSVM performs classification tasks by constructing the optimal separating hyper-plane (OSH). OSH maximizes the margin between the two nearest data points belonging to the two separate classes.

Suppose the training set $(x_1, y_1), \ldots, (x_n, y_n)$, $x_i \in R^m$, $i = 1,2,\ldots,n$, and every $x_i$ belongs to Class 1 or Class 2, where $y_i \in \{-1,+1\}$ is the label of $x_i$. The key idea of kernel-based methods is mapping data instances from the input space $R^m$ into high dimensional Hilbert space $H$, called feature space. Usually the nonlinear mapping is performed implicitly by means of a kernel function $K$, The feature map $\varphi(\cdot)$ associates with every $x_i$ from the input space to the image $\varphi(x_i)$ in the feature space.

The kernel function corresponds to inner product in the feature space: $K(x_i, x_j) = \varphi(x_i)^T \varphi(x_i)$. The SVM aim is to find the optimal maximum margin hyper-plane separating in the feature space Class 1 from Class 2. The geometrical margin for the point $(\varphi(x_i), y_i)$ is defined as a distance to the hyper-plane

$$\rho(x, y) = y(\omega^T \varphi(x) + b)/\|\omega\|. \tag{6}$$

For canonical hyper-plane the margin is $1/\|\omega\|$. Because outliers may present the optimal soft margin hyper-plane should be found. It guarantees some reasonable tradeoff between the generalization ability and minimization of the classification error. It leads to the following optimization problem

$$\min_{\omega, \xi} \frac{1}{2}\|\omega\|^2 + C\sum_{i=1}^{n} \xi_i,$$

$$\text{subject to: } y_i(\omega^T \varphi(x_i) + b) \geq 1 - \xi_i, \tag{7}$$

where $\xi_i \geq 0$ is a slack variable; parameter C controls the tradeoff between the margin maximization and classification error minimization. Inequalities (7) are relaxed form of the conditions that all images of objects from Class 1 lie on the positive side of the canonical hyper-plane and all images of objects from Class 2 lie on the negative side. The solution of Equation (7) defines the optimal separating hyper-plane in H and depends on some subset of the training set only. This subset consists of supports vectors (SVs). Thus, SVM decision function is defined as follows:

$$f(x) = \omega^T \varphi(x) + b = \sum_{i \in SV_s} \alpha_i K(x_i, x) + b . \tag{8}$$

where $\{\alpha_i\}$ is a solution of dual optimization problem for Equation (7). In 2-class problem, $x$ is assigned to Class 1 if $f(x) \geq 0$ and, otherwise $x$ is assigned to Class 2.

In the nonlinearly separable cases, the SVM map the training points, nonlinearly, to a high dimensional feature space using kernel function $K(x_i, x)$, where linear separation may be possible. Some of the kernel functions of SVM are: Linear, Gaussian radial basis function (GRBF), Polynomial and Sigmoid. The performance of a SVM depends on penalty parameter(C) and the kernel parameters, which are called hyper-parameters.

## 3.2   Multi-class SVM-Based Classifier

There are two widely used methods to extend binary SVMs to multi-class problems. One of them is called the one-against-all (OAA) method and the other is called one-against-one (OAO) method. OAA method is the earliest approach for multi-class SVM, and had been used to recognize the patterns of control chart.

For $k$-class problem, it constructs $k$ binary SVMs. The $i^{th}$ SVM is trained with all the samples from the $i^{th}$ class against all the samples from the rest classes. Let the $i^{th}$ decision function that classifies class $i$ and the remaining classes be

$$D_i(x) = \omega_i^T \varphi(x) + b_i = \sum_{l \in SV_{si}} \alpha_{i,l} K(x_l, x) + b_i . \tag{10}$$

The hyper-plane $D_i(x) = 0$ forms the optimal separating hyper-plane and the support vectors belonging to class $i$ satisfy $D_i(x) = 1$ and to those belonging to the remaining classes satisfy $D_i(x) = -1$. For conventional support vector machine, given a sample $x$ to classify if

$$D_i(x) \geq 0 \tag{11}$$

is satisfied for one $i$, $x$ is classified into class $i$.

But if (11) is satisfied for plural $i$'s, or there is no $i$ that satisfies (11), $x$ is unclassifiable (Shown as the shadows in Figure 2).

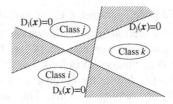

**Fig. 2.** Unclassifiable region by the OAA method

In this paper, we consider that the six patterns mentioned above can be re-described with the four types of four patterns, including normal pattern, up pattern, down pattern and cycle pattern. Since the sequences in the downtrend (D) and downshift (F) are some kind of similar to each other, as shown in figure 1(d) and (f), can we group them together as a bigger category {D, F} and match it against {C, E}?

After a test sample is decided to lie in {D, F}, we can go further to see which class it should be labeled. In this way, we have a recursive binary classification problem, which of course can be implemented by regular SVM. The structure of the classifier is same as a decision tree with each node as binary SVM. The classification procedure goes from root to the leaf guided by the binary SVM, same as traveling a decision tree.

Figure 3 shows a possible SVM decision tree for the control chart patterns data and the intuitive divisions of the 6 classes.

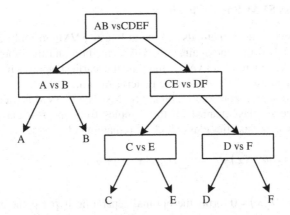

**Fig. 3.** Multi-class SVM for control charts

## 4    Training Data Set Preparation

Before the Multi-class SVM is used to work as a control chart pattern recognizer, it must be trained through enough numbers of data to get the pattern recognition ability. And after the training, the Multi-class SVM is used to monitor the abnormal patterns and further associate the abnormal patterns with the specific process factors. In the process of training Multi-class SVM, we had better use the pattern data (natural and unnatural) collected from a real process to train and test the Multi-class SVM, but it is difficult and costly for us to collect the large amount of various pattern data. To economically generate the data to train and test the Multi-class SVM used for control charts pattern recognition, this paper use Monte-Carlo simulation method to produce the required data sets. In this research, the six different pattern data sets for training and testing are simulated using Equation (1).

(1) For normal pattern,

$$y(t) = x(t),$$

where $x(t) \sim N(0,1)$.

(2) For cyclic pattern,

$$y(t) = a \times sin(2\pi t / \Omega) + x(t).$$

where, $a$ is cycle amplitude in terms of σ, $\sigma \le a \le 3\sigma$, in this paper. Random value $\Omega$ is the cycle period, and $\Omega = 8$ in this paper.

(3) For up-trend and downtrend trend pattern,

$$y(t) = \pm \rho \times t + x(t).$$

where $\rho$ is trend slope in terms of σ, 0.1σ≤k ≤0.3σ, σ is an random value in this paper.

(4) For upward and downward shift pattern,

$$y(t) = \pm(t - t_0) \times s + x(t).$$

where $(t - t_0) = 1$, when $t > t_0$, and $(t - t_0) = 0$, when $t < t_0$, and $t_0$ is a parameter to determine the shift position ($t_0 = 24, 25$, or 26, random integer value in this paper). And s is the shift magnitude in terms of σ, and σ≤s≤3σ, random value.

In this study, all together 12000 (2000 × 6) pattern data were used for training and testing the designed decision tree, and there into 6000 (1000 × 6) pattern data were used to train these multi-class SVM randomly, and the other 1000 data are used to test the decision tree.

## 5  Experimental Results

The radial basis function (RBF) was selected as kernel function, and SVM algorithm was offered from "SSVM toolbox", which can be downloaded from "http://dmlab1. csie.ntust.edu.tw/downloads". In the process of simulation experiment through MATLAB software, we could find that the parameter C of SVM and the parameter $\delta$ of RBF kernel function had a great impact on the simulation outcomes. When the experiment was been doing, the regular factor C , and RBF parameter $\delta$ were determined by 5-fold cross validation (CV) method.

Take out 200 data from every basic class randomly and make up the training sample sets for the five SVMs. Train BP, OAA, and decision tree of SVM-based model with training sets until all the data were classified correctly.

The rest of the data (each class included 1000 samples) were used to set up the testing sample sets. Test BP, OAA and decision tree of SVM based model respectively, and the recognition results are showed in Table 1.

**Table 1.** The recognition results of different model

| Pattern | BP | | OAA | | Decision tree | |
|---|---|---|---|---|---|---|
| | Mis | Rate | Mis | Rate | Mis | Rate |
| Normal | 149 | 85.1% | 81 | 91.9% | 68 | 93.2% |
| Cycle | 158 | 84.2% | 86 | 91.4% | 53 | 94.7% |
| Up-trend | 113 | 88.7% | 75 | 92.5% | 56 | 94.4% |
| Down-trend | 78 | 92.2% | 47 | 95.3% | 35 | 96.5% |
| Up-shift | 136 | 86.4% | 49 | 95.1% | 47 | 95.3% |
| Down-shift | 123 | 87.8% | 94 | 90.6% | 54 | 94.6% |

P.S. MIS. here means misclassification; Rate is recognition rate.

From the Table 1, the average recognition ratio of BP is the lowest, only about 87.4%. The recognition rate of OAA is higher comparison with that of BP-based model, because it avoids some disadvantages, such as over-training and weak normalization capability, which BP has. Without the influence of unclassifiable region, the decision tree of multi-class SVM can get the highest recognition accuracy (the average recognition rate is 94.8%).

## 6    Conclusions

Control chart patterns are important SPC tools for determining whether a process is run in its intended mode or in the presence of unnatural patterns. In this paper we propose a decision tree of multi-class SVM for recognition of the CCP. Simulation was used to evaluate the performance of the proposed model. The numerical results show that the problem of false recognition has been addressed effectively by the proposed method. From the testing results, the proposed decision tree is more effective in detecting unnatural patterns on control charts than the traditional OAA approaches.

## References

1. Yang, J.H., Yang, M.S.: A control chart pattern recognition system using a statistical correlation coefficient method. Computers and Industrial Engineering 48(2), 205–221 (2005)
2. Shewhart, W.A.: Economic control quality manufactured products. Van Nostrand, New York (1931)
3. Grant, E.L., Leavenworth, R.S.: Statistical Qualify Control, 6th edn. McGraw-Hill Book Company, New York (1988)

4. Shewhart, M.: Interpreting statistical process control (SPC) charts using machine learning and expert system techniques. In: Aerospace and Electronics Conference, Proceedings of the IEEE 1992 National, vol. 3, pp. 1001–1006 (1992)
5. Swift, J.A., Mize, J.H.: Out-of-control pattern recognition and analysis for quality control charts using lisp-based systems. Computers & Industrial Engineering 28, 81–91 (1995)
6. Guh, R.S., Hsieh, Y.C.: A neural network based model for abnormal pattern recognition of control charts. Computers & Industrial Engineering 36, 97–108 (1999)
7. Perry, M.B., Spoerre, J.K., Velasco, T.: Control chart pattern recognition using back propagation artificial neural networks. International Journal of Production Research 39, 3399–3418 (2001)
8. Jiang, P.Y., Liu, D.Y., Zeng, Z.J.: Recognizing control chart patterns with neural network and numerical fitting. Journal of Intelligent Manufacturing 20(6), 625–635 (2009)

# Robust Adaptive Neural Network Control for Nonlinear Time-Delay Systems

Geng Ji

School of Mathematics and Information Engineering,
Taizhou University, Linhai, China
gjiwust@yahoo.com.cn

**Abstract.** This paper presents a robust adaptive neural network control scheme for a class of strict-feedback nonlinear time-delay systems, with both unknown nonlinearities and uncertain disturbances. Unknown smooth function vectors and unknown time-delay functions are approximated by two neural networks, respectively, such that the requirement on the unknown time-delay functions is relaxed. The uncertain disturbances are solved by nonlinear damping technique. In addition, by theoretical analysis, closed loop signals is proved to be semi-globally uniformly ultimately bounded, and the output of the system converges to a small neighborhood of the desired trajectory. Finally, the feasibility is investigated by a simulation example.

**Keywords:** Nonlinear time-delay systems, robust adaptive control, neural control, backstepping.

## 1 Introduction

In recent years, by using the idea of backstepping design [1], adaptive neural network control (ANNC) has received considerable attention and become an active research area [2-4]. In adaptive neural control design, neural networks are mostly used as approximators for unknown nonlinear functions in system models. In [3], a robust adaptive neural control scheme was presented for a class of uncertain chaotic systems in the disturbed strict-feedback form, with both unknown nonlinearities and uncertain disturbances. Two different backstepping neural network control approaches were presented for a class of affine nonlinear systems in [4]. The proposed controller made the neural network approximation computationally feasible.

Stabilization of nonlinear systems with time delay is receiving much attention[5-7]. In [6], an adaptive output feedback neural network tracking controller was designed for a class of unknown output feedback nonlinear time-delay systems. The time-delay exists in output variable. In [7], an adaptive neural control design approach was proposed for a class of strict-feedback nonlinear systems with unknown time delays. The uncertainties of unknown time delays were compensated for through proper Lyapunov-Krasovskii functionals. The unknown time-delay functions were not approximated by neural networks.

In this paper, we extend the adaptive neural network control approaches to a class of nonlinear time-delay systems. The main contributions of this paper lie in: neural

G. Lee (Ed.): Advances in Intelligent Systems, AISC 138, pp. 43–51.

networks are used to approximate the completely unknown time delay functions. The uncertain disturbances are solved by nonlinear damping technique. Simulation study is conducted to verify the effectiveness of the approach.

## 2    Problem Formulation

Consider a class of single-input-single-output (SISO) nonlinear time-delay systems

$$\dot{x}_i(t) = x_{i+1}(t) + f_i(\bar{x}_i(t)) + h_i(\bar{x}_i(t-\tau)) + \phi_i(\bar{x}_i(t))d_i$$
$$(1 \le i \le n-1)$$
$$\dot{x}_n(t) = u(t) + f_n(\bar{x}_n(t)) + h_n(\bar{x}_n(t-\tau)) + \phi_n(\bar{x}_n(t))d_n \tag{1}$$
$$y = x_1$$

where $\bar{x}_i = [x_1, x_2, \cdots, x_i]^T \in R^i, i = 1, 2 \cdots, n$ , $u \in R$ , $y \in R$ are state variables, system input and output, respectively. $f_i(\cdot)$ , $h_i(\cdot)$ $(i = 1, 2, \cdots, n)$ are unknown smooth functions, and $\tau$ is known time delay constant of the states. $\phi_i(\cdot)$ is an unknown smooth function of $(x_1, x_2, \cdots, x_i)$ , and $d_i$ is the uniformly bounded disturbance input with its bound not necessarily known. Here, the disturbance input $d_i$ is allowed to take any form of uniformly bounded terms, including un-modeled states and external disturbances. We have the following assumptions for the system (1):

*Assumption 1:* $|\phi_i(\bar{x}_i)| \le \varphi_i(\bar{x}_i)$ , where $\varphi_i(\bar{x}_i)$ are known smooth nonlinear functions.

*Assumption 2:* $|d_i| \le d_i^*$ , with $d_i^*$ is not necessarily known.

In the following, we let $\| \cdot \|$ denote that 2-norm, $\lambda_{max}(B)$ and $\lambda_{min}(B)$ denote the largest and smallest eigenvalues of a square matrix $B$ , respectively.

## 3    Robust Adaptive Neural Network Control Design

**Step 1:** Defined $z_1 = x_1 - y_d$ . Its derivative is

$$\dot{z}_1(t) = x_2(t) + f_1(x_1(t)) + h_1(x_1(t-\tau))$$
$$+ \phi_1(x_1(t))d_1 - \dot{y}_d(t) \tag{2}$$

Two neural networks are adopted to approximate the unknown smooth functions $f_1(x_1(t))$ and $h_1(x_1(t-\tau))$ , i.e.

$$f_1(x_1) = W_{11}^{*T}S_{11}(Z_{11}) + \varepsilon_{11},$$
$$h_1(x_1(t-\tau)) = W_{12}^{*T}S_{12}(Z_{12}) + \varepsilon_{12} \tag{3}$$

where $Z_{11} = [x_1]^T \subset R^1$ , $Z_{12} = [x_1(t-\tau)]^T \subset R^1$ . $W_{11}^*$ and $W_{12}^*$ are the optimal weight vectors of $f_1(x_1)$ and $h_1(x_1(t-\tau))$ , respectively. The neural reconstruction error

$e_1 = \varepsilon_{11} + \varepsilon_{12}$ is bounded, i.e., there exists a constant $\varepsilon_1^* > 0$ such that $|e_1| < \varepsilon_1^*$. Throughout the paper, we shall define the reconstruction error as $e_i = \varepsilon_{i1} + \varepsilon_{i2}$, where $i = 1, 2, \cdots, n$. Like in the case of $e_1$, $e_i$ is bounded, i.e., $|e_i| < \varepsilon_i^*$. Since $W_{11}^*$ and $W_{12}^*$ are unknown, let $\hat{W}_{11}$ and $\hat{W}_{12}$ be the estimate of $W_{11}^*$ and $W_{12}^*$, respectively. The other uncertainty is the disturbance term $\phi_1(x_1) d_1$, where the $d_1$ is bounded but unavailable through measurement. This term cannot be approximated by neural networks. To cope with this uncertainty, a nonlinear damping term $-p_1 \varphi_1^2(x_1) z_1$ $(p_1 > 0)$ is introduced to counteract the disturbance term.

Defining error variable $z_2 = x_2 - \alpha_1$ and choosing the virtual control

$$\alpha_1 = -c_1 z_1 - \hat{W}_{11}^T S_{11}(Z_{11}) - \hat{W}_{12}^T S_{12}(Z_{12}) - p_1 \varphi_1^2(x_1) z_1 + \dot{y}_d(t) \tag{4}$$

$\dot{z}_1$ can be obtained as

$$\begin{aligned}
\dot{z}_1 = z_2 - c_1 z_1 - \tilde{W}_{11}^T S_{11}(Z_{11}) - \tilde{W}_{12}^T S_{12}(Z_{12}) \\
- p_1 \varphi_1^2(x_1) z_1 + \phi_1(x_1) d_1 + e_1
\end{aligned} \tag{5}$$

where $\tilde{W}_{11} = \hat{W}_{11} - W_{11}^*$, $\tilde{W}_{12} = \hat{W}_{12} - W_{12}^*$. Through out this paper, we shall define $(\tilde{\cdot}) = (\hat{\cdot}) - (\cdot)^*$.

Consider the following Lyapunov function candidate:

$$V_1 = \frac{1}{2} z_1^2 + \frac{1}{2} \tilde{W}_{11}^T \Gamma_{11}^{-1} \tilde{W}_{11} + \frac{1}{2} \tilde{W}_{12}^T \Gamma_{12}^{-1} \tilde{W}_{12} \tag{6}$$

where $\Gamma_{11} = \Gamma_{11}^T > 0$, $\Gamma_{12} = \Gamma_{12}^T > 0$ are adaptation gain matrices.

The derivative of $V_1$ is

$$\begin{aligned}
\dot{V}_1 = z_1 z_2 - c_1 z_1^2 + z_1 e_1 + \tilde{W}_{11}^T \Gamma_{11}^{-1} \left[ \dot{\hat{W}}_{11} - \Gamma_{11} S_{11}(Z_{11}) z_1 \right] \\
+ \tilde{W}_{12}^T \Gamma_{12}^{-1} \left[ \dot{\hat{W}}_{12} - \Gamma_{12} S_{12}(Z_{12}) z_1 \right] \\
- p_1 \varphi_1^2(x_1) z_1^2 + \phi_1(x_1) d_1 z_1
\end{aligned} \tag{7}$$

Consider the following adaptation laws:

$$\dot{\hat{W}}_{11} = \dot{\tilde{W}}_{11} = \Gamma_{11} \left[ S_{11}(Z_{11}) z_1 - \sigma_{11} \hat{W}_{11} \right]$$

$$\dot{\hat{W}}_{12} = \dot{\tilde{W}}_{12} = \Gamma_{12} \left[ S_{12}(Z_{12}) z_1 - \sigma_{12} \hat{W}_{12} \right] \tag{8}$$

where $\sigma_{11} > 0$, $\sigma_{12} > 0$ are small constants. Formula (8) is so-called $\sigma$-modification, introduced to improve the robustness in the presence of the NN approximation error $e_1$, and avoid the weight parameters to drift to very large values.

Let $c_1 = c_{10} + c_{11}$, with $c_{10}$ and $c_{11} > 0$. Then, (7) become

$$\dot{V}_1 = z_1 z_2 - c_{10} z_1^2 - c_{11} z_1^2 + z_1 e_1 - \sigma_{11} \tilde{W}_{11}^T \hat{W}_{11}$$
$$- \sigma_{12} \tilde{W}_{12}^T \hat{W}_{12} - p_1 \varphi_1^2(x_1) z_1^2 + \phi_1(x_1) d_1 z_1 \qquad (9)$$

By completion of squares, we have

$$-\sigma_{11} \tilde{W}_{11}^T \hat{W}_{11} = -\sigma_{11} \tilde{W}_{11}^T \left( \tilde{W}_{11} + W_{11}^* \right)$$
$$\leq -\frac{\sigma_{11} \left\| \tilde{W}_{11} \right\|^2}{2} + \frac{\sigma_{11} \left\| W_{11}^* \right\|^2}{2} \qquad (10)$$

$$-\sigma_{12} \tilde{W}_{12}^T \hat{W}_{12} = -\sigma_{12} \tilde{W}_{12}^T \left( \tilde{W}_{12} + W_{12}^* \right)$$
$$\leq -\frac{\sigma_{12} \left\| \tilde{W}_{12} \right\|^2}{2} + \frac{\sigma_{12} \left\| W_{12}^* \right\|^2}{2} \qquad (11)$$

$$-c_{11} z_1^2 + z_1 e_1 \leq -c_{11} z_1^2 + z_1 |e_1| \leq \frac{e_1^2}{4c_{11}} \leq \frac{\varepsilon_1^{*2}}{4c_{11}} \qquad (12)$$

$$-p_1 \varphi_1^2(x_1) z_1^2 + \phi_1(x_1) d_1 z_1 \leq -p_1 \varphi_1^2(x_1) z_1^2 + \phi_1(x_1) d_1 z_1$$
$$\leq \frac{d_1^2}{4p_1} \leq \frac{d_1^{*2}}{4p_1} \qquad (13)$$

Substituting (10) (11) (12) (13) into (9), we have the following inequality:

$$\dot{V}_1 \leq z_1 z_2 - c_{10} z_1^2 - \frac{\sigma_{11} \left\| \tilde{W}_{11} \right\|^2}{2} - \frac{\sigma_{12} \left\| \tilde{W}_{12} \right\|^2}{2}$$
$$+ \frac{\sigma_{11} \left\| W_{11}^* \right\|^2}{2} + \frac{\sigma_{12} \left\| W_{12}^* \right\|^2}{2} + \frac{\varepsilon_1^{*2}}{4c_{11}} + \frac{d_1^{*2}}{4p_1} \qquad (14)$$

where the coupling term $z_1 z_2$ will be canceled in the next step.

Step $i$ ($2 \leq i \leq n-1$): The derivative of $z_i = x_i - \alpha_{i-1}$ is

$$\dot{z}_i = x_{i+1}(t) + f_i \left( \overline{x}_i(t) \right) + h_i \left( \overline{x}_i(t-\tau) \right) + \phi_i \left( \overline{x}_i(t) \right) d_i - \dot{\alpha}_{i-1}$$

Similarly, choose the virtual control

$$\alpha_i = -z_{i-1} - c_i z_i - \hat{W}_{i1}^T S_{i1}(Z_{i1}) - \hat{W}_{i2}^T S_{i2}(Z_{i2})$$
$$- p_i \varphi_i^2(\overline{x}_i) z_i + \dot{\alpha}_{i-1} \qquad (15)$$

where $c_i > 0$, $Z_{i1} = [x_1, x_2, \cdots, x_i]^T \subset R^i$,

$$Z_{i2} = [x_1(t-\tau), x_2(t-\tau), \cdots, x_i(t-\tau)]^T \subset R^i.$$

Then, we have

$$\dot{z}_i = z_{i+1} - z_{i-1} - c_i z_i - \tilde{W}_{i1}^T S_{i1}(Z_{i1}) - \tilde{W}_{i2}^T S_{i2}(Z_{i2})$$
$$- p_i \varphi_i^2(\overline{x}_i) z_i + \phi_i \left( \overline{x}_i(t) \right) d_i + e_i \qquad (16)$$

where $z_{i+1} = x_{i+1} - \alpha_i$.

Consider the Lyapunov function candidate

$$V_i = V_{i-1} + \frac{1}{2}z_i^2 + \frac{1}{2}\tilde{W}_{i1}^T \Gamma_{i1}^{-1} \tilde{W}_{i1} + \frac{1}{2}\tilde{W}_{i2}^T \Gamma_{i2}^{-1} \tilde{W}_{i2} \qquad (17)$$

Consider the following adaptation laws:

$$\dot{\hat{W}}_{i1} = \dot{\tilde{W}}_{i1} = \Gamma_{i1}\left[ S_{i1}(Z_{i1})z_i - \sigma_{i1}\hat{W}_{i1} \right]$$

$$\dot{\hat{W}}_{i2} = \dot{\tilde{W}}_{i2} = \Gamma_{i2}\left[ S_{i2}(Z_{i2})z_i - \sigma_{i2}\hat{W}_{i2} \right] \qquad (18)$$

where $\sigma_{i1} > 0$, $\sigma_{i2} > 0$ are small constants. Let $c_i = c_{i0} + c_{i1}$, where $c_{i0}$ and $c_{i1} > 0$. By using (14), (16), and (18), and with some completion of squares and straightforward derivation similar to those employed in the former steps, the derivative of $V_i$ becomes

$$\dot{V}_i < z_i z_{i+1} - \sum_{k=1}^{i} c_{k0} z_k^2 - \sum_{k=1}^{i} \frac{\sigma_{k1}\left\|\tilde{W}_{k1}\right\|^2}{2} - \sum_{k=1}^{i} \frac{\sigma_{k2}\left\|\tilde{W}_{k2}\right\|^2}{2}$$
$$+ \sum_{k=1}^{i} \frac{\sigma_{k1}\left\|W_{k1}^*\right\|^2}{2} + \sum_{k=1}^{i} \frac{\sigma_{k2}\left\|W_{k2}^*\right\|^2}{2} + \sum_{k=1}^{i} \frac{\varepsilon_k^{*2}}{4c_{k1}} + \sum_{k=1}^{i} \frac{d_k^{*2}}{4p_k} \qquad (19)$$

Step $n$: This is the final step. The derivative of $z_n = x_n - \alpha_{n-1}$ is

$$\dot{z}_n = u + f_n\left(\bar{x}_n(t)\right) + h_n\left(\bar{x}_n(t-\tau)\right) + \phi_n\left(\bar{x}_n(t)\right)d_n - \dot{\alpha}_{n-1}$$

Similarly, choosing the practical control law as

$$u = -z_{n-1} - c_n z_n - \hat{W}_{n1}^T S_{n1}(Z_{n1}) - \hat{W}_{n2}^T S_{n2}(Z_{n2})$$
$$- p_n \varphi_n^2(\bar{x}_n)z_n + \dot{\alpha}_{n-1} \qquad (20)$$

where $c_n > 0$, $Z_{n1} = [x_1, x_2, \cdots, x_n]^T \subset R^n$,

$$Z_{n2} = [x_1(t-\tau), x_2(t-\tau), \cdots, x_n(t-\tau)]^T \subset R^n.$$

We have

$$\dot{z}_n = -z_{n-1} - c_n z_n - \tilde{W}_{n1}^T S_{n1}(Z_{n1}) - \tilde{W}_{n2}^T S_{n2}(Z_{n2})$$
$$- p_n \varphi_n^2(\bar{x}_n)z_n + \phi_n\left(\bar{x}_n(t)\right)d_n + e_n \qquad (21)$$

Consider the overall Lyapunov function candidate

$$V_n = V_{n-1} + \frac{1}{2}z_n^2 + \frac{1}{2}\tilde{W}_{n1}^T \Gamma_{n1}^{-1} \tilde{W}_{n1} + \frac{1}{2}\tilde{W}_{n2}^T \Gamma_{n2}^{-1} \tilde{W}_{n2} \qquad (22)$$

Consider the following adaptation laws:

$$\dot{\hat{W}}_{n1} = \dot{\tilde{W}}_{n1} = \Gamma_{n1}\left[ S_{n1}(Z_{n1})z_n - \sigma_{n1}\hat{W}_{n1} \right]$$

$$\dot{\hat{W}}_{n2} = \dot{\tilde{W}}_{n2} = \Gamma_{n2}\left[ S_{n2}(Z_{n2})z_n - \sigma_{n2}\hat{W}_{n2} \right] \qquad (23)$$

where $\sigma_{n1} > 0$, $\sigma_{n2} > 0$ are small constants. Let $c_n = c_{n0} + c_{n1}$, where $c_{n0}$ and $c_{n1} > 0$. By using (19), (21), and (23), and with some completion of squares and straightforward derivation similar to those employed in the former steps, the derivative of $\dot{V}_n$ becomes

$$\dot{V}_n < -\sum_{k=1}^{n} c_{k0} z_k^2 - \sum_{k=1}^{n} \frac{\sigma_{k1} \|\tilde{W}_{k1}\|^2}{2} - \sum_{k=1}^{n} \frac{\sigma_{k2} \|\tilde{W}_{k2}\|^2}{2}$$

$$+ \sum_{k=1}^{n} \frac{\sigma_{k1} \|W_{k1}^*\|^2}{2} + \sum_{k=1}^{n} \frac{\sigma_{k2} \|W_{k2}^*\|^2}{2} + \sum_{k=1}^{n} \frac{\varepsilon_k^{*2}}{4c_{k1}} + \sum_{k=1}^{n} \frac{d_k^{*2}}{4p_k} \quad (24)$$

Let $\delta \triangleq \sum_{k=1}^{n} \frac{\sigma_{k1} \|W_{k1}^*\|^2}{2} + \sum_{k=1}^{n} \frac{\sigma_{k2} \|W_{k2}^*\|^2}{2} + \sum_{k=1}^{n} \frac{\varepsilon_k^{*2}}{4c_{k1}} + \sum_{k=1}^{n} \frac{d_k^{*2}}{4p_k}$. If we choose $c_{k0}$ such that

$c_{k0} > \dfrac{\gamma}{2}, k = 1, 2, \cdots, n$, where $\gamma$ is a positive constant, and choose $\sigma_{k1}$, $\sigma_{k2}$, $\Gamma_{k1}$

and $\Gamma_{k2}$ such that $\sigma_{k1} \geq \gamma\lambda_{\max}\{\Gamma_{k1}^{-1}\}$, $\sigma_{k2} \geq \gamma\lambda_{\max}\{\Gamma_{k2}^{-1}\}$, $k = 1, 2, \cdots, n$, then from (24) we have the following inequality:

$$\dot{V}_n < -\sum_{k=1}^{n} c_{k0} z_k^2 - \sum_{k=1}^{n} \frac{\sigma_{k1} \|\tilde{W}_{k1}\|^2}{2} - \sum_{k=1}^{n} \frac{\sigma_{k2} \|\tilde{W}_{k2}\|^2}{2} + \delta$$

$$< -\sum_{k=1}^{n} \frac{\gamma}{2} z_k^2 - \sum_{k=1}^{n} \frac{\gamma \tilde{W}_{k1}^T \Gamma_{k1}^{-1} \tilde{W}_{k1}}{2} - \sum_{k=1}^{n} \frac{\gamma \tilde{W}_{k2}^T \Gamma_{k2}^{-1} \tilde{W}_{k2}}{2} + \delta$$

$$= -\gamma V_n + \delta \quad (25)$$

The following theorem shows the stability and control performance of the closed-loop adaptive system.

*Theorem 1:* Consider the closed-loop system consisting of the plant (1), the controller (20), and the NN weight updating laws (8) (18) and (23). Assume that there exists sufficiently large compact sets $\Omega_i \in R^i$, $i = 1, 2, \cdots, n$ such that $Z_{i1} \in \Omega_i$, $Z_{i2} \in \Omega_i$ for all $t \geq 0$. Then, for bounded initial conditions, we have the following:

1) All signals in the closed-loop system remain semiglobally uniformly ultimately bounded;
2) The output tracking error $y(t) - y_d(t)$ converges to a small neighborhood around zero by appropriately choosing design parameters.

*Proof:* 1) From (25), using the boundedness theorem (e.g., [8]), we have that all $z_i$, $\hat{W}_{i1}$ and $\hat{W}_{i2}$ are uniformly ultimately bounded. Since $z_1 = x_1 - y_d$ and $y_d$ are bounded, we have that $x_1$ is bounded. From $z_i = x_i - \alpha_{i-1}, i = 1, 2, \cdots, n$, and the definitions of virtual controls (4), (15) we have that $x_i, i = 2, 3, \cdots, n$ remain bounded. Using (20), we conclude that control $u$ is also bounded. Thus, all the signals in the closed-loop system remain bounded.

2) Let $\rho = \dfrac{\delta}{\gamma} > 0$, then (25) satisfies

$$0 \leq V_n(t) \leq \rho + (V_n(0) - \rho)\exp(-\gamma t) \quad (26)$$

From (26), we have

$$\sum_{i=1}^{n}\frac{1}{2}z_k^2 < \rho + \left(V_n(0)-\rho\right)\exp(-\gamma t) < \rho + V_n(0)\exp(-\gamma t) \qquad (27)$$

That is

$$\sum_{i=1}^{n}z_k^2 < 2\rho + 2V_n(0)\exp(-\gamma t) \qquad (28)$$

which implies that given $\mu > \sqrt{2\rho}$, there exists $T$ such that for all $t \geq T$, the tracking error satisfies

$$|z_1(t)| = |x_1(t) - y_d(t)| < \mu \qquad (29)$$

where $\mu$ is the size of a small residual set which depends on the NN approximation error $e_i$ and controller parameters $c_i$, $\sigma_{i1}$, $\sigma_{i2}$, $\Gamma_{i1}$ and $\Gamma_{i2}$. It is easily seen that the increase in the control gain $c_i$, adaptive gain $\Gamma_{i1}$, $\Gamma_{i2}$ and NN node number $l_j$ will result in a better tracking performance.

*Remark 1:* Compared with the works in [2][3] and [4], the proposed adaptive neural network controller in this paper can cope with nonlinear time-delay systems.

*Remark 2:* Compare with reference [6], the system model presented in this paper is more general. The time-delay exists in state variable other than output variable. In addition, the practical systems may contain many types of uncertainties, such as unknown nonlinearities, exogenous disturbances, unmodeled dynamics, etc. Thus, it is significant and necessary to design robust controller for nonlinear systems with uncertain disturbances.

*Remark 3:* Compared with the work in [7], the unknown time-delay functions in this paper are approximated by neural networks. However, in [7], by constructing proper Lyapunov-Krasovskii functionals, the uncertainties of unknown time delays are compensated. So, the requirement on the unknown time delay functions is relaxed in our paper.

## 4    Simulation

Consider the following strict-feedback system:

$$\dot{x}_1(t) = x_2(t) + 0.5x_1(t) + 0.5x_1^{2}(t)\cos(1.5t)$$
$$\dot{x}_2(t) = u(t) + x_1(t)\cdot x_2(t) + \sin\left(x_1(t-\tau)\cdot x_2(t-\tau)\right) \qquad (30)$$
$$y = x_1$$

where $\tau = 5$. $0.5x_1^{2}(t)\cos(1.5t)$ is the disturbance term with $d_1 = \cos(1.5t)$ and $\phi_1 = 0.5x_1^{2}$. The initial condition $\left[x_1(0), x_2(0)\right]^T = [0.5, 0]^T$ and the desired reference signal of system is $y_d(t) = \cos(t)$. For clarity of presentation, we assume that $d_2 = 0$ and $\varphi_1 = x_1^{2}$.

The adaptive neural network controller is chosen according to (26) as follows:

$$u = -z_1 - c_2 z_2 - \hat{W}_{21}^T S_{21}(Z_{21}) - \hat{W}_{22}^T S_{22}(Z_{22}) + \dot{\alpha}_1 \qquad (31)$$

where $z_1 = x_1 - y_d$, $z_2 = x_2 - \alpha_1$,

$Z_{11} = [x_1]^T$, $Z_{21} = [x_1, x_2]^T$, $Z_{22} = [x_1(t-\tau), x_2(t-\tau)]^T$,

$\alpha_1 = -c_1 z_1 - \hat{W}_{11}^T S_{11}(Z_{11}) - p_1 \varphi_1^2(x_1) z_1 + \dot{y}_d(t)$, and neural network weights $\hat{W}_{11}$, $\hat{W}_{21}$ and $\hat{W}_{22}$ are updated by (8) and (18) correspondingly.

Neural networks $\hat{W}_{11}^T S_{11}(Z_{11})$ contains 13 nodes (i.e., $l_1 = 13$), with centers $\mu_l (l = 1, 2, \cdots, l_1)$ evenly spaced in $[-6, 6]$, and widths $\eta_l = 1(l = 1, 2, \cdots, l_1)$. Neural network $\hat{W}_{21}^T S_{21}(Z_{21})$ and $\hat{W}_{22}^T S_{22}(Z_{22})$ contains 169 nodes (i.e., $l_2 = 169$), with centers $\mu_l (l = 1, 2, \cdots, l_2)$ evenly spaced in $[-6, 6] \times [-6, 6]$, and widths $\eta_l = 1(l = 1, 2, \cdots, l_2)$. The design parameters of above controller are $c_1 = 4$, $c_2 = 4$, $\Gamma_{11} = \Gamma_{21} = \Gamma_{22} = diag\{2.0\}$, $\sigma_{11} = \sigma_{21} = \sigma_{22} = 0.2$. The initial weights $\hat{W}_{11} = 0.5$, $\hat{W}_{21} = 0$, $\hat{W}_{22} = 0$.

Figure 1 shows the simulation result of applying controller (31) to system (30) for tracking desired signal $y_d$. From Figure 1, we can see that good tracking performance is obtained.

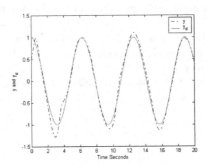

**Fig. 1.** Output tracking performance

## 5    Conclusion

In this paper, an adaptive neural network control approach is proposed for a class of strict-feedback nonlinear time-delay systems, with both unknown nonlinearities and uncertain disturbances. The unknown time delay functions are approximated by neural networks, such that the requirement on unknown time delay functions is relaxed. The uncertain disturbances are solved by nonlinear damping technique. Finally, a numerical simulation is given to show the effectiveness of the approach.

# References

1. Krstic, M., Kanellakopoulos, I., Kokotovic, P.: Nonlinear and Adaptive Control Design. Wiley, New York (1995)
2. Ge, S.S., Wang, C.: Direct adaptive NN control of a class of nonlinear systems. IEEE Transactions on Neural Networks 13, 214–221 (2002)
3. Ge, S.S., Wang, C.: Uncertain chaotic system control via adaptive neural design. International Journal of Bifurcation and Chaos 12, 1097–1109 (2002)
4. Li, Y.H., Qiang, S., Zhuang, X.Y., Kaynak, O.: Robust and adaptive backstepping control for nonlinear systems using RBF neural networks. IEEE Transactions on Neural Networks 15, 693–701 (2004)
5. Zhang, T.P., Zhou, C.Y., Zhu, Q.: Adaptive variable structure control of MIMO nonlinear systems with time-varying delays and unknown dead-zones. International Journal Automation and Computing 6, 124–136 (2009)
6. Chen, W.S., Li, J.M.: Adaptive output feedback control for nonlinear time-delay systems using neural network. Journal of Control Theory and Applications 4, 313–320 (2006)
7. Ge, S.S., Hong, F., Lee, T.H.: Adaptive neural network control of nonlinear systems with unknown time delays. IEEE Transactions on Automatic Control 48, 2004–2010 (2003)
8. Qu, Z.: Robust Control of Nonlinear Uncertain Systems. Wiley, New York (1998)

# The Analysis of Thermal Field and Thermal Deformation of a Water-Cooling Radiator by Finite Element Simulation

Yang Lianfa, Wang Qin, and Zhang Zhen

Faculty of mechanical & Electrical Engineering,
Guilin University of Electronic Technology, GUET,
Guilin, P.R. China
y-lianfa@163.com, wangqin252686193@126.com,
zhangzhenzdh@yahoo.com.cn

**Abstract.** A water-cooling radiator is the effective heat dissipation device to ensure the high-power electrical equipment work properly and the heat transfer process and thermal deformation of the water-cooling radiator become hot research areas currently. Based on a water-cooling radiator of a company, the simulation model for its single water-cooling flow channel is established and the influence of the flow velocity of the inlet fluid and thermal convection of air on the outlet temperature of fluid is analyzed by the commercial software ANSYS. In addition, the thermal deformation of the water-cooling flow channel is discussed. The simulation results show that the outlet temperature declines progressively as the inlet velocity increases, the outlet temperature goes downlinearly with the increase of coefficient of air convection. And the thermal deformation of the brass water-cooling radiator is obviously non-uniform and becomes warp inevitably.

**Keywords:** Water-cooling radiator, Heat transfer, Thermal deformation, Finite element simulation.

## 1 Introduction

Because of its widespread application and severe operating environment e.g. military environment, an excellent heat dissipation and cooling performance is required for high-power electrical equipments. The forced water cooling, of which the dissipation efficiency is about 20 times that of the natural heat, is a good alternative for radiator design [1]. The relationships among the flow field, thermal field and thermal deformation within a water-cooling radiator are too complicated to be determined exclusively via experimental approach; nevertheless, the finite element simulation provides a feasible scheme to solve the problem. Some researches have been performed on the heat exchanging process and optimize the structure of radiators via finite element simulation. Partankar et al have initially established the theory foundation for the finite element simulation of a radiator [2]. WU Shuquan et al have carried out finite element simulation on a forced air cooling process of an electronic supply and to verify the simulation results by experiments [3]. LIU Yanping has conducted finite element

G. Lee (Ed.): Advances in Intelligent Systems, AISC 138, pp. 53–59.
springerlink.com     © Springer-Verlag Berlin Heidelberg 2012

simulation on the water-cooling radiator of power electronics equipments, aiming to investigate the influence of the ambient temperature, the fluid temperature, the inlet velocity and the radiator structure on the thermal performance of the radiator [4]. GUO Chongzhi has investigated the thermal stress distribution on the sheet, the tube and the shell joints of a fixed tube-sheet exchanger via finite element simulation [5].

This paper investigates the thermal field and thermal deformation of a water-cooling radiator via finite element simulation. The purpose of the present work is to gain some worthy clue for the thermal design and the structural optimization of such radiators.

## 2    The Water-Cooling Radiator and Its Simulation Model

### 2.1    The Water-Cooling Radiator

The main body of the brass water-cooling radiator is comprised of 16 parallel and equidistant water-cooling flow channels. The flow channel is 520 mm in length and 1 mm in wall thickness, as shown in Fig. 1. The main part of the flow channel is at the middle with 40 mm×40 mm square cross section and 480 mm in length, of which the heat from a high-power electrical equipment is loaded on the superface. At the two ends of the square main part are two hollow cylinder noses with diameter of 20 mm and is connected with the inlet and outlet fluid pipes, respectively (not illustrated in the Fig.1). The running water inside the flow channel works as cooling medium and carries the heat away from the high-power electrical equipment and hence guarantees it work normally.

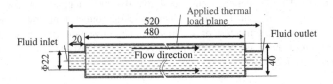

**Fig. 1.** Schematic diagram of a water-cooling flow channel

### 2.2    The Simulation Model for Flow Channel

The commercial software ANSYS finds wide applications, e.g. structures, electromagnetism thermofluids and thermotics. Besides, it provides flexible multi-physics coupling approach and interact function with popular computer-aided design (CAD) systems [6].

Since all the water-cooling flow channels of a water-cooling radiator are identical in structure and dimensions, and even approximate in heat exchanging way. Simulation model of a water-cooling flow channel of is established to simulate the thermal field and thermal deformation of the water-cooling radiator. The water-cooling flow channel is freely meshed by FLOTRAN 142 into 10453 nodes and 55199 elements, as shown in Fig.2. Some assumptions are made to simplify the simulations model as follows: (1) the fluid is incompressible and its flow is laminar and steady; (2) the physical parameters of the model, as listed in Table 1, are constants; (3) the interior wall surface of the flow channel is extremely smooth and the thermal resistance and slide velocity are ignored; (4) the thermal radiation instead of thermal convection and thermal conduction is ignored.

**Fig. 2.** Simulation model of a water-cooling flow channel

The Reynolds number $Re = \rho v Dh/\mu$ is used to determine the state of the fluid: in laminar state when $Re < 2300$ and in turbulence state when $Re > 2300$, where $\rho$ is the medium density (kg/m³), $v$ is the fluid velocity (m/s), $\mu$ is the medium viscosity (Pa·s) and $Dh$ is the hydraulic diameter (m).

**Table 1.** The physical parameters of the simulations model

| Material | Density (kg/ m³) | Thermal conductivity (W/m·K) | Specific heat ( J/kg·K) | Viscosity (m²/s) |
|---|---|---|---|---|
| Cooling Medium (Water) | 997 | 0.613 | 4179 | 1e-6 |
| Fluid Channel (Brass) | 8500 | 93 | 380 | — |

# 3    The Analysis of Fluid Outlet Temperature

Thermal convection and thermal conduction play important roles in heat transfer of the water-cooling radiator. The inlet velocity of the fluid and thermal convection of air may greatly affect the heat transfer. The outlet temperature of fluid, which can be easily measured, is a practical index to indicate the effectiveness of heat transfer of the water-cooling radiator at certain inlet velocity of fluid. The influence of the flow velocity of the inlet fluid and thermal convection of air on the outlet temperature of fluid is analyzed in this section.

## 3.1    The Influence of Inlet Velocity of Fluid

The coefficient of air convection is generally within 5-25 W/(m²·k) [8]. In present analysis, it is set to be 10 W/(m²·k). Heat power 500 W is converted to heat flux 2.5e4 W/m², which is loaded on the superface of the flow channel. The ambient temperature and the inlet temperature of the fluid are both set to be 25 degrees centigrade, and the outlet pressure to be 0 Pa.

The simulation reasults under six velocities of inlet fluid $v=0.01$ m/s, 0.02 m/s, 0.05 m/s, 0.1 m/s, 0.2 m/s, 0.5 m/s are shown in Fig.3. The outlet temperature of fluid is, in fact, the average on the cross section of the fluid. As can be seen in Fig. 3, the outlet temperature, which is above ambient temperature or 25 degrees centigrade, declines progressively as the inlet velocity increases; the higher the inlet velocity is, the less the temperature rise (the difference between outlet temperature and inlet temperature) is and vice versa. The reason for this is that the increasing flow velocity decreases the contact time between the fluid and the interior surface of the flow channel and takes less heat away. Moreover, the downtrend of the outlet temperature becomes moderate with the increasing of the inlet velocity especially beyond 0.2 m/s. Consequently, 0.2 m/s is a reasonable inlet velocity for effective heat transfer performance.

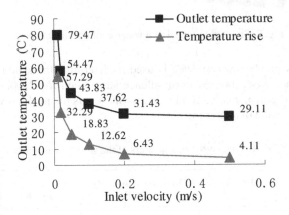

**Fig. 3.** The influence of inlet velocity on outlet temperature of fluid

## 3.2    The Influence of Air Convection

Forced air cooling is usually adopted in working practice to assist the heat dissipation of the water-cooling radiator. The coefficient of air convection at the range of 10-60 W/(m²·k) is loaded on simulation model as boundary condition of the forced air cooling effect. The influence of air convection on outlet temperature at the inlet velocities of 0.01 m/s, 0.05 m/s and 0.1 m/s is as shown in Fig.4. As can be seen from the figure, the outlet temperature goes down linearly with the increase of the coefficient of air convection; and the lower the inlet velocity is, the more quickly the outlet temperature decreases. That is to say, the coefficient of air convection has greatly effect on the outlet temperature of the fluid when its inlet velocity is low. Therefore, the performance of heat dissipation can't be improved effectively in too high inlet velocity.

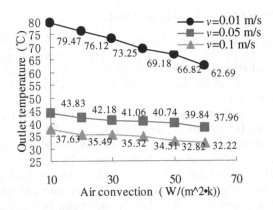

**Fig. 4.** The influence of air convection on outlet temperature of fluid

# 4    The Thermal Deformation of the Water-Cooling Radiator

Although a brass water-cooling radiator has a good thermal conductivity and also an excellent heat dissipation performance, thermal deformation inevitably occurs in brass and severe non-uniform thermal deformation which may lead to a poor heat dissipation performance and even cause cracking on the radiator. So it is make sense to analyze the thermal deformation phenomenon of the brass water-cooling radiator. A low inlet velocity causes a high outlet temperature of the fluid, as mentioned in the previous section. So the thermal deformation simulation of single water-cooling flow channel is presented at the inlet velocity $v$=0.01 m/s in present section.

The simulations are conducted by means of indirect fluid-solid coupling or sequential coupling method [9]. Suppose that the water-cooling radiator is under free deformation without displacement constraint in present simulation model. Some parameters of the flow channel are set as follows: elastic module EX=1.17e11 Pa, Major Poisson's ratios PRXY=0.324, thermal coefficient of expansion ALPX=1.9e-5 K-1. The structure element SOLID 45 is applied to replace the FLOTRAN 142 element and remesh the simulation model established in section 2.*B*. The temperature field at the inlet velocity of 0.01 m/s obtained in the previous section, as shown in Fig.5, is taken as initial temperature load in the simulation of the thermal deformation. It is necessary to state that the origin of coordinates in Fig.5 is located at the center of the cross section of the inlet fluid and X denotes the flow direction of the fluid.

The water-cooling flow channel before and after thermal deformation is shown in Fig.6. As can be seen from the figure, the magnitude and direction of the thermal deformation at each node are in diversity—in particularly the locations of maximal and minimal thermal deformation in the X-, Y- and Z- directions are diversified, as list in table 2. The generated non-uniform thermal deformation may cause some unfavorable results such as expansion, warping and even cracking in the radiator. Hence, in thermal design of radiator close attention should be given to the locations of maximal thermal deformation. From both Fig.5 and Fig.6, it can be seen that both the total thermal deformation and the temperature of the fluid go up along the flow direction or X-direction of the flow channel and reach their maximum values at node H, which are 40.3 mm and 185.487 degrees centigrade, respectively.

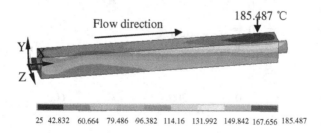

25  42.832   60.664   79.486   ;96.382   114.16   131.992   149.842  167.656  185.487

**Fig. 5.** Temperature field of the flow channel at the inlet velocity of 0.01 m/s (degrees centigrade)

To consider the displacements on the top and bottom contours of XY-section of the flow channel to demonstrate distinctly the thermal deformation. As shown in Fig.7, The largest thermal displacement occurs at the outlet of the flow channel and smallest one occurs at $X=150$ mm from the inlet. This implies that the water-cooling radiator suffers irregular thermal deformation or warp phenomenon -- two ends are upwarp and the middle part is sunken.

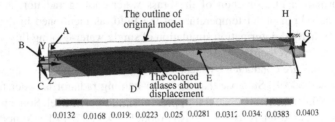

0.0132   0.0168   0.019:   0.0223   0.025   0.0281   0.031:   0.034:   0.0383   0.0403

**Fig. 6.** The water-cooling flow channel before and after deformation(The data denotes the total displacement, m)

**Table 2.** The maximal and minimal thermal displacements of flow channel and their locations (MM)

| Displace-ment | X-direction | Y-direction | Z-direction | Total-dir-ection |
|---|---|---|---|---|
| Maximum | 3.610 (F) | -12.185 (E) | 15.376 (C) | 40.285 (H) |
| Minimum | -2.096 (A) | -20.905 (B) | -35.404 (G) | 12.969 (D) |

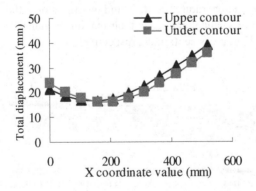

**Fig. 7.** The displacements on the top and bottom contours of XY-section of the flow channel

# 5    Conclusions

On the basis of a water-cooling radiator of a company, the 3-D simulation model for its flow channel is established and simulation analysis is accomplished by the commercial software ANSYS. The following conclusions can be drawn from the present research work:

(1) The outlet temperature declines progressively as the inlet velocity of the fluid increases; the higher the inlet velocity is, the smaller the temperature rise. In addition, the downtrend of the outlet temperature becomes moderate with the increasing of the inlet velocity especially beyond 0.2 m/s. Therefore, inlet velocity 0.2 m/s may be a reasonable value for effective heat transfer performance.

(2) The outlet temperature goes down linearly with the increase of coefficient of air convection; and the lower the inlet velocity is, the more quickly the outlet temperature decreases. The performance of heat dissipation could not be improved effectively in high inlet velocity

(3) Through the analysis of thermal deformation of the flow channel, it is found that the thermal deformation field is consistent with temperature field. The thermal deformation of the brass water-cooling radiator is obviously non-uniform and becomes warp inevitably.

**Acknowledgment.** The authors would like to acknowledge the support of Guangxi Key Laboratory of Manufacturing System & Advanced Manufacturing Technology (Grant No. 09-007-05s007), Guangxi Science Foundation (Grant No.0832243) and the Science Foundation of Guilin University of Electronic Technology (Grant No.Z20703).

# References

1. Liu, R.: Analytical and experimental study on the comprehensive performance of plate fin and staggered heat sinks. Tianjin University of Science and Technology (2006)
2. Sun, S., Zhang, H.: Analysis of CFD simulation with experiment of heat transfer and pressure drop for heat pipe heat exchanger. Journal of Nanjing University of Technology, Science 2, 62–65 (2004)
3. Zhou, M., Wu, S.: Forced Air Cooling Simulation of an Electronic Supply. Computer Engineering, Science 9, 171–172 (2003)
4. Zhi, M., Liu, Y.: Numerical simulation of electric power and electronic equipment radiator. North China Electric Power University (2006)
5. Guo, C., Zhou, J.: Numerical analysis of the thermal stress induced by temperature differences of fixed tube-sheet heat exchangers. Chemical Machinery, Science 1, 41–46 (2009)
6. Xiao, S.S.: The latest classic ANSYS and Workbench tutorial, p. 1. Publishing house of electronic industry, China (2004)
7. Xu, Y., Dang, S.: Example Guide to flow analysis by ANSYS 11.0/FLOTRAN, p. 82. China machine press (2009)
8. Zhang, X., Ren, Z.: Heat transfer (fourth version), p. 102. China Building Industry Press, Beijing (2001)
9. ANSYS, Realse 10.0 Documentation for ANSYS

# Incremental Mining Algorithm of Sequential Patterns Based on Sequence Tree

Jiaxin Liu[1], Shuting Yan[1], Yanyan Wang[1], and Jiadong Ren[1,2]

[1] College of Information Science and Engineering Yanshan University
Qinhuangdao City, P.R. China
[2] School of Computer Science and Technology Beijing Institute of Technology
Beijing City, P.R. China
{ljx,jdren}@ysu.edu.cn

**Abstract.** At present, the existing incremental mining algorithms of sequential patterns can not make full use of the mining information in original database, the updated database is scanned many times and the projected databases for frequent items are constructed. In this paper, we propose an incremental mining algorithm of sequential patterns based on Sequence tree, called ISPBS. ISPBS uses the sequence tree (Stree) to store the mining information in the original database. Sequence tree is a novel data structure, it is similar in structure to the prefix tree. But the sequence tree stores all the sequences in the original database. When the database is updated, ISPBS only deals with the incremental sequences and adds the sequences to the sequence tree without constructing the projected database. It can find all the sequential patterns in the updated database by scanning the sequence tree. Experiments show that ISPBS outperforms PrefixSpan and IncSpan in time cost.

**Keywords:** Sequential patterns, Incremental mining, Sequence tree.

## 1 Introduction

Sequential patterns mining is an important research in data mining[1]. It was first introduced by Agrawal and Srikant in [2], and they presented three mining algorithms of sequential patterns, including AprioriAll, AprioriSome and DynamicSome. In the implementation process of AprioriAll, database was scanned many times, and the huge candidate sequences were generated. Considering the shortages of AprioriAll, Srikant proposed GSP in [3]. The technologies of time constraints, sliding time windows and taxonomies were introduced in GSP to improve the performance of the algorithm. GSP need repeat database scans, and used complex hash structures which have poor locality. Zaki[4] presented SPADE which utilized combinatorial properties to decompose the original problem into smaller sub-problems, that can be independently solved in main-memory using efficient lattice search techniques, and using simple join operations. The weakness of Apriori-based mining algorithms of sequential patterns is that the huge candidate sequences needed to be stored. In order to reduce the huge set of candidate sequences, another approach that mining sequential patterns by pattern-growth was proposed by J. Pei, called PrefixSpan[5]. Using the prefix projection

G. Lee (Ed.): Advances in Intelligent Systems, AISC 138, pp. 61–67.
springerlink.com    © Springer-Verlag Berlin Heidelberg 2012

presented in PrefixSpan can reduce the size of the projected database and improve the efficiency of the algorithm.

In many applications, databases are updated incrementally. In order to improve mining efficiency, incremental algorithm is generally used to mine incremental sequences in database. When some new sequences are added to the original database, incremental algorithm makes full use of the set of frequent item mined from original database. It can reduce the overhead of scanning the original database. At present, incremental mining algorithms of sequential patterns include IncSpan[6], IncSpan+[7], PBIncSpan[8], ISC[9], BSPinc[10] and so on. Projection-based incremental mining algorithms of sequential pattern include IncSpan, IncSpan+ and PBIncSpan. Projection-based incremental mining algorithms avoid generate huge candidate sequences.

Cheng[6] proposed an incremental mining algorithm, called IncSpan. In order to reduce the using time in mining incremental sequences, the algorithm used semi-frequent patterns for incremental mining. IncSpan had some weakness, that is, it can not find the complete set of sequential patterns in the updated database. Nguyen[7] clarified this weakness by proving the incorrectness of the basic properties in the IncSpan and proposed a new algorithm called IncSpan+. IncSpan+ rectified the shortcomings in generating the set of frequent sequential pattern and the set of semi-frequent sequential pattern.

PBIncSpan constructed a prefix tree to represent the sequential patterns. It need maintain pSeqId of the node α, that pSeqId was a set of seqID in α-projected database. And the algorithm had to check weather pSeqId of each node α was included in IASIDS (insert and append sequence id). Therefore it required extra time consuming.

In order to make full use of the mining information in original database, to avoid constructing the projected database for frequent item and reduce the runtime of the algorithm, we use the structure of sequence tree to store all sequences and its support in the original database. Two kinds of data-processing strategies that are presented in this paper are used to add the sequences in incremental database to the sequence tree and maintain the sequence tree dynamically. We can find all the sequential patterns in the updated database through traversing the sequence tree.

The remaining of the paper is organized as follows. In section 2, we introduce the structure of sequence tree. ISPBS is formulated in section 3. The experimental results and performance analysis are presented in Section 4. In Section 5, we summarize our study.

## 2    The Structure of Sequence Tree

Sequence tree is a prefix tree, it is similar in structure to the prefix tree proposed by Chen in [8]. However, the sequence tree not only stores the frequent sequences in the original database, but also stores the non-frequent sequences in the original database, and the support of each sequence is stored in the sequence tree. Constructing the sequence tree has the same procedure as mining sequential patterns in a sequence database using PrefixSpan algorithm. All of the 1-sequence mined in the projected database as a child node is inserted into the sequence tree, the last one in the prefix of the projected database is its parent node. We give the definition of the sequence tree.

**Definition 2.1:** The root node of the sequence tree is an empty node. In addition to the root node, each node in the sequence tree contains two attributes. One stores 1-sequence in the projected database, and the other stores support of the sequence. The path from the root node to any leaf node represents a sequence in the database, and its support is the support of the leaf node. The support of any node is not smaller than the support of its child nodes.

We give an example to illustrate the sequence tree. For brevity, we assume an element has only one item. When an element has multiple items, the results may be deduced by analogy. A sequence database D is shown in table1.

**Table 1.** A Sequence Database D

| SeqID | Sequence |
|-------|----------|
| 1 | a b c d |
| 2 | d e |
| 3 | a c d |
| 4 | b d |

The sequence tree of the sequence database D is shown in Figure 1.

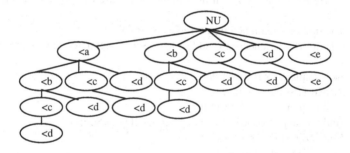

**Fig. 1.** The sequence tree of the sequence database D shown in table 1.

From the sequence tree we can see that the set of all sequences in the sequence database D with its support is {<a>:2, <b>:2, <c>:2, <d>:4, <e>:1, <ab>:1, <ac>:2, <ad>:2, <bc>:1, <bd>:2, <cd>:2, <de>:1, <abc>:1, <abd>:1, <acd>:2, <bcd>:1, <abcd>:1}.

## 3    Incremental Mining Algorithm of Sequential Patterns Based on Sequence Tree

IncSpan uses FS and SFS to store the frequent sequences and semi-frequent sequences in the original database, PBIncSpan uses the prefix tree to store the frequent sequences in the original database. These two algorithms only store the sequences that meet the support threshold. When the database is updated, both algorithms must take into account such a possibility that the sequences that are not frequent in the original database become frequent sequences in the updated database because of the incremental sequences. Therefore both algorithms need to construct the projected database for frequent item to find the sequential patterns in the updated database. In order to avoid constructing the projected database for frequent item and reduce occupied space and runtime in constructing the projected database, we propose an incremental mining algorithm of sequential patterns based on sequence tree using the structural characteristic of the sequence tree. As the sequence tree stores all the sequences and its support in the original database, when the database is updated, ISPBS does not need to construct the projected database, just adds the new sequences to the sequence tree and maintains the sequence tree dynamically. Finally, it can get all the sequences and its support in the updated database and find all the sequential patterns by scanning the sequence tree.

We give two kinds of data-processing strategies in ISPBS. One is INSERT operation, the other is APPEND operation. The INSERT operation means that the new sequences are inserted into the original database. These old sequences are still unchanged, but the total number of the sequences in the updated database is increased. The APPEND operation means that some old sequences are appended with new sequences. And the total number of the sequences in the updated database is unchanged.

**Algorithm 1.** ISPBS(D, d, min_sup, Stree)

Input: An original database D, an incremental database d, min_sup, sequence tree Stree

Output: The updated sequence tree Stree', FS' in D'

1:  Scan d, insert sequences are added to InSet, append sequences are added to Appset;
2:  For each sequence s in Inset do
3:       INSERT(Stree', root, s);
4:  For each sequence s in Appset do
5:       Find the sequence $\alpha$ in D, making $\alpha$.SeqID = s.SeqID;
6:       APPEND(Stree', root, $\alpha$, s);
7:       INSERT(Stree', root, s);
8:  Scan the Stree', find FS';
9:  Return;

**Algorithm 2.** INSERT(Stree, root, s)

Input: A sequence tree Stree, the root node, an insert sequence s
Output: The updated sequence tree Stree'

1:  For each item $e_i$ in the sequence s do
2:       Given s', s' is the postfix of sequence s w.r.t. prefix $e_i$;
3:       If $e_i$ is the child node of root and $e_i$ appears the first time in s
4:            support($e_i$) = support($e_i$) + 1;
5:       ElseIf $e_i$ is not the child node of root and $e_i$ appears the first time in s
6:            add a new child node $e_i$, its parent node is root, support($e_i$)=1;
7:       INSERT(Stree', $e_i$, s');
8:  Return;

**Algorithm 3.** APPEND(Stree, root, α, s)

Input: A sequence tree Stree, the root node, a sequence α in the database D that it has the same SeqID with the sequence s, an append sequence s

Output: The updated sequence tree Stree'

1: For each item $e_i$ in the sequence α do
2:    Given α', α' is the postfix of sequence α w.r.t. prefix $e_i$;
3:    If α' ≠ Φ
4:        For each item $e_j$' in the sequence s do
5:            If $e_j$' is not the child node of $e_i$
6:                add a new child node $e_j$', its parent node is $e_i$, support($e_j$') = 1;
7:            Else
8:                support($e_j$') = support($e_j$') + 1;
9:            APPEND(Stree', $e_i$, α', s);
10:    Else
11:        INSERT(Stree', $e_i$, s);
12:        Do_append(Stree', $e_i$, s); // Do_append() function is used to handle new added brother nodes of $e_i$
13: Return;

When the database is updated, ISPBS scans the incremental database once. The new sequences in the incremental database are added to InSet, the append sequences are added to AppSet. Each sequence in InSet and AppSet respectively performs INSERT operation and APPEND operation. ISPBS algorithm only maintains the sequence tree without constructing the projected database. The sequence tree stores all the sequences in the updated database, so ISPBS can find all the sequential patterns in the updated database through using depth-first search strategy to traverse the sequence tree.

## 4    Experimental Results and Performance Analysis

In the experiment, we compare and analyze the performance of ISPBS with IncSpan and PrefixSpan. All the experiments are performed on a 1.5GHz Pentium PC machine with 1 gigabytes main memory, running Microsoft Windows XP. All the algorithms are implemented using Microsoft Visual Studio 2005 and Microsoft SQL Server 2005. The dataset we used for our experiments is synthetic data, with letters to mark the item. The item set contains 26 elements, the size of the dataset is 200K.

Fig. 2 shows the running time of the three algorithms in different support threshold on condition that the number of incremental sequences is 1% of the total number of the original database sequences. The running time of ISPBS is mainly affected by the number of incremental sequences, so when the support threshold is varied, the change in the running time of ISPBS is not obvious. When the support threshold is high, the numbers of sequential patterns are limited, and the length of sequential patterns is short, the three algorithms are close in terms of running time. However, when the support threshold decreases, the gaps become clear. PrefixSpan and IncSpan have to spend much time in constructing the projected database, so ISPBS is the fastest.

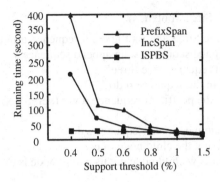

**Fig. 2.** Varying support threshold

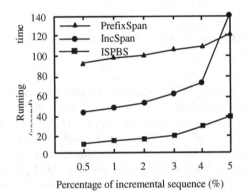

**Fig. 3.** Varying percentage of incremental sequence

Fig. 3 shows how the three algorithms can be affected when min_sup = 0.5% and the percentage of incremental sequences is varied. The curves show that the time increases as the incremental portion of the database increases. PrefixSpan and IncSpan spend much time in constructing the projected database. ISPBS only deals with the incremental sequences, so ISPBS is more efficient than PrefixSpan and IncSpan. When the incremental part exceeds 5% of the database, PrefixSpan outperforms IncSpan. This is because if the incremental sequences is not very small, the number of sequential patterns brought by it increases, increasing the overhead of IncSpan.

In summary, our experimental results show that ISPBS outperforms PrefixSpan and IncSpan in time performance.

## 5    Conclusions

In this paper, we propose an incremental mining algorithm of sequential patterns based on Sequence tree, called ISPBS. ISPBS uses a sequence tree to store all the sequences and its support in the original database. When the database is updated, the incremental sequences are added to InsSet and AppSet in terms of sequence ID. Each sequence in

InSet and AppSet respectively performs INSERT operation and APPEND operation that the incremental sequences can be added to the sequence tree. The updated sequence tree stores all the sequences in the updated database, so ISPBS can find all the sequential patterns in the updated database through using depth-first search strategy to traverse the sequence tree. Experimental results show that ISPBS outperforms PrefixSpan and IncSpan in time cost.

**Acknowledgment.** This work is supported by the National High Technology Research and Development Program ("863"Program) of China No. 2009AA01Z433 and the Natural Science Foundation of Hebei Province P.R. China No.F2008000888.

# References

1. Zou, X., Zhao, L., Guo, J., Chen, X.: An advanced algorithm of frequent subgraph mining based on ADI. ICIC Express Letters 3, 639–644 (2009)
2. Agrawal, R., Srikant, R.: Mining sequential patterns. In: Proc. Int. Conf. Data Engineering, pp. 3–14. IEEE Press, Taipei (1995)
3. Srikant, R., Agrawal, R.: Mining sequential patterns: generalization and performance improvements. In: Lect. Notes Comput. Science, vol. 1057, pp. 3–17 (1996)
4. Zaki, M.J.: SPADE: an efficient algorithm for mining frequent sequences. Mach. Learning 42, 31–60 (2001)
5. Pei, J., Han, J., Mortazavi-Asl, B., Pinto, H., Chen, Q., et al.: PrefixSpan: mining sequential patterns efficiently by prefix-projected pattern growth. In: Proc. Int. Conf. Data Engineering, pp. 215–224. IEEE Press, Heidelberg (2001)
6. Cheng, H., Yan, X., Han, J.: IncSpan: incremental mining of sequential patterns in large database. In: KDD Proc. Tenth ACM SIGKDD Int. Conf. Knowl. Discov. Data Mining, pp. 527–532. Association for Computing Machinery Press, Seattle (2004)
7. Nguyen, S.N., Sun, X., Orlowska, M.E.: Improvements of IncSpan: Incremental Mining of Sequential Patterns in Large Database. In: Ho, T.-B., Cheung, D., Liu, H. (eds.) PAKDD 2005. LNCS (LNAI), vol. 3518, pp. 442–451. Springer, Heidelberg (2005)
8. Chen, Y., Guo, J., Wang, Y., Xiong, Y., Zhu, Y.: Incremental Mining of Sequential Patterns Using Prefix Tree. In: Zhou, Z.-H., Li, H., Yang, Q. (eds.) PAKDD 2007. LNCS (LNAI), vol. 4426, pp. 433–440. Springer, Heidelberg (2007)
9. Ren, J., Sun, Y., Guo, S.: Incremental sequential pattern mining based on constraints. J. Comput. Inf. Systems 4, 571–576 (2008)
10. Lin, M.-Y., Hsueh, S.-C., Chan, C.-C.: Incremental discovery of sequential patterns using a backward mining approach. In: Proc. IEEE Int. Conf. Comput. Sci. Engineering, pp. 64–70. IEEE Press, Vancouver (2009)

# Micropower HF and SHF Operational Amplifiers

Sergey Krutchinsky, Vasiliy Bespyatov, Alexander Korolev,
Eugeniy Zhebrun, and Grigoriy Svizev

Educational Research Center of System Design Technologies
of Southern Federal University, Russian Federation,
Rostov region, Taganrog, 347928, Shevchenko 2
{sgkrutch,jackjk}@mail.ru,
{bespyatov,korolev}@nocstp.ru,
virpil07@gmail.com

**Abstract.** Single-stage OA schematic circuit a based on the functional-topological processes of self-compensation and cancellation of field-effect and bipolar transistors small-signal parameters influence was proposed. A distinctive feature of the scheme is the equality of p-MOS and n-p-n transistor with heterojunction contributions. Simulation results of schematic circuit based on components of SGB25VD technical process were presented. The advantages of the circuit solution were shown.

**Keywords:** OAs, IP blocks, self-compensation and cancellation, parametric sensitivity, compensating feedback loops.

## 1 Introduction

Designing of integral project IP blocks of mixed systems on a chip (SoC) HF and SHF ranges for a number of objective reasons requires relatively universal operational amplifiers (OAs). To increase their gain coefficient and to reduce the "electrical" length in input differential stages, as a rule, dynamic loads based on transistors with the opposite type of electronic conductivity are applied [1]. From the perspective of maximizing the operating frequency range of these OAs with the practical constraints on power consumption the bipolar transistors with heterojunction based on SiGe technology are preferred [2]. However, in some of the cheapest processes - SGB25VD, SGB25H1 it is possible to use only n-p-n transistors and CMOS transistors with significantly more low-frequency bands. Thus, for the SGB25VD technical process consuming current of 1mA the frequency properties of the bipolar transistor almost by an order of magnitude greater than the same characteristic of the field-effect transistor.

To extend the operating frequency range of amplifying stages additional loops of self-compensation connecting inverting and non-inverting inputs of the bipolar transistor can be used [3]. However, as shown in [4], in this case, the operating frequency range of resulting amplifier stage is limited to cutoff frequency of voltage follower based on FET, which is insufficient to a wide range of practical problems in the IP-blocks. That is why further improvement of circuit design in BiCMOS basis is the relevant problem of modern analog microelectronics.

G. Lee (Ed.): Advances in Intelligent Systems, AISC 138, pp. 69–78.

## 2    Statement of the Problem

As a rule, to solve the general problem of OA constructing, structural compromise associated with the creation of dynamic loads based on the p-MOS transistors for the main n-p-n transistor collector circuit of with a heterojunction was used. In this case, while deteriorating the stability in static mode, coordination of the currents is executed using autonomous current source, and the gain coefficient is increased by increasing current consumption of differential input stage. A standard solution of this problem is a single-stage OA circuit, considered in [5]. It can be stated that such decisions are not optimal either structurally, or parametrically.

From the viewpoint of differential parameters of bipolar and field-effect transistors influence on the gain of each stage we have the following ratio

$$K_0 = \frac{S}{g_l + g_i + h_{22}} \ , \tag{1}$$

where $S = \alpha / h_{11}$ – the steepness of the bipolar transistor; $g_i$ – the field-effect transistor drain-source conductivity; $h_{11}, h_{22}$ – input resistance and output conductivity of the bipolar transistor; $g_l$ – the conductivity of the stage load.

In a relatively wide range of frequencies on the same stage in the first approximation, transfer function of the first order:

$$K(p) = \frac{K_0}{1 + pK_0 \tau_s} \tag{2}$$

where $\tau_s = (c_{cb} + c_{ds} + c_{dg}) / S$ - the equivalent time constant; $c_{cb}, c_{ds}, c_{dg}$ - interelement capacities of transistors.

From these ratios we can identify the dominant factors determining the maximum gain coefficient and cutoff frequency of stage and formulate a number of particular problems, which solution allows achieving this goal. So, the direction of structural optimization of the designed stage is determined by comparison of $g_i$ and $h_{22}$ and the necessity to use loops of self-compensation and cancellation justified by analysis of $\tau_s$ components. For example, in SGB25VD process regime dependences of the input n-p-n and p-MOS transistors, presented in Fig. 1, show that the dominant factor for $K_0$ is $R_i = 1/g_i$ that almost by an order of magnitude smaller than $1/h_{22}$.

In addition, the increasing of bipolar transistor $S$ by increasing the current consumption leads to $K_0$ increase only until the specified dominant factor remains. This is the approach used in [5] and allowed for single-stage OA following parameters: $K_0 = 32dB$, $f_1 = 16GHz$ at $I_0 = 10mA$.

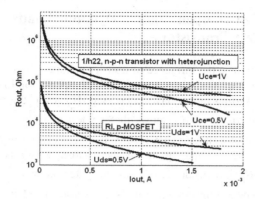

**Fig. 1.** Dependence of output resistance of transistors on operation mode

Note that the unity gain frequency of SiGe bipolar transistor is defined by the expression [6]:

$$f_{t.GBT} = \left[ 2\pi \left( \frac{1}{g_m}(C_{eb} + C_{cb}) + \tau_b + \tau_c + \tau_e \right) \right]^{-1} \qquad (3)$$

analogous for a MOSFET:

$$f_{t.MOS} = \left[ 2\pi \left( \frac{1}{g_m}(C_{gs} + C_{gd}) + \tau_{ps} \right) \right]^{-1} \qquad (4)$$

where $\tau_{ps}$ – the time required for the passage of charge carrier from the source to the drain, $\tau_b, \tau_c, \tau_e$ – transit times of charge carriers across the base region, the collector, emitter, respectively.

However, at moderate collector currents, the frequency of unity gain of SiGe GBT is found from a simplified expression:

$$f_{t.GBT} = \frac{1}{2\pi(\tau_b + \tau_c + \tau_e)} \qquad (5)$$

i.e. it is determined purely by a temporary delay. Whereas the frequency of unity gain of submicron field-effect transistors is determined mainly by a parasitic capacitances $C_{gs}$ and $C_{gd}$.

That is why, in the structure of the OA the main functions of signal transformation has to be performed on n-p-n transistor.

Thus, to further increase the gain coefficient it is required another principle of dynamic load constructing which would provide lower influence of $g_i$ on $K_0$. As it will be shown below this provides a significant improving $K_0$ and broadbandness of OA with decreasing current consumption.

## 3    Feature of Self-compensation of Dynamic Loads on Field-Effect Transistors

As shown in [3], to reduce the influence of transfer conductivity of transistors on the characteristics of the amplifier stages an extra self-compensation loop, connecting its inverting and non-inverting inputs have to be used. However, the nature of physical processes in field-effect transistors shows that numerical value of $R_i = 1/g_i$ does not influence its control voltage. That is why making compensating feedback loop requires finite resistance in the circuit of the source, which is the indicator of numerical value of $R_i$. Independence of transistor gate voltage on this factor changes the structure of compensating circuit. This issue is considered in "Design circuitry feature of stages with high-gain coefficient on field-effect transistors". From the perspective of the problem, we note that the increasing of additional resistance in the circuit of the source is equivalent to increasing feedback depth of the compensating $R_i$ influence, so its implementation as additional dynamic load on the transistor of the same type can significantly reduce the influence of dynamic load on realized gain coefficient. Schematic circuit of the dynamic load for the differential stage OA is shown in Fig. 2.

In this case, the equivalent output conductivity

$$g_i \approx g_{i4}\, g_{i3} \big/ g S_4 \qquad (6)$$

is reduced by the static gain of T4 ( $\mu = S_4 / g_{i4}$ ). From ratios (1) and (6) it follows that such a feedback loop reduces or, at least, does not increase the sensitivity of the gain coefficient to the small-signal parameters of MOSFET

$$S_{g_{i3}}^{K} = S_{g_{i4}}^{K} = -S_{S_4}^{K} = -K\frac{g_{i3}}{\mu S} \qquad (7)$$

and with maximization does not increase its instability.

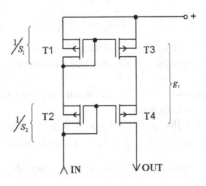

**Fig. 2.** Schematic circuit of asymmetric differential stage dynamic load

According to [3], this feedback loop reduces the influence of output parasitic capacities T3 and T4 on an equivalent time constant of the stage

$$\tau_\theta = \frac{1}{S \cdot S_4}(C_3 g_{i4} + C_4 g_{i3}) \approx \frac{1}{S\mu}(C_3 + C_4) \tag{8}$$

where $C_3$ and $C_4$ – output parasitic capacities of T3 and T4.

Thus, even in the extreme case, where due to the action of this circuit $K_0$ is increased in $\mu$ times, according to ratio (2) it does not decrease its cutoff frequency under this dynamic load.

A favorable factor aimed at expanding operating frequency range of the differential stage is the increment of its transfer function

$$\Delta K(p) = \frac{S_4}{g_{i4} + h_{22} + g_l} \cdot \frac{1}{1 + S_4/g_{i3}} \cdot \frac{p\,C_3/g_{i3} + 1}{p\,C_3/S_4 + 1}, \tag{9}$$

associated with increased transmission of T4 on the circuit gate-drain. It will be shown below that this factor together with a loop of cancellation improves the stability margin for OA phase.

In terms of the general principle of signal transformation, such dynamic load is appropriate to consider "cascode".

Further improvement of quality values of cascode load with additional compensating feedback connecting T4 source with its gate through the inverter stage.

## 4    Principles of Cancellation of Parasitic Capacities Influence on Operating Frequency Range of Operational Amplifier

Obtained results show that the dominant factors limiting the operating frequency range of OA with a minimal electrical length are transfer capacities and the capacities of transistors to substrate ($C_s$). In [7] it is shown that minimizing their influence on $f_1$ leads to the principle of cancellation, when additional capacity $C_c$ reduces the equivalent time constant of entire amplifier.

To implement the principle of cancellation of load equivalent capacity influence on i-th stage it is required connect the output of j-th stage to the output of additional (in this case the compensating) capacitor $C_c$.

If series connection of cascades is used in the structure of the amplifier

$$K_0 = \prod_i K_{oi} \text{æ}_i, \quad H_i(p) = \prod_i K_{oi} \text{æ}_i, \quad F_i(p) = \prod_{j=1}^{i} K_{oj} \text{æ}_j \tag{10}$$

then this condition can be specified in numerical values of the additional capacitor

$$C_c = \left( C_s + \sum_o C_{in\,j} \right) \cdot \left[ \prod_{l=i+1}^{m} K_{0l} - 1 \right]. \tag{11}$$

where $æ_i$, $æ_j$ – transfer coefficient of interstage connection; $H_i(p)$ – transfer function of the amplifier when connecting the input signal source to the non-inverting input of the i-th transistor; $F_i(p)$ – transfer function at the input of the i-th stage.

Current ratio shows that the effectiveness of this method of solving the general problem depends on the identity of the processes in those components, which models characterize these capacities. Thereupon, it is appropriate to use one of the active components as $C_c$ in the proper mode of operation.

Direct use of the found principle is shown in Fig. 3.

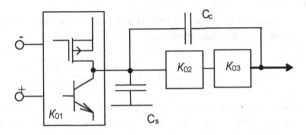

**Fig. 3.** Cancellation of the influence of $C_s$ и $C_c$ on frequency characteristics of the amplifier

Here, if $K_0 \approx K_{01}$ conductions of $g_{in2}$ and $g_{out1}$ are sufficiently small, and the influence of $C_s$ is maximal, so defines its dominant role. In accordance with (10) - (11) the implementation of $C_c$ satisfying condition

$$C_c = C_s \cdot (K_{02}K_{03} - 1) \tag{12}$$

influence of $C_c$ and $C_s$ is excluded.

The disadvantage of cancellation is relatively high sensitivity of the condition to the instability of $C_s$ and $C_c$. So, for a given in Fig. 3 case the relative sensitivity of the time constant of the amplifier and, consequently, its cutoff frequency

$$S_{C_s}^{\tau} = \frac{C_s(1 - K_{22}K_{03})}{C_c + C_s(1 - K_{02}K_{03})}, \tag{13}$$

$$S_{C_c}^{\tau} = \frac{C_c}{C_c + C_s(1 - K_{02}K_{03})} \tag{14}$$

directly determined by desired (achieved) level of compensation. That is why there is device unity gain frequency regime dependence.

In single-stage amplifiers satisfying of the above conditions leads to a large value of $C_c$, and therefore require an increase in current of *output* stage $K_{03}$. From ratios (12), (13) and (14) implies that compensating feedback signal is appropriate to inject into a node, which provides numerical values $H_i(p)$ and $F_i(p)$ close to realized gain coefficient. As can be seen from the circuit of the dynamic load (Fig. 2) such a node is the source of transistor T4. However, in general, the connecting node of $C_c$ output pin requires detailed analysis of the aggregate circuit of the designed amplifier.

## 5    Micropower Single-Stage Amplifier in BiMOS Basis

As seen from Fig. 1, n-p-n transistor has a higher output resistance and it is non-dominant element limiting the gain coefficient (1). Thus, for the technology SGB25VD (p-MOS: width 50um, length 250nm, number of gates 1; n-p-n bipolar: emitter length 840nm, emitter width 420nm, y-multiplier 2, x-multiplier 8) in microregime ( $I_c = I_d \approx 0.1mA$ ) $R_i = 15kOhm$ , $1/h_{22} = 500kOhm$ , so

$$S_i = g_{i4}\, g_{i3}/S_4 \gg h_{22} \tag{15}$$

As seen from Fig. 1 and (1), this reserve can be used to increase equivalent steepness of n-p-n transistor and hence, the gain coefficient by increasing current consumption until the contributions of bipolar and field-effect transistor become equivalent. ( $I_c = I_d \approx 0.2mA$ )

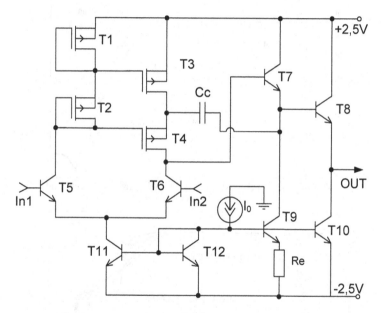

**Fig. 4.** Schematic circuit of micropower single-stage OA

Schematic circuit of single-stage OA that meets the restrictions is shown in Fig. 4. An additional solution feature is using of the loop of cancellation of n-p-n and MOS transistors transfer capacities influence on the OA cutoff frequency. According to [7], such a reactive feedback decrease equivalent time constant of the amplifier by an amount

$$\Delta \tau_\theta = 2C_c/S. \tag{16}$$

Besides, according to (9) ( $C_3 = C_c$ ), there is a correction of phase-frequency characteristics in a range of high frequencies. Indeed, in the neighborhood of frequency

$$\omega_0 = \sqrt{g_{i3}S} / (C_c + C_3) \qquad (17)$$

there is phase difference decrease on

$$\Delta \varphi = arctg \sqrt{\frac{S_4}{g_{i3}}} - arctg \sqrt{\frac{g_{i3}}{S_4}} \qquad (18)$$

This feature of the characteristics is shown in fig.5 and may be exploited to increase stability margin by connecting additional $C_3$. Shown in Fig. 5 amplitude-frequency characteristic of OA shows that the loop of cancellation allows to expand cutoff frequency perceptibly for given stability margin and required total negative feedback depth.

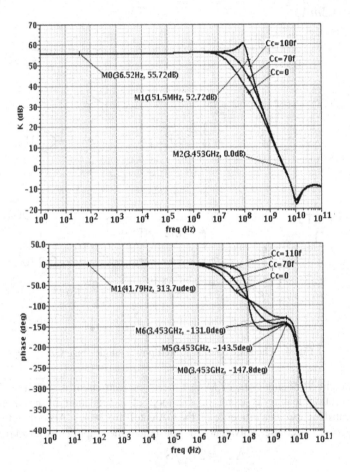

**Fig. 5.** Amplitude (left) and phase (right) frequency response of the OA where $C_c = 0$, $C_c = 70 fF$ and $C_{c\,opt} = 110 fF$

**Table 1.** Results of OA simulation

| parameter / scheme | $K^a_{cmrr}$ [dB] | $f^b_{br\_cmrr}$ [MHz] | $G^c$ [dB] | $f^d_{br\_G}$ [MHz] | $f^e_1$ [GHz] | $\vartheta^{+\,f}$ [kV/us] | $\vartheta^{-\,g}$ [kV/us] | $U^{+\,h}_{out.max}$ [V] | $U^{-\,i}_{out.max}$ [V] | $E^j_{off}$ [mV] | $I^k_{sup}$ [mA] |
|---|---|---|---|---|---|---|---|---|---|---|---|
| Fig. 4 microregime Cc=0fF | 91 | 139.5 | 56 | 18.5 | 5.6 | 12.1 | 1.5 | 0.83 | -1.84 | -0.8 | 1.6 |
| Fig. 4 microregime Cc=110fF | 91 | 15.6 | 56 | 151.5 | 6.1 | 21.9 | 1.6 | 0.83 | -1.84 | -0.8 | 1.6 |
| Analogous OA[5] microregime | 58 | 29.5 | 30 | 88.91 | 3 | 1.7 | 0.7 | 1.6 | -1.2 | 4.1 | 1.2 |
| Fig. 4 macroregime Cc=0fF | 74 | 0.822 | 37 | 444 | 26.2 | 15.2 | 8.7 | 0.56 | -1.75 | -6.4 | 14.5 |
| Fig. 4 macroregime Cc=20fF | 74 | 0.821 | 37 | 450 | 26.9 | 14 | 9.1 | 0.56 | -1.75 | -6.4 | 14.5 |
| Analogous OA[5] macroregime | 42 | 48.5 | 35 | 477 | 23.84 | 13.7 | 11.6 | 1.6 | -1.5 | -24.6 | 14.8 |

Here: a. Common-mode rejection ratio; b. Boundary frequency of $K_{cmrr}$; c. Differential gain coefficient; d. Boundary frequency of G; e. Unity gain frequency; f. Rate of front rise; g. Rate of front fall; h. Maximal positive output voltage; i. Maximal negative output voltage; j. offset EMF; k. Circuit current consumption.

The results of a comparative analysis of the proposed SHF OA scheme and its analogue [5] listed in Table 1 show that the composition of loops of self-compensation [3] and cancellation [7] of the influence of small-signal parameters of transistors can significantly improve OA basic quality values. In addition, such loops of compensation create additional parametric circuits of freedom that can be used to reduce the OA current consumption.

For example, the above principle of parametric optimization by the criterion of equal contributions of p-MOS and n-p-n transistors to maximize gain coefficient can reduce OA power consumption almost by an order of magnitude. This assertion is confirmed by simulation of the schematic circuit of the OA with dynamic load, while implementation error of $K_0$ does not exceed 2.2dB, and its cutoff frequency 15%.

## 6    Conclusion

Obtained results confirm the initial theoretical position [3], [7], associated with the ability to track circuitry innovations in the amplifying devices. Indeed, the loops of self-compensation of transistors small-signal parameters influence on the attainable OA gain coefficient while maintaining low parametric sensitivity and unchanged cutoff frequency. The loop of cancellation of the influence of active elements parasitic capacitances not only extends the operating frequency range of the OA, but also makes it possible to save necessary amount of phase stability.

The minimum electrical length of these OAs creates additional parametric freedom that can be used for maximizing operating frequency range of functional devices based on these OAs. For example, when creating scaled or instrumentation amplifiers when the amount of feedback required remains unchanged, it is possible to increase several times the OA's threshold frequency by means of increasing balancing capacitance (see Fig. 5 and 6). At that required phase margin remains.

Represented results may be used in the course of integrated circuit design, e.g. using IC Flow produced by Mentor Graphics which provides the opportunity to consider not only semiconductor elements SPICE-models features, but also design and technological constraints when modeling.

# References

1. Heinemann, B.: Complementary SiGe BiCMOS. Electrochemical Society Proceeding 07, 25–31 (2004)
2. http://www.ihp-microelectronics.com
3. Krutchinsky, S.G., Prokopenko, N.N., Starchenko, E.I.: Structurally topological principles of self-compensation in electronic devices. In: Proceeding ICCSC 2004, Moscow, Russia, pp. 26–30 (2004)
4. Krutchinsky, S.G.: Modern microcircuitry and competitiveness of the domestic analog IC and mixed IP blocks. Scientific and Technical Journal "Electronic Components" (1), 6–10 (2009) (in Russia)
5. Annual Report of the Institute of Innovations for High Performance Microelectronics (IHP) (2006), http://www.ihp-microelectronics.com/lileadmin/ihp-template/about-ihp/publications/IHP_AR_2006.pdf
6. Ashburn, P.: SiGe Heterojunction Bipolar Transistors, p. 286. Wiley & Sons (2003)
7. Krutchinsky, S.G., Prokopenko, N.N., Budyakov, A.S.: Compensation Methods of Basic Transistors Output Capacitance Components in Analog Integrated Circuits. In: Proceeding ICCSC 2006, Bucharest, Romania, pp. 44–49 (2006)

# Influence of Irrigation Water-Saving on Groundwater Table in the Downstream Irrigation Districts of Yellow River

Zhou Zhen-min, Zhou Ke, and Wang Xuechao

North China University of Water Conservancy and Hydropower,
Zhengzhou 450011,China
{Zhouzhenmin,wangxuechao}@ncwu.edu.cn,
zkzhouke@yahoo.com.cn

**Abstract.** Selecting the People's Victory Canal Irrigation District as a typical example, firstly, underground water kinetic situation was analyzed considering the different irrigation water saving measures. Then, kinetic simulation model was set up for the typical irrigation district. Based on the simulation model, influence of different irrigation water saving methods on underground water and eco-environment were studied. It is proved by the study findings that irrigation water saving measures play important role in overall enhancement of irrigation water use efficiency, decreasing groundwater table and alleviation of irrigation water shortage in the irrigation district. But, if irrigation water saving technology cannot be used rationally, it would result in reverse influence on crop growth and agricultural eco-environment. It is believed that the study results would play certain active values in irrigation water saving structures rehabilitation and eco-environment protection in the irrigation districts.

**Keywords:** Irrigation Water-Saving, Simulation Model, Groundwater, Influence.

## 1 Background[1]

The People's Victory Canal Irrigation District (PVCID) is the first large-scale irrigation area set up in the lower reach of the Yellow River after the new China founded[1]. Located in the north part of Henan Province, the PVCID covers 3 cities and 8 counties with the command area 1835 km$^2$. Since starting irrigation in 1952, the real irrigation area maintains 3~4×10$^4$ ha, with the highest irrigated area 6.1×10$^4$ha. In 1999, the PVCID carried out new water-saving and facilities rehabilitation layout according to requirement in the middle-long term national economic development plan and based on the detail irrigation water saving regulation. The command area was redesigned. The total irrigated area reached to 12.17×10$^4$ ha, in which water-saving irrigated area amounts to 7.89×10$^4$ hm$^2$, compensable water source irrigated area amounts to 4.28×10$^4$ ha. Up to 2002, the total water diversion volume from the Yellow River amounted to 30.5billion cm$^3$, which made great contribution in agricultural irrigation, urban and industrial water supply, Yellow River water supply to Tianjin, salinity-alkaline farmland reclamation, grain production security and eco-environment improvement, and created important social and economic benefits.

G. Lee (Ed.): Advances in Intelligent Systems, AISC 138, pp. 79–83.
springerlink.com        © Springer-Verlag Berlin Heidelberg 2012

## 2    Ground Water Dynamic Analysis for Pre-Post--Water Saving

The hydraulic slop in the aquifer of the typical area is slight [1]. The relation between lateral recharge and drainage of ground water is not obvious. The dynamic situation of ground water clearly shows the type of infiltration and evaporation, in which the lateral influential factor dominates the process, i.e, the main recharge source for ground water is from irrigation infiltration. Shallow water evaporation is the main outlet. Since 1999, water saving structures rehabilitation, in which canal lining is the main measures, have been carried out. Project water saving rehabilitation raised irrigation efficiency, achieved obvious water saving benefits [2], greatly reduced leakage lost from the Yellow River water diversion and irrigation canals, and decreased water table. It can be seed from Fig.1 that water table decreased obviously in the typical area after realization of water saving projects.

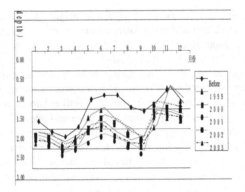

**Fig. 1.** Water table before and after water saving project construction

## 3    Model Analysis

Typical parameter model was studied to simulate groundwater dynamic situation in the typical area [2]. Groundwater balance model was used in PVCID, as Fig.2.

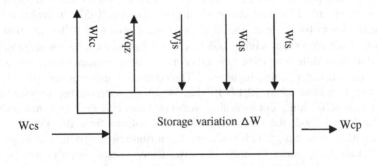

**Fig. 2.** Typical parameter model of underground water balance

Underground water balance model in the typical area can be written as,

$$\Delta W = \Delta H \cdot F \cdot \mu = W_{ts} + W_{qs} + W_{js} + W_{cs} - W_{qz} - W_{cp} - W_{kc} \qquad (1)$$

Where, $\Delta H$ , groundwater table storage variation (m); $\mu$ , groundwater specific yield; $F$ , groundwater area (ha.), $W_{ts}$ , infiltration recharge of farmland (×104); $W_{qs}$ , infiltration recharge of canal systems (×104); $W_{cs}$ , groundwater lateral recharge (×104); $W_{cp}$ , groundwater lateral drainage (×104); $W_{js}$ , rainfall infiltration recharge (×104); $W_{qz}$ ,groundwater evaporation (×104); $W_{kc}$ ,groundwater exploitation (×104).

## 3.1    Dynamic Influence of Canal System Lining on Groundwater

Canal system lining can largely reduce infiltration recharge, and cause obvious influence on groundwater. Different canal lining location can result in different influence degree. In order to simulate influence of canal system seepage prevention on groundwater, several canal lining schedules were assigned.

Scheme0, scheme1, scheme2 and scheme3 simulated four schemes, i.e., non-water-saving measures, only secondary canals lining, secondary and tertiary canals lining, all secondary and tertiary and farmland canals lining, to study groundwater table at the end of different periods under different flow.

**Table 1.** Groundwater table variation under different canal lining    unit:   m

| Water diversion (×10⁴m³) | Water table at the beginning of periods | Scheme0 non-water-saving measures | Scheme1 secondary canal lining | Scheme2 secondary and tertiary canals lining | Scheme3 all secondary, tertiary and farmland canal lining |
|---|---|---|---|---|---|
| 1799.89 | 2.13 | | 1.18 | 1.2 | 1.26 |
| 1855.55 | 2.13 | | 1.14 | 1.17 | 1.22 |
| 1911.21 | 2.13 | | 1.1 | 1.13 | 1.19 |
| 1966.87 | 2.13 | | 1.07 | 1.1 | 1.15 |
| 2022.53 | 2.13 | 0.95 | | | |

It can be seen from Table1 that under certain water diversion volume, the larger the canal lining area, the deeper the groundwater table. It means that canal lining system reduced water leakage recharge of groundwater.

## 3.2    Influence of Farmland Water Saving on Groundwater

Besides canal lining in typical irrigation area, water saving project rehabilitation also includes land leveling, small land plot preparation, sprinkle and tube-well irrigation, etc. which realized obvious water saving results. Scheme 4 is the simulation of influence of different water diversion on groundwater under the condition of farmland water saving measures only. The simulated results can be seen in Table2.

**Table 2.** Groundwater table under farmland water saving measures only    Unit: m

| Water diversion ($\times 10^4 m^3$) | Groundwater table at the beginning of periods | Scheme 4 |
|---|---|---|
| | | Farmland water saving only |
| 1799.89 | 2.13 | 1.45 |
| 1855.55 | 2.13 | 1.41 |
| 1911.21 | 2.13 | 1.38 |
| 1966.87 | 2.13 | 1.35 |

It can be seen from Table2 and Fig4 that farmland water saving measures reduced water leakage recharge on groundwater to cause groundwater deterioration.

### 3.3    Water Saving Irrigation Influence on Groundwater Table

In the typical area, water saving project rehabilitation includes not only canal lining, but also sprinkle irrigation, drip irrigation and preparation of smaller irrigation basin, which raised water use efficiency of canal systems from 0.614 to 0.921. Water saving projects reduced greatly water leakage loss of canal systems and ineffective water loss of farmland. Scheme 5, scheme 6 and scheme 7 simulated groundwater variation under different water diversion with combination measures of different canal lining area and farmland water saving. See Table3 and Fig4.

**Table 3.** Groundwater variation under different irrigation water saving schemes    Unit: m

| Water diversion ($\times 10^4 m^3$) | Groundwater table at the beginning of periods | S. 5 | S. 6 | S. 7 |
|---|---|---|---|---|
| | | Secondary canal lined, farmland water saving | Secondary and tertiary canal lined, farmland water saving | Secondary and tertiary and farmland canals lined, farmland water saving |
| 1799.89 | 2.13 | 1.49 | 1.58 | 1.69 |
| 1855.55 | 2.13 | 1.46 | 1.54 | 1.66 |
| 1911.21 | 2.13 | 1.43 | 1.51 | 1.64 |
| 1966.87 | 2.13 | 1.4 | 1.48 | 1.61 |

It can be seen from Table3 that under certain water diversion volume, with the increase of water-saving measures, groundwater infiltration recharge was reduced accordingly, groundwater table was reduced as well.

# 4  Conclusions

Influence of water-saving irrigation on groundwater table was studied through water balance analysis in typical groundwater and relevant models. It is shown by research findings that water saving irrigation reduced water diversion requirement from the Yellow River, decreased groundwater infiltration recharge, raised water resources effective utilization, reduced ineffective water loss, decreased groundwater table and improved effectively soil saline-alkaline in irrigated area. It can be concluded through study that it may not obtain the best results with the absolute water saving measures. There is a threshold that irrigation should meet crop water requirement. If irrigation water is lower than the threshold, the crop yield would be reduced. The present findings should have the reference values to irrigation water saving and irrigation project rehabilitation.

# References

1. Cheng, M.-J., Shen, L.-G., et al.: Comprehensive technology experiment on water saving project rehabilitation in the Inner Mongolia Hetao Irrigation District. China Water and Hydropower press, Beijing
2. Yue, W.-F., Yang, J.-Z., et al.: Analysis on water consumption principle of Hetai Irrigation District. China Rural Water and Hydropower (8) (2004)
3. Li, J.-S., et al.: Study on water saving efficiency of small basin irrigation. China Rural Water and Hydro-Power (9) (2006)
4. Xu, D., Li, Y.-N.: Study and application on farmland water saving irrigation technology. China Agricultural press, Beijing
5. Wuhan University of Water Conservancy and Hydropower. Farmland Hydraulics. Water Conservancy and Hydropower Press (1980)
6. Water Conservancy Research Institute of Gansu Province. Precise level basin irrigation in the Hexi Inland area, vol. 11 (2005)

# Chaining OWS Services with Graphic-Workflow

Zhang Jianbo[1], Liu Jiping[2], and Wang Bei[3]

[1] School of Resource and Environmental Science,
Wuhan University, Wuhan,430079, China
finecho@163.com
[2] Chinese Academy of Surveying and Mapping, Beijing, 100830, China
[3] Institute of Geographic Sciences and Natural Resources Research,
CAS, Beijing, 100101, China

**Abstract.** We propose an aggregation model of OWS chaining. By analyzing the main problems between the interface and data control about OWS services used in the workflow, we present the corresponding methods of the interface improvements, and GML-compressed scheduling strategy. Finally, combined with Kepler graphic-workflow, processing flow modeling with OWS services chain and solutions are presented in the paper. By dispatching the Concrete OWS chaining with extended interface in Kepler, the implementation is proved feasible and efficient.

**Keywords:** OWS, combining, services chaining, GML data flow dispatch, graphic-workflow, Kepler.

## 1 Introduction

Chaining spatial information services based on the workflow provides an effective solution for the sharing and the interoperability about distributed spatial information over the network[1]. But now, there are two major problems with the method of OWS (OGC Web Service) chaining: (1)Interface level. The services Running In the Service-oriented workflow engine are adopted by OASIS[2] standard, but spatial information services are adopted by OGC (Open Geospatial Consortium) standard which leads to the spatial information services not be aggregated by the workflow. (2)Spatial data level. The OGC standardizes spatial information services like WFS (Web Feature Service) and WPS (Web Processing Service) exchanged with the GML format, and the intensive operation with GML makes the efficiency more lowly in chaining spatial information services by workflow.

By now, The research of spatial information services chain based on the workflow focuses on two aspects: one aspect is about the services chain ordered by BPEL (workflow description language), and then be sent into the workflow engine[3-4]. The other aspect is that the using of ontology or semantic approach to build a service-driven workflow chain[5-6]. But these methods are mainly discussed about workflow in the conceptual level, and not for the modeling of spatial information services from the field of GI (Geographic information). Therefore, in this paper, we research how to change the OWS service interface, and improve the execution strategies in services chain so that a seamless services chain can be used in OWS graphics workflow.

G. Lee (Ed.): Advances in Intelligent Systems, AISC 138, pp. 85–93.
springerlink.com     © Springer-Verlag Berlin Heidelberg 2012

## 2    OWS Services Chain and Graphical Workflow

**The choice of spatial information services.** OGC (Open Geospatial Consortium) standardizes multi-spatial information services: the cartography services like (WMS、WCS)、 the features services  like(WFS) and the processing services (WPS). In this paper, we use the WFS to be the spatial date resource, and the WPS to be the services chain nodes to build the OWS services chain.

**Graphics workflow.** The kind of Workflow engine in accordance with the process modeling can be divided into BPEL process modeling and graphical workflow modeling. In the processing modeling of BPEL-based workflow, the principle is the use of BPEL language to order service chain, and then to sent into the workflow engine to implement. But because of lacking of visual tools to support, the workflow wrote in the BPEL is more difficult.

The graphical modeling workflow should be drag in the manual way to customize the service chain through graphics proxy. Compared with BPEL workflow process modeling it has the following advantages: The graphical modeling workflow can provide a visual process tool which users do not need to manually write a process modeling language. The client applications only need to implement the visualization of WMS and WFS, and the services chain can be executied by the graphics workflow which guarantee The architecture of application client、 graphical workflow engine be the maximum degree of loosely coupled. The service-oriented agents are able to dynamically discover、 aggregate OWS services through the interface transformation, and it is easy to implement the scheduling policies of GML data (3.2 explained). The typical graphical workflow is the Kepler[7] made by University of California. In this paper we choice the Kepler workflow to chain the OWS services.

## 3    The Model of Graphical Workflow Drive OWS Services Chain

**Traditional WPS services chain pattern.**WPS[8] standard defined a simple services chain which required individual WPS should be in KVP edcoding to submit the GET request. This can be described:

$$GetURL=req\ (req_1(req_2(req_i(...))))\ \ \ \ \ \ \ \ \ \ \ \ (1)$$

In the model, the "req" is the service request for the current WPS, "req$_i$"  is the OWS request participated in the service chain, including the operation of GetFeature in the WFS and the operation of Execute in the WPS. From the formulate(1) we can seen, the service chain of traditional WPS defined by the way of nested URL service request in the aggregation, because of the limiting of URL encoded character for this service chain which only suit for a small number of OWS service chain; And the nested services in the aggregation bring a great deal of difficulty to the service monitoring and exception capture.

**Descripton of graphical workflow model.** Using the graphics workflow engine to aggregate the OWS services is no longer adopted KVP encoding interface. Seen from the Figure 1. We assume there are four OWS services like WFS$_1$, WFS$_2$, WPS$_1$, WPS$_2$. First, We add the WSDL description to the services above, and the message exchanged among the services which be wrapped with SOAP binding; and then we change the WFS interface to enable to output the GML address reference, as well as add the parser for WPS output which be easy to pass the compressed GML data like GML ZIP; Finally, we use the visual interface of the  workflow to build OWS service chain, then implement and monitor the processing of services chain through the graphical workflow engine.

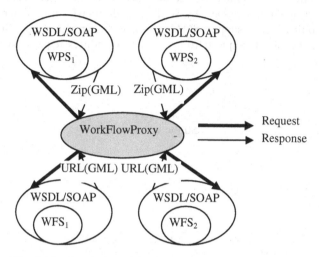

**Fig. 1.** The description of OWS chaining model with Workflow

The graphics workflow engine can be a control center to control each OWS services. The aggregation follows (2):

$$V = WorkFlow (OWS_1, OWS_2, ..., OWS_n) \tag{2}$$

In this model, WPS is no longer to be a service chain entrance uniquely, but to be split into multiple nested WFS, WPS services. OWS services in the Workflow aggregate each other by way of multi-point value, rather than the single point of WPS service aggregation described in formulae (1).

## 4   The Implementation of Graphical Workflow Drive OWS Services Chain

According to the characteristic between Kepler workflow and the spatial information services, we propose the OWS aggregation strategy though graphic workflow engine.

**OWS interface wrapper.** There is a big difference in the internal interface and external interfaces between OWS services and graphic workflow:

1) External Interface: ①Kepler workflow oriented standard Web services. The services are controlled by workflow Node which be described with the WSDL, and the message for the communication between services is adopted in SOAP protocols. But the OGC

did not give the relevant specifications; ②While OGC proposed that the WPS would
support WSDL、SOAP protocol requirements in the file of OWS-5, but did not clearly
support the WFS wrapped with WSDL so that integration between WFS and WPS can
not be reconciled though graphical workflow.

To the question about the inconsistencies in external interface, we need package the
interface between WFS and WPS. Take the WPS for example, it contains three interfaces :GetCapabilities、Execute and DescribeProcess. We only need to encapsulate the
interface of GetCapabilities request by WSDL, while transform the interface of Execute request so that it can   support SOAP protocol, shown in Figure 2. So the kepler
engine can get the SOAP request from the GetCapabilities request with packaged
interface.

```
<wsdl:definitions name="SpatialAnalysis" >
<wsdl:types>
<xsd:schema elementFormDefault="qualified" >
<xsd:element name="ExecuteProcess_GMLBufferResponse">
<xsd:complexType>
<xsd:sequence>
<xsd:element maxOccurs="1" minOccurs="0" name="ExecuteProcess_GMLBufferResult" />
</xsd:sequence>
</xsd:complexType>
</xsd:element>
</xsd:schema>
</wsdl:types>
<wsdl:message name="ExecuteProcess_GMLBufferSoapRequest">
<wsdl:part name="parameters" element="tns:ExecuteProcess_GMLBuffer" />
</wsdl:message>
```

**Fig. 2.** WPS with WSDL

2) Internal Interface: ① While we add the WSDL description to the   GetCapabilities
request in WPS, and the type of input parameters in WPS is not described in the Get-
Capabilities interface, but in the DescribeProcess (DescribeFeatureType in WFS)
interface, kepler workflow can not extract the required parameter type   which is necessary for SOAP request from the GetCapabilities interface.② There are not OGC
standards to do peremptory specialization for the input and output parameters in WPS,
thus, the parameters could be a separate type of argument, Either it can be complex
multi-parameter, so the WPS can not dynamically identify the complex type parameters
form the entrance.

**Fig. 3.** WPS output interface definition

For example, seen the depiction of the OWS service in the Fig.1, the type of Re-
sponse1 of WPS1 may be the GML entity data types also be the reference type in the
URL address. So, in the service chain driven by Kepler, the input type of $WPS_2$ can not

be confirmed, because the two types of parameters are the weakly typed language like AnyURL. The essence of the problem is which the graphics workflow can not automatically resolve non-entity of GML data.

To Solve this problem, it requires add the GML parser for WFS and WPS services in the output interface. Seen from the Fig.3. Supposing the output type of WPS1 is GML contains the URL address, non LineString type, then the GML parser can parse out the correct type like URL address for next node called WPS₂, instead of GML data. Should be noted that, the OWS service chain does not need to be design, Kepler provides XML parsing component, users need to use this component to extract the required parameters and types.

**Scheduling strategy in OWS service chain.** OWS service chain exchange each other with GML for data sharing, but the GML-intensive operation bring lowly efficiency to OWS service chain. Because large data volume of GML increased the congestion overhead in network, there need more transmission time.

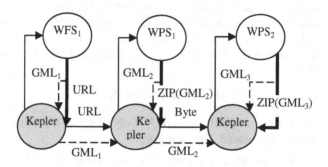

**Fig. 4.** Data flow of OWS services chaining

Figure 4 depictures the two strategies among the data scheduling. Thin solid line means the service requests using the kepler agent, and the dashed and thick solid lines represent two ways of data transmission. Data request, transmission are controlled   by each graphical node.

The methods of traditional data Scheduling are described in dotted lines, all of the OWS data services return the data of GML, which is a high-density data operation mode. the data transmission using the following method:

$$GML_1= KeplerProxy(WFS_1:GetFeature) \tag{3}$$

$$GML_2=KeplerProxy(WPS_1:Execute(GML_1)) \tag{4}$$

$$GML_3=KeplerProxy(WPS_2:Execute(GML_2)) \tag{5}$$

There are two steps include service response and data transfer from one OWS service to next OWS service, the overall time consumption is:

$$T=2O(N_1)+ 2O(N_2) +O(N_3) \tag{6}$$

therein,  N1、 N2、 N3 represent the quantity respectively about $GML_1$, $GML_2$, $GML_3$.

In order to reduce the number of GML transmission in the network, this paper changes the  scheduling strategy in OWS service chain, as shown in thick solid line above, the processing include three step: ①Change the output type of WFS to the complexType contains the URL. ②Change output type of WPS to the binary stream transmitted over the network.③ Add the Interface parser (3.1) for WFS and WPS, and ensure the parameters type can be resolved correctly. The process can be expressed by following formula.

$$URL=KeplerProxy(WFS1:GetFeature) \tag{7}$$

$$ZIP(GML_2)=KeplerProxy(WPS_1:Execute(URL)) \tag{8}$$

$$ZIP(GML_3)=KeplerProxy(WPS_2:Execute(ZIP(GML_2))) \tag{9}$$

In the data stream Scheduling, because of the parser for the interface of Kepler, $WPS_1$, $WFS_1$ could accept the URL from GML. The workflow engine should not have to first download GML locally, and then transmit to the $WPS_1$, but could respond the  results as a URL handling directly to the $WPS_1$, which   avoid the GML response cost to the transfer agent client; at the same time, WPS1 could output GML in ZIP format, decrease the time consumption. The entire time consumption of service chain transmission is expressed as:

$$T= O(N_1)+ 2O(K*N_2) +O(K*N_3) \tag{10}$$

K is the compression rate of GML. Since K * N <N (0 <K <1), the time consumption significantly reduced Compared with formula (6).

**The process of Kepler schedules OWS services chain.** Under the conditions of corrective interface and the improved scheduling strategies, we propose the OWS service agregation process by the kepler workflow.

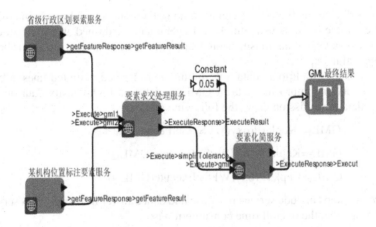

**Fig. 5.** Kepler workflow model

Fig.5 describes the flow diagram using kepler engine to chain the OWS service. In kepler, the nodes oriented Web services called Actors, which are essentially service proxy client, and execute any WSDL-defined Web services through a simple plug-in mechanism, at same time, links to other Actors through its port. Using the customized graphical interface supported by kepler workflow, firstly, we need to initialize the service agent Actors, then configure the proxy port in accordance with the service parameters required to by WFS、WPS respectively, finally, connect the Actors following the business of OWS services chain. The mapping between data and Actors is described in Fig.4.

1) Interface reconstruction ①Firstly, change the interface of OWS services.②Transform WFS output interface which contains the URL type reference.③ Add the parser for the WPS interface.

2) Service discovery: kepler engine provides search tools as agent Actors for Web services which could automatically send SOAP requests described by WSDL, and capture the output or abnormal.

3) Service Aggregation: ①Configure agents of Actor1 and Actor2. Send GetFeature request (SOAP protocol) to server and capture data of $GML_1$, $GML_2$ returned.②Put out the URL of address reference by the GML Parser.③Agent actor3 receive URL reference and call to the WFS's GetFeature request for the GML data; then invoke intersection operation algorithm to get the data $GML_3$.Finally, referred to the feature simplification services to get the data of $GML_4$, and then display the $GML_4$ by the client application.

# 5    Experimentation and Analysis

Experiment showed the effect of Kelper workflow aggregate OWS services chain, tested the performance of the aggregation strategy, and would use the address reference, compressed GML and GML-intensive operations strategy to do a comparative analysis.

**The data and environment of experiment.** Here, the instance of OWS services chain participating the experiment was proposed. Firstly, experiment accesses the point features of the national institution, and then Point Features make the intersection with polygons of the provinces. After Douglas-Pcucker polygon simplified, the results returned to the client to display. Spatial information services related to experiments are used OGC standards, for example:

1) China's provincial-level administrative division factor services ($WFS_1$)
2) Point features of the national institution services ($WFS_2$)
3) Processing services that make the intersection between the elements ($WPS_1$)
4) Simplify elements services ($WPS_2$)

In the experiment, WFS involved in aggregation were used Geoserver (Open Source mapping service engine) for distribute, and the WPS were distributed by spatial information processing components called 52oNorth which is open source software. The GML data generated by aggregation was displayed by the client called Openlayers.

**The result and the analysis.** The execution of the workflow has been shown in Figure 5. Figure 6 is the result of aggregation. Figure(a)、Figure(b) and Figure(c) shown the atomic service by the client of openlayers. Figure(d) displayed the results of GML which is be simplified by WPS (Douglas-Pcucker Simplification).

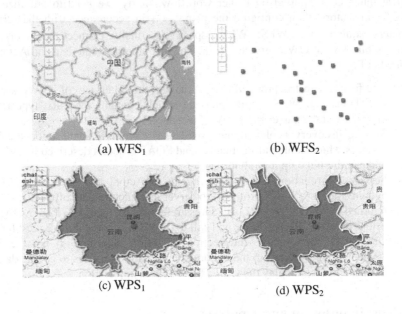

(a) WFS$_1$　　　　　　　　　　　(b) WFS$_2$

(c) WPS$_1$　　　　　　　　　　　(d) WPS$_2$

**Fig. 6.** OWS services

In order to verify different effects of the implementation using the different approach as address reference、compressed GML and GML-intensive strategies for different size of data, the experiment offers three different particle size data at the provincial level administrative divisions: 400W、100W and 25W vector data (mainly simplify administrative divisions). After The GML data distributed by Geoserver, the data size were 1.47 MB and 11.2 MB and 23.5 MB. The table.1 and table.2 show that the time cost in different strategies and different environments. The experimental results show that when the amount of GML data is small, the time advantage after optimal scheduling strategy is not obvious; But when the greater amount of GML data processed by the service chain, the optimal scheduling of the aggregate effect is more evident.

The time costs in implementation of the entire service chain aggregation can be divided into three parts: ①the costs of GML transported.②the costs of GML parsed and processed by WPS$_1$. ③the costs of GML parsed and processed by WPS$_2$. Actually, the spatial processing algorithm depends on the superior performance of the algorithm on the service side. If two strategies using the same algorithms of WPS, the time spent was no difference. Therefore, the overall performance depends on the capabilities of transmission and serialization and parsing about GML.

If a small amount of GML data transmitted, there was little difference between GML parsing and transmitting in network. However, when the GML data is large, the cost of GML transmission in network doubled. Compared to formula (6) and formula (10), we can find that using the GML strategy of address reference, and compressed GML avoided the time consumption of download GML to the local, while GML by way of compression reduces costs in network transmission, the overall time consumed nearly was half of the time of GML-intensive strategy.

**Table 1.** Time costs of Strategies with larger GML/s

| GML | $T_{WFS}$ | $T_{WPS1}$ | $T_{WPS2}$ |
|---|---|---|---|
| 1.47MB | 1.2 | 2 | 3.5 |
| 11.2 MB | 17 | 19 | 11 |
| 23.5 MB | 52 | 61 | 29 |

**Table 2.** Time costs of optimizing Strategies/s

| GML | $T_{WFS}$ | $T_{WPS1}$ | $T_{WPS2}$ |
|---|---|---|---|
| 1.47 MB | 1 | 2.5 | 1 |
| 11.2 MB | 9 | 16 | 5 |
| 23.5 MB | 28 | 37 | 10 |

## 6    Conclusion

This paper provides a new solution for the aggregation among OWS services using graphical workflow. The experiment shows that this approach not only breaks the interface bottlenecks, but also improves efficiency of the OWS services chain by re-building the scheduling strategies about the flow of data.

## References

1. Alameh, N.: Chaining Geographic Information Web Services. IEEE Internet Computing 7(5), 22–29 (2003)
2. OASIS, Reference model for service oriented architecture 1.0. Technical report, Organization for the Advancement of Structured Information Standards (2006)
3. Jia, W.-Y., Li, B., Gong, J.-Y.: Research on Dynamic GIS Chain Based on Workflow Technology. Geomatics and Information Science of Wuhan University 30(011), 982–985 (2005)
4. Kiehle, C., Greve, K., Heier, C.: Standardized geoprocessing-taking spatial data infrastructures one step further. Citeseer (2006)
5. Lemmens, R., Wytzisk, A., de By, R., et al.: Integrating semantic and syntactic descriptions to chain geographic services. IEEE Internet Computing, 42–52 (2006)
6. Friis-Christensen, A., Ostländer, N., Lutz, M., et al.: Designing service architectures for distributed geoprocessing: Challenges and future directions. Transactions in GIS 11(6), 799–818 (2007)
7. Kepler: An Extensible System for Scientific Workflows,
   http://kepler.ecoinformatics.org
8. OGC, Web processing service v1.0.0. Implementation specification OGC 05-007r7, Open Geospatial Consortium Inc. (2007b)

# Complex-Valued Neural Network Based Detector for MIMO-OFDM Systems

Kai Ma, Fengye Hu, and Peng Zhang

College of Communication Engineering, Jilin University, Changchun 130025, China
mk710@sina.com

**Abstract.** The MIMO-OFDM detector with near-optimum performance and low complexity is valuable. This paper uses the complex-valued neural network to implement a detector for MIMO-OFDM systems according to the channel matrix in frequency domain which is composed of complex number. The parameters related to the network stability are given. The output of proposed detector is just the output of all neurons. This detector has advantage of low complexity because neuro-computing is a computational process with low-complexity. The simulations show that the proposed detector is one which can obtain a near-optimum performance, but with a low-complexity.

**Keywords:** Complex-Valued Neural Network, MIMO-OFDM, maximum likelihood detector, computational complexity.

## 1 Introduction

Next-generation mobile communication (B3G/4G) has become a research focus. It is well known that the combination of MIMO and OFDM is considered as the most suitable technology for 4G [1]. In MIMO-OFDM systems, the difficulty of signal detector research is the balance of contradictions between computational complexity of detector and its performance. Among those typical detector, maximum likelihood detector (MLD) has the best performance than any others, but also has the unacceptable computational complexity which is $O(M^{Nt})$, it means the solution space of maximum likelihood (ML) function exponential expands when the modulation order $M$ or the number of transmit antennas $N_t$ increases. So the ML algorithm has no practicality, and the research of detector with low-complexity and near-optimum performance is absolutely meaningful [2].

In recent years, the research techniques on near-optimum detector with lower complexity can be probably divided into two areas. On the one hand, reducing the search space of MLD or excluding the unreliable candidate vectors is used for MIMO systems [3,4]. On the other hand, by taking the advantage of intelligent optimization algorithms which can quickly find the optimum solution, the times of searching the whole candidate vectors space is reduced: The Hopfield neural network was first employed in CDMA multi-user detector [5]. The genetic algorithms can provide a near-optimum MIMO-OFDM multiuser detector with significantly lower computation complexity [6]. A reduced complexity MIMO detector employed Hopfield neural

G. Lee (Ed.): Advances in Intelligent Systems, AISC 138, pp. 95–101.
springerlink.com      © Springer-Verlag Berlin Heidelberg 2012

network was proposed in [7]. Those applications mean that the intelligent optimization algorithms can provide us a compromised detection algorithm between complexity and performance.

In this paper, we proposed a novel detector based on complex-valued neural network (CVNN) [8] to give a solution to hard combinatorial optimization problem which comes from maximum likelihood detector of MIMO-OFDM systems. And it also can balance the complexity and performance.

## 2    MIMO-OFDM Model

The mathematical relationship of signals between transmitter and receiver in an $N_t \times N_r$ MIMO-OFDM systems can be simply described as:

$$y = Hx + n. \tag{1}$$

Here $y$ represents the ($N_r \times 1$)-dimensional received signal, $x$ is the ($N_t \times 1$)-dimensional transmitted modulated symbol, $n$ is the ($N_r \times 1$)-dimensional noise vector, which is a complex Gaussian noise with zero-mean and variance of $\sigma^2$, $H$ is the ($N_r \times N_t$)-dimensional matrix which describes the fading channel in frequency-domain, $h_{ij}$, is independent, stationary, complex Gaussian distributed processes with zero-mean and unit variance, represents the channel frequency response form transmit antenna $j$ to receive antenna $i$. Specifically, the vector $y, x, n$ are given by:

$$y = [y_1, y_2, \cdots, y_{Nr}]^T, \; x = [x_1, x_2, \cdots, x_{Nt}]^T, \; n = [n_1, n_2, \cdots, n_{Nr}]^T. \tag{2}$$

Here $[\cdots]^T$ denotes transpose. The frequency domain subcarrier index is omitted for the sake of efficient notation.

## 3    ML Detctor for MIMO-OFDM

The optimal maximum likelihood detection uses a global search to find the most likely transmitted symbol minimizing the Euclidean distance. It is given by:

$$\hat{x} = \arg \min_{x \in M^{Nt}} \| y - Hx \|^2. \tag{3}$$

Here we also omitted the frequency domain index. The $y, x$ and $H$ are defined by (2), respectively. The set $M^{Nt}$ in (3) is a set of vector, contains all the possible candidates of ($N_t \times 1$)-dimensional transmitted symbol vector, is formulated as:

$$M^{Nt} = \left\{ x = \left( x^{(1)}, x^{(2)}, \cdots x^{(Nt)} \right) \middle| x^{(1)}, x^{(2)}, \cdots x^{(Nt)} \in M_c \right\}. \tag{4}$$

Where $M_c$ is the constellation points set that includes $2^M$ complex constellation points associated with the specific modulation scheme employed. For the reason of vector search space $M^{Nt}$ expanding exponentially when the system employs numerous transmit antennas and high modulation scheme, so that the MLD is quite constricted in practical application.

# 4    Complex Valued Neural Network Based Detector

**Complex-Valued Neural Network(CVNN).** Complex-valued neural networks has an architecture of Hopfield network that is a network with iterative feedback neurons [8].The neuron is a complex-signum function, computes an output $s_i(t)$ from $h_i(t)z^{1/2}$. Specifically, the input $h_i(t)z^{1/2}$ and output $s_i(t)$ are given by:

$$s_i(t) = CSIGN\left(h_i(t-1) \cdot z^{1/2}\right). \tag{5}$$

Here, $z = e^{i\phi_0}$, $\phi_0 = 2\pi / K, (K = 2,3\cdots, N)$, $K$ is called the resolution factor, separates complex plane into $K$ equal sectors.

$$h_i(t) = Ws(t) + I = \sum_j w_{ij} s_j(t) + \theta_i. \tag{6}$$

Where $w_{ij}$ represents the synaptic weight between every two neurons $i, j$, it can be complex-valued. $\theta_i$ is a complex-valued threshold. *CSIGN*, the complex-signum function, is defined as follows:

$$CSIGN(u) \overset{df}{=} \begin{cases} z^0, & 0 \le \arg(u) < \varphi_0 \\ z^1, & \varphi_0 \le \arg(u) < 2\varphi_0 \\ \vdots & \vdots \\ z^{K-1}, & (K-1)\varphi_0 \le \arg(u) < K\varphi_0 \end{cases}. \tag{7}$$

$N$ neurons generate a $(N \times 1)$-dimensional output vector:

$$s(t) = [s_1(t), s_2(t), \cdots, s_N(t)]. \tag{8}$$

Network stability can be measured by energy function for the reason of the network is Hopfield structured. The function decreasing to an unchanged value means the networks stable after updating network several times. The energy function of the complex-valued network is postulated as:

$$E(s) \overset{df}{=} -\frac{1}{2} \sum_i \sum_j w_{ij} s_i s_j + \sum_{i=1} s_i \theta_i. \tag{9}$$

Here $s_i$, $s_j$ are the output from different neurons respectively. When the synaptic weight matrix is Hermitian ( $w_{ij} = \overline{w}_{ji}$ ), diagonal entries are nonnegative ( $w_{ii} \ge 0$ ) and $s_i \theta_i$ is constant bias, the energy $E$ is real-valued, is properly defined and it will not increase during the evolution (iteration) of network.

**Extract the Network Parameters.** Utilizing CVNN to perform ML detector can be done by transforming ML equation into the form of energy function equation and extracting the corresponding network parameters in energy function. Derivation for the equation transmutation is as follows:

$$
\begin{aligned}
&\|y - Hx\|^2 \\
&= (y - Hx)^T \overline{(y - Hx)} \\
&= (y^T - x^T H^T)\overline{(y - Hx)} \\
&= y^T \overline{y} - y^T \overline{Hx} - x^T H^T \overline{y} + x^T H^T \overline{Hx}
\end{aligned}
\qquad (10)
$$

Where $\overline{(\cdots)}$ denotes conjugate. By removing the constant term $y^T \overline{y}$, and for a BPSK modulation scheme, $x$ is a vector which is composed of -1 or 1, hence, $x = \overline{x}$, the equation can be simplified as follows:

$$
-x^T \overline{H}^T y - x^T H^T \overline{y} + x^T H^T \overline{H} x = x^T (H^T \overline{H}) x - x^T (\overline{H}^T y + H^T \overline{y}). \qquad (11)
$$

It is easy to see $\overline{H}^T y = \overline{H^T \overline{y}}$, so that $\overline{H}^T y + H^T \overline{y}$ is a vector which is composed of real number, thus the later term in (11) equals to a real number. In order to satisfy other conditions for network stability, the first term needs to be expanded as follows:

$$
x^T (H^T \overline{H}) x = \sum_{i=1}^{N}\sum_{j=1}^{N} h_i^T \overline{h_j} x_i x_j - \underbrace{\sum_{i=1}^{n} h_i^T \overline{h_i}(x_i^2 - 1)}_{0} = \sum_{i=1}^{N}\sum_{j=1}^{N} h_i^T \overline{h_j} x_i x_j - \underbrace{\sum_{i=1}^{n} h_i^T \overline{h_i}}_{\substack{\text{diagonal}\\\text{entry}}} x_i^2 + \underbrace{\sum_{i=1}^{n} h_i^T \overline{h_i}}_{\text{constant}}
$$
$$
\Leftrightarrow \sum_{i=1}^{N}\sum_{j \neq i}^{N} h_i^T \overline{h_j} x_i x_j
\qquad (12)
$$

Where $h_i$ represents the $i$th column of $H$, and $(H^T \overline{H})$ is a $(N \times N)$-dimensional matrix, the diagonal entry is $h_i^T \overline{h_i}$, otherwise $h_i^T \overline{h_i}$ is constant which has no relationship with variable $x$, so it is removed in the above derivation, which will not influence minimizing the $\|y - Hx\|^2$. After a series of derivation and simplification, we obtain:

$$
\hat{x} = \arg\min \|y - Hx\|^2
$$
$$
\Leftrightarrow \arg\min\left\{ \sum_{i=1}^{N}\sum_{j \neq i}^{N} h_i^T \overline{h_j} x_i x_j + \sum_{i=1}^{N} x_i (\overline{h_i}^T y + h_i^T \overline{y}) \right\}. \qquad (13)
$$

Now we can extract the parameters from (12) by comparing with (8). Note that the $s_i$ in (8) is equivalent to $x_i$. Hence, the network parameters are defined as:

$$
W_{ij} = \begin{cases} -2h_i^T \overline{h_j}, & i \neq j \\ 0, & i = j \end{cases}, \quad I = \overline{h}^T y + h^T \overline{y}. \qquad (14)
$$

Where matrix $W$ is Hermitian, and the mathematical justification is given below:

$$\overline{W_{ji}} = \overline{-2h_j^{T}\overline{h_i}} = \overline{-2h_j^{T}}h_i = -2h_i^{T}\overline{h_j} = W_{ij}, \quad i \neq j. \qquad (15)$$

When $x=s$, Equation (5) can be equally substituted by energy function form equation:

$$\hat{x} = \arg\min\|y - Hx\|^2 = \arg\min\left(-\frac{1}{2}\sum_{i=1}^{N}\sum_{j\neq i}^{N}w_{ij}x_ix_j + \sum_{i=1}^{N}x_i\theta_i\right) = \arg\min\left(-\frac{1}{2}x^{T}Wx + x^{T}I\right). \qquad (16)$$

When employing BPSK modulation scheme, the $x_i$ is 1 or -1. When resolution factor $K$ is 2, the output of each neuron is given by:

$$CSIGN(u) \stackrel{df}{=} \begin{cases} z^0 = 1, & 0 \le \arg(u) < \pi \\ z^1 = -1, & \pi \le \arg(u) < 2\pi \end{cases}. \qquad (17)$$

Thus, when the output of neurons is stable, which means the CVNN has converged and the network energy reaches its minimum. In fact, this minimum is the minimum of MLD formulation (5), so that the output which minimizes the network energy is just the solution of MLD.

**Algorithm Steps.** MIMO-OFDM detector is achieved by detecting every single subcarrier separately. CVNN based detector detecting steps is as follows:

a) Set the network parameters according to (14).
b) Choose a initialization vector from candidate vector search space $M^{Nt}$, then use it to calculate the initial input $s(0)$ according to (6) and (7).
c) Update the network with initial input $s(0)$ for $n$ times.
d) When the iterations is reached, the output $s(n)$ is just considered as the output of CVNN based detector. If the network has been already stable when the iterations has not reached, the network is not going to evolve toward next time, and the output of this time $s(t)$ is just what we want.

## 5    Simulations and Results Analysis

The simulations were performed on the slow fading typical urban channel 6 (TU6) channel. The MIMO-OFDM systems simulation parameters and CVNN parameters can be found in Table 1. Channel parameters are given in Table 2, where $\tau$ is tap delay and $P$ is fading coefficient.

**Table 1.** Simulation Parameters

| MIMO-OFDM Parameters | | CVNN Parameters | |
|---|---|---|---|
| Transmit antennas | 4 | Numbers of neuron | 4 |
| Receive antennas | 4 | Resolution Factor | 2 |
| FFT size | 128 | Iterations | 8 and 12 |
| Subcarrier Bandwidth | 15KHz | | |
| Modulation Scheme | BPSK | | |
| Simulation Frames | 80 | | |

Performance comparison between purposed detector and conventional detector is shown in Fig. 1. Performance of purposed detector on different iterations is shown in Fig. 2.

**Table 2.** Channel Parameters

| $\tau$ [ns] | 0 | 200 | 500 | 1600 | 2300 | 5000 |
|---|---|---|---|---|---|---|
| $P$ [dB] | -3 | 0 | -2 | -6 | -8 | -10 |

Four detectors were included in simulation in Fig. 1. It is easy to see that CVNN based detector with 8 iterations has clearly lower bit-error-rate (ber) than MMSE and ZF detector. Because of more iterations of network employing, more likely stable the network will be, and the energy value is more likely to reach the minimum that minimizes the Euclidean distance of MLD. The CVNN based detector with 12 iterations has a lower ber than the one with 8 iterations for the reason of a lot more of subcarriers are perfectly detected by completely stable network. CVNN based detector doesn't have a better ber than ML detector, because all subcarriers are not fully detected by a stable network or the network may converge to a local stable state.

The complexity of CVNN based detector is much lower than MLD because of the initialization vector we chose is MMSE solution which is a linear detector with low-complexity, and the CVNN based detector judge whether the MMSE solution is a bit error by network stability. The MMSE solution which causes bit error will activate the network. Then the network will iterate certain times. The complexity of per network iteration is a matrix multiplying a vector, and then pulse a vector, which is much lower than ML detector.

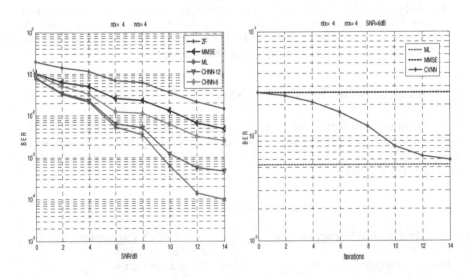

**Fig. 1.** BER versus SNR performance comparison between CVNN based detector and conventional detector

**Fig. 2.** BER versus Iterations performance of CVNN based detector while SNR=6dB

# 6   Conclusion

This paper represents a novel detector for MIMO-OFDM systems, which can judge the outcome of linear detector, and keep on detecting the outcome which cannot stabilize the network when it is a initialization vector. And this further detection is able to find a solution in the manner of maximum likelihood detector by transforming ML equation into energy-function-like equation. Neuro-computing has a very low complexity, so that the CVNN based detector has clearly lower complexity than ML detector, and simulation results show there are two advantages than conventional detector. First, it is near-optimum detector comparing with ML detector. Second, the performance of this detector can be actively changed by choosing different iterations when we need a detector with low ber or with a low complexity.

# References

1. Paulraj, A.J., Gore, D.A., Nabar, R.U., Bölcskei, H.: An Overview of MIMO Communications—A Key to Gigabit Wireless. Proceedings of the IEEE 92(2), 198–218 (2004)
2. Akhtman, J., Wolfgang, A., Chen, S., Hanzo, L.: An Optimized-Hierarchy-Aided Approximate Log-MAP Detector for MIMO Systems. IEEE Transactions on Communication 6(5), 1900–1909 (2007)
3. Kim, J.S., Moon, S.H., Lee, I.: A new reduced complexity ML detection scheme for MIMO systems. IEEE Transactions on Communication 58(4), 1302–1310 (2010)
4. Woo, H.M., Kim, J., Yi, J.H., Cho, Y.S.: Reduced-Complexity ML Signal Detection for Spatially Multiplexed Signal Transmission Over MIMO Systems With Two Transmit Antennas. IEEE Transactions on Vehcular Thechnology 59(2), 1036–1041 (2010)
5. Kechriotis, G.I., Manolakos, E.S.: Hopfield neural network implementation of the optimal CDMA multiuser detector. IEEE Transactions on Neural Networks 7(1), 131–141 (1996)
6. Jiang, M., Hanzo, L.: Multiuser MIMO-OFDM for next-generation wireless systems. Proc. IEEE 95(7), 1430–1469 (2007)
7. Louw, D.J., Botha, P.R., Maharaj, B.T.: A low complexity soft-input soft-output MIMO detector which combines a Sphere Decoder with a Hopfield network. In: 15th IEEE Mediterranean Electrotechnical Conference, pp. 521–526 (2010)
8. Jankowski, S., Lozowski, A., Zurada, J.M.: Complex-Valued Mulitstate Neural Associative Memory. IEEE Transactions on Neural Networks 7(6), 1491–1496 (1996)

# Innovation System of Agricultural Industrial Cluster: A Perspective from Dynamical Structure Model

Xiaotao Li

School of Economics & Management, Wuhan Polytechnic University, Wuhan, China
xiaotaowhu@163.com

**Abstract.** Applying the theories of Innovation and System Dynamics, We analysis quantitatively and research the element structures, functions and dynamical mechanism of the Cluster Innovation System by means of setting up the Cluster Innovation System's dynamical structure model. Finally we make analysis on the driving force of economic development and development path in clusters, which supplies the theory and decision support for the formation and implementation of clusters' innovative strategies.

**Keywords:** innovation system, agricultural industrial cluster, dynamical structure.

## 1   Introduction

At present, many provincial and municipal governments in China have formulated policies for development of agricultural industrial clusters as one of crucial measures to strengthen the basic position of agriculture in the national economy and improve the competitiveness of agricultural industry. And some localities laid down corresponding regulations to promote the development of agricultural industrial clusters. However, the academic field has failed to pay due attention to the cultivation of agricultural industrial clusters. Most studies focus on the macro qualitative study of the nature of agricultural industrial clusters, while only a few addresses the evolution of agricultural industrial clusters. Because the development of agricultural industrial clusters, restricted by historic and institutional factors (such as the promotion of type –transitional  economy), has its own endogeny and self-organization,  small-world networks model,  a theory addressing self-organization problems,  provides us with a way  to study non-balanced growth, which tallies with the formation process of agricultural industrial clusters. Therefore, the study of the formation mechanism of agricultural industrial clusters with Dynamical Structure model may generate some fascinating ideas.

After the appearance of "Innovation system" concept, the dynamic and interactive essence in the innovation process has been seriously paid attention to, and the innovation has been regarded as the social process of knowledge creation, transference and proliferation. The integration of innovative system paradigm and cluster's innovative behavior, which forms a new research subject-- Innovation System of clusters (ISC). ISC is an overlay between a cluster Innovation Network and

G. Lee (Ed.): Advances in Intelligent Systems, AISC 138, pp. 103–108.
springerlink.com        © Springer-Verlag Berlin Heidelberg 2012

clusters of innovative environment, It's a cluster carrier for the innovative system. The research on the system is built on innovation in the region, which is inheritance and development of the existing innovation theories and belongs to the domain of Socio-economic system. The System is is a higher-order, nonlinear, time-varying multiple and complex feedback system. We have introduced System Dynamics (System Dynamics referred to "SD") that are "strategy and tactics laboratories" to make discussion on structure features and operation mechanism of ISC in the socio-economic system simulation in order that we could make a deep analysis of system structures and grasp the system operations quantitatively. We analysis quantitatively and research the element structures, functions and dynamical mechanism of the Cluster Innovation System by means of setting up the Cluster Innovation System's dynamical structure model to make sure of development path in ISC, which supplies the theory and decision support for the formation and implementation of clusters' innovative strategies.

The establishment of ISC dynamical structure model is on the purpose of optimization between innovation output and innovation performance. For the one hand, we need to consider the final requirement for innovative strategies, on the other hand we should make good use of innovation to adjust the economic structure, so as to achieve economic development being a virtuous circle. Based on the characteristics of ISC and the main point of modeling, we could find the dominant feedback loop to acquire the further grasp about the cluster's innovative features. Then we could find the key points to constrain clusters' innovations and developments, which provides the basic policy for promoting clusters' innovation.

## 2    System Dynamics Theory

The following are several basic principles in our research based on the System Dynamics Theory. :

**Micro-Structures and Functional Principles.** We research on ISC from micro-structure based on the System Dynamics Theory. And we construct the ISC model in terms with the relation between system structures and functions. The combination of the inner elements of the clusters' innovation is the unity of structures and functions. The opposition between structures and functions is relative, and structures and functions are complementary and reciprocal causations. They can be transformed into each other under certain conditions. Therefore, we should set up a structure model that better reflect the ISC model by inspecting structures and functions of ISC interactively. System Dynamics believes that only through research and analysis on the phenomenon of surface features can not be fully reflected the internal structure of the system. To truly construct ISC structure model we must go to those causal relationships to feedback which can not be fully measured, understand the interrelated elements and the interaction of information. We should relate measurably dynamical and changeable trend to the internal factors by acquiring a correct understanding for ISC, then integrate the understanding into the model structures.

**Movement and Information Feedback Principles.** System Dynamics thinks that ISC is composed of units, the movement of units and information, and information is the foundation of a system structure. Unit is the basis of system, and the information plays a key role in the system. The movement of units could form behaviors and functions of the system and the units of system could form structures depending on the information systems unit to form the structure of the unit movement to form and function of the system. The information feedback loop is the basic structure of ISC, the interaction and couple of these feedback loops form the ISC.

**Dominant Circuit Principles.** There are one or a few dominant circuits in ISC, which is a part of the system. The circuits decide the main structure and dynamics of change in trend. The principle is the dominant circuit principle. System Dynamics is based on this principle in the ISC to find the dominant loop circuit through the analysis of system structures and function behaviors, so that we could further understand the system structure and dynamical characteristics to classify and simplify the system.

## 3    Overall Dynamic Mechanism Description of Innovation System of Agricultural Industrial Clusters

Cluster is a kind of specialized "gather" and inner relationship among related enterprises or institutions which are based on specialized division of labor and cooperations. The process of cluster innovation means to create, store, transfer and apply knowledge by means of formal and informal methods, and then creates agglomeration effect so as to gain advantage of cluster innovation. The paper divides cluster innovation system into external subsystem and inner subsystem, which is divided into meddle-view sub-sub-system and micro sub-sub-system. The operation of cluster innovation system performs as the interaction between subsystems and sub-sub-systems.

We grasp the formation process of the dynamic mechanism from the general view, and provide foundation for dynamical model. The external innovation subsystem includes related policies, environment (regional, country and global level), external clients, external providers, relevant institutions and competitors outside of clusters. This subsystem mainly provides political support with innovation ability and cluster innovation environment for country and region from governmental level (including innovation platform and rules of cluster system, economic and social cultural environment, technical development and market conditions of clusters). The middle-view innovation sub-subsystem includes social institution (profession associations, subject committee, entrepreneurs and leadership associations), government organization, public service agencies, cultural environments, etc. It can promote innovation ability and resources allocation ability through interaction and cooperation between innovation subjects. It can also promote structure adjustment and upgrading, so the economic of country and surrounding areas will develop persistently, steadily and healthily. The enterprise micro sub-subsystem innovation system includes knowledge, information, technology communications and interactions among enterprises and inner enterprises. There are formal economic connections and informal technology contact ions which create, store and transfer knowledge.

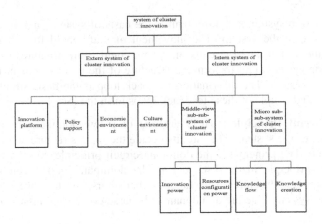

**Fig. 1.** The General Diagram of ISC Dynamical Mechanism

# 4    Structural Dynamics of ISC Model

According to the entire description to the clusters of innovate system dynamical mechanism and the structural characteristics of the system itself, we consider the innovate activities of the different objects in-and-out of clusters(including enterprises, governments, research institutions and intermediary organizations) as the principle series, and aim at improving the innovate capability and efficiency. At the same time, we conform the interactions among these variables to establish a new structural model for clusters of innovation system dynamical mechanism. In the term of system principles, several factors has been involved in the innovation process, such as technology invention, achievement conversion, enterprise technology innovations, market demand inductions, structural changes of the market, diffusion of innovation and the support of colleges, institution, government and enterprises, all of which are related with each other and sometimes need to cooperate in the way of parallel, cycle or overlap. From the dynamical point of view, the cluster innovation is a process including communication of information, and converts the knowledge, skills and substances into the products meeting, the customers' needs and then into clusters.

In the aim of modeling our system, the limitation of cluster innovation system mainly includes the aspects such as Innovation consciousness of government, Innovation encouragement policy of government, Innovation impetus of sci-tech institution, System innovation of higher education, Transfer of sci-tech production, Construction of innovation infrastructure, Collaboration of industry, academics & research, Innovation impetus of enterprise, Market environment, Finance environment, Risk investment of sci-tech innovation, Information & knowledge flow of enterprise, Innovation pervasion of knowledge, Extern customer requirement, Innovation of material&supply of extern supplier.In addition, there are many other elements, such as level of education, scientific research institutions, new products' price, etc., which are excluded out of the limitation of our system.

In the system dynamics, the relationship among elements can be generalized as a causal relations, which help to form the function and activities of the system. So the analysis to the causal relations is the basis of our research of system dynamical modeling. For the many factors in clusters' innovation, we establish the model on the basis of analysis to the factors. According to the description of the innovation characteristics and running procedure, we refine the variables of the system and subsystems, and then combine them to form the causal-relation diagram for the cluster innovation system:

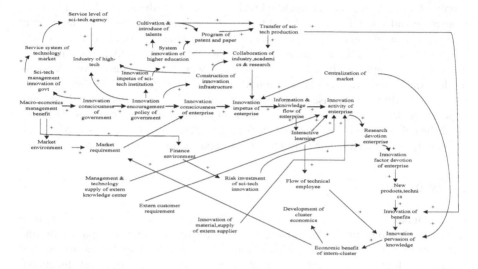

**Fig. 2.** The Causal-relation Structure Diagram for the Sluster Innovation System

From the above causal links between the diagram, it is not difficult to find several development feedback loops of cluster economics:

(1) Macro-economics management benefit→Innovation consciousness of government→Innovation of encouragement policy of government→Innovation impetus of sci-tech institution→System innovation of higher education→Program of patent and paper→Transfer of sci-tech production→Collaboration of industry, academics & research→Innovation impetus of enterprise→Information & knowledge flow of enterprise→Innovation activity of enterprise→Research devotion enterprise→Innovation factor devotion of enterprise→New product, technics →Innovation of benefits→Innovation pervasion of kowledge→Economic benefit of intern-cluster→Development of cluster economics.

(2) Macro-economics management benefit→Innovation consciousness of government →Innovation of encouragement policy of government→Innovation consciousness of enterprise→Innovation impetus of enterprise→Information & knowledge flow of enterprise→Interactive learning→Innovation activity of enterprise→Research devotion enterprise→Innovation factor devotion of enterprise→New product, technics→Innovation of benefits→Innovation pervasion of kowledge→Economic benefit of intern-cluster→Development of cluster economics.

(3) Macro-economics management benefit→Innovation consciousness of government→Innovation of encouragement policy of government→I Innovation

consciousness of enterprise→Innovation impetus of enterprise→Information&
knowledge flow of enterprise→Interactive learning→Flow of technical employee→
Innovation pervasion of kowledge→Economic benefit of intern-cluster→Development of cluster economics.

These positive feedback loops indicate that the macro-level of the driving force for innovation clusters is come from the cluster of external innovation system in the management of external knowledge and technical supply, demand and supply, the supply of innovation, and the middle-level is come from the level of awareness of innovation and scientific research institutions in the innovative ability of micro-level enterprises from the micro-entrepreneurs in the innovation of systems, the enterprise's sense of innovation. From the feedback loops, it is easy to find that clustering innovative strength is based on the skilled workers. By interactive learning mechanism collective enterprises could enhance the flow of knowledge and technology within enterprises and between enterprises to provide the basis and impetus for the business engaged in the technological innovation activities. The flow of technical staff in the clusters makes proliferation of the knowledge and technological innovation in the cluster, and promotes innovation in the entire cluster.

## 5    Summary

By continuous deepening of understanding on the basis of the innovation and regional economic development in the cluster, seeing a lot of literatures on innovation and system dynamics and building a dynamics model of innovation system of agricultural industrial  cluster , which defines the elements of the system structural, functional and dynamic mechanism. Finally this paper put an outline of the analysis on the force and path of clusters economic development based on the causal link between the structure of the clusters. Introducing the system dynamics theory in the research of cluster innovation system, which  provides an effective ways both on qualitative and quantitative analysis and has some reference.

## References

1. OECD. National Innovation Systems. Internet Paper (1997)
2. Yao, X.-J.: Strength of Clusters. Chinese Business Joint Press, Beijing (2006)
3. Lin, Y.-X.: Regional Innovation Advantage. Economic Management Publishing House, Beijing (2006)
4. Zheng, Y.: Analysis and Construction of the Cluster Innovation System's Dynamical Structure Model. CSSE (1), 645–649 (2008)
5. Zhao, L.-M.: The research on the Urban Innovation Dynamical Mechanism. Scientific Research Journal 21(1) (February 2003)
6. Xie, L.-J.: The Knowledge Flow Analysis in the Industrial Clusters Innovation System flow. The Journal of Entrepreneurs World 12 (2006)
7. Wei, J.: The evolution and construction of innovative clusters. Review of Natural Dialectics 01, 51–53 (2004)
8. Li, F.: A study of evolution from industrial clusters to innovative clusters. doctoral dissertation of Wuhan University (2007)

# Improved Ensemble Empirical Mode Decomposition Method and Its Simulation

Jinshan Lin

School of Mechanical and Electronic Engineering, Weifang University,
Weifang 261061, P.R. China
jslinmec@yahoo.cn

**Abstract.** Ensemble empirical mode decomposition (EEMD) is a powerful tool for processing signals with intermittency. However, a problem existing in the EEMD method is the absent guide to how much amplitude of the added white noise should be appropriate for the researched signal. To begin with, the problem was investigated using a noiseless simulated signal. Moreover, based on the conclusions obtained in the above step, the improved EEMD (IEEMD) method was proposed to deal with the noisy signals. Then, a noisy simulated signal was used to measure the performance of the IEEMD method. The results showed that the IEEMD method could greatly alleviate the problem concerning the EEMD method. Additionally, the paper indicates that the IEEMD method may be an improvement on the EEMD method.

**Keywords:** Ensemble empirical mode decomposition (EEMD), Improved ensemble empirical mode decomposition (IEEMD), mode mixing, signal processing.

## 1 Introduction

It is a topic issue to extend signal processing techniques for non-stationary and noisy signals, which has aroused wide concerns in recent decades [1]. Many methods, such as short time frequency transform [2] and wavelet transform [3] , have been addressed for solving the problem and showed useful in some applications. Unfortunately, these methods usually need a priori knowledge about the researched signals, so they are naturally short of the self-adaption for the researched signals. In addition, the Wigner-Ville distribution has high time-frequency resolution, but its cross terms is unbearable [4]. Empirical mode decomposition (EMD) is a self-adaptive method and suitable for analyzing the non-stationary and nonlinear signals [4], successful in various fields [4]. Nonetheless, when the EMD algorithm is employed to explore an intermittent signal, the mode mixing constantly occurs as a vexing problem [5-7]. To alleviate the mode mixing problem, ensemble empirical mode decomposition (EEMD) is proposed as a substitute for the EMD method [8]. The EEMD method adds some white noise with finite amplitude to the researched signals, adequately utilizing the uniform statistical characteristics of white noise throughout the frequency domain, then projects the signal components onto the proper frequency bands and, finally, removes

G. Lee (Ed.): Advances in Intelligent Systems, AISC 138, pp. 109–115.
springerlink.com

the added white noise by ensemble mean of enough corresponding components [8]. Hence, the EEMD method is thought of as a significant improvement on the EMD method and recommended as a substitute for the EMD method [8]. Indeed, the EEMD method has shown its superiority over the EMD method in some applications [9].

However, the EEMD method lacks a guide to how much amplitude of the added noise should be appropriate for the researched signals. Although the reference [8] suggested that the amplitude of the added white noise should be about 0.2 times of the standard deviation of the investigated signal, notwithstanding, with the suggested value, the decomposition results with the EEMD method frequently deviate from the realistic contents of the signals [10]. Additionally, if the researched signal is a naturally noisy signal, its intrinsic noise will inevitably interact with the noise added through the EEMD method, which may further complicate the above problem in regard with the EEMD method. In particular, when the researched signals become very noisy, the above problem remains untouched.

The paper investigated the problem and, as a result, the improved EEMD (IEEMD) method was presented. The application to a noisy simulated signal indicated that the IEEMD method was an improvement on the EEMD method.

## 2    The EMD Method and the EEMD Method

**The EMD method.** The EMD method can self-adaptively decompose any a non-stationary and nonlinear signal into a set of intrinsic mode functions (IMFs) from high frequency to low frequency, which may be written as

$$x(t) = \sum_{i=1}^{N} c_i(t) + r_N(t), \tag{1}$$

where $c_i(t)$ indicates the $i$th IMF and $r_N(t)$ represents the residual of the signal $x(t)$. An IMF is a function which must satisfy the following two conditions: (1) the number of extrema and the number of zero crossings either equal to each other or differ at most by one, and (2) at any point, the local average of the upper envelope and the lower envelope is zero [4]. The residual $r_N(t)$ usually is a monotonic function or a constant.

**The EEMD method.** An annoying problem occurring in the EMD method is the mode mixing due to intermittency, defined as either a single IMF consisting of widely disparate scales, or a signal residing in different IMF components [8]. To alleviate the imperfection of the EMD method, the ensemble EMD (EEMD) [8], a noise-assisted method, is proposed. The EEMD method can be stated as follows:

$$x_m(t) = x(t) + w_m(t), \quad m = 1, 2, \cdots, N, \tag{2}$$

$$x_m(t) = \sum_{i=1}^{L} c_{m,i}(t) + r_{m,L}(t), \quad m = 1, 2, \cdots, N, \tag{3}$$

$$x(t) = \frac{1}{N} \sum_{i=1}^{L} \sum_{m=1}^{N} c_{m,i}(t) + \frac{1}{N} \sum_{m=1}^{N} r_{m,L}(t), \tag{4}$$

where $x(t)$ is the original signal, $w_m(t)$ is the $m$th added white noise, $x_m(t)$ is the noisy signal of the $m$th trial, $c_{m,i}(t)$ is the $i$th IMF of the $m$th trial, $L$ is the number of IMFs from the EMD method, and $N$ is the ensemble number of the EEMD method.

## 3    Improved EEMD Method

**The choice of the amplitude of the added noise for the EEMD method.** The amplitude of the added white noise is essential for the EEMD method, which will have a decisive effect on whether or not the EEMD method can produce the reasonable decomposition results. If the added noise is too faint to cause the changes of extrema of the original signal, the EEMD method will degenerate into the EMD method. Instead, if the added noise is too strong to expose the original signal, the EEMD method will achieve worthless results which mainly reflect the added noise and are barely related to the original signal [10], whether or not the ensemble number is large enough. The issue was carefully investigated in the paper. Successively, the Pearson's correlation coefficient (PCC) was used as a parameter to assess the performance of the EEMD method with different amplitude of the added white noise. Here, the ensemble number of the EEMD method was set as 100.

The problem was examined using a noiseless simulated signal. The signal holding four components models realistic vibration signals of a rolling bearing, shown in Fig. 1, and its formula can refer to [10]. The relationship between the PCCs and the amplitude of the added white noise is represented in Fig. 2. As shown in Fig. 2, when the amplitude of the added white noise is located in the range of 0.0085-0.0138, the four PCCs almost simultaneously arrive at their maximum values. The average powers of four components

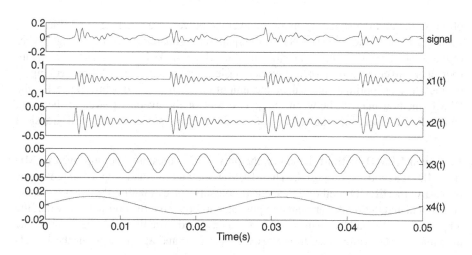

**Fig. 1.** A noiseless simulated signal and its four components

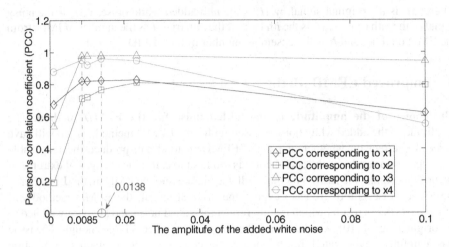

**Fig. 2.** The relationship between the Pearson's correlation coefficient (PCC) and the amplitude of the added white noise for a noiseless simulated signal

**Table 1.** The average powers of four components of the noiseless simulated signal and the square roots of the average powers

| Parameter | The four components of the noiseless simulated signal | | | |
| --- | --- | --- | --- | --- |
| | $x_1(t)$ | $x_2(t)$ | $x_3(t)$ | $x_4(t)$ |
| The average power | $1.9032\times10^{-4}$ | $1.9377\times10^{-4}$ | $5.445\times10^{-4}$ | $7.2\times10^{-5}$ |
| The square root of the average power | 0.0138 | 0.0139 | 0.0233 | 0.0085 |

of the signal and the square roots of the average powers are described in Table 1. As shown in Table 1, the value 0.0085 is just equal to the square root of the average power of the weak sinusoid component $x_4(t)$ and the value 0.0138 is just equal to the square root of the average power of the weak transient component $x_1(t)$. Moreover, Fig. 2 denotes that an optimal internal of the amplitude of the added white noise for the EEMD method may lie between the square root of the average power of the weak sinusoid component and the square root of the average power of the weak transient component.

**Improved EEMD method.** Based on the previous conclusions, the improved EEMD (IEEMD) method is proposed. The IEEMD method is made up of three parts: primarily, the EEMD method is executed coursely for evaluating the weak sinusoid component and the weak transient component; subsequently, the optimal interval of the amplitude of the added white noise is determined; finally, add the appropriate amplitude of the white noise to the researched signal and again carry out the EEMD method.

## 4   Numerical Evaluation

A noisy signal with SNR of 40dB, shown in Fig. 3, was constituted by adding some noises to the preceding simulated signal. The noisy signal was decomposed using the EEMD method and the IEEMD method, and the decomposition results are depicted in Fig. 4 and Fig. 5, respectively. As shown in Fig. 4, there exists the mode spiting in the low-frequency components from the EEMD method. By contrast, as shown in Fig. 5, the IEEMD method is successful in relieving the mode splitting of the EEMD method.

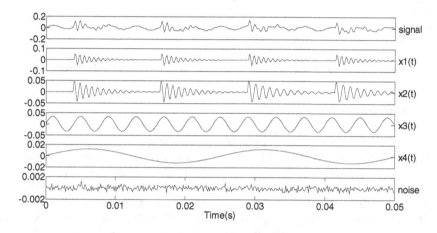

**Fig. 3.** The noisy signal and its components

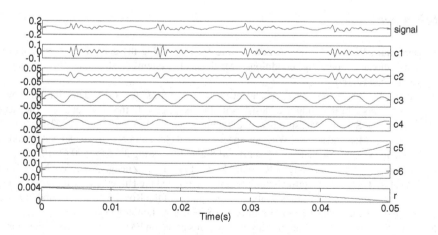

**Fig. 4.** The decomposition results of the noisy signal with the EEMD method

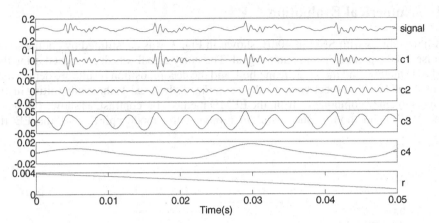

**Fig. 5.** The decomposition results of the noisy signal with the IEEMD method

## 5  Summary

The paper aimed to provide guidance on how much amplitude of the added white noise should be appropriate for the EEMD method. First, a noiseless simulated signal was used to probe the problem. Consequently, the IEEMD method was addressed for analyzing the noisy signals in this paper. Besides, the numerical example has tested the ability of the IEEMD method. The comparisons with the EEMD method showed that the IEEMD method seemed to outperform the EEMD method in alleviating the mode mixing and splitting. Moreover, the paper displays that the IEEMD method is apparently an improvement on the EEMD method.

## References

1. Bartelmus, W., Zimroz, R.: A new feature for monitoring the condition of gearboxes in non-stationary operating conditions. Mechanical Systems and Signal Processing 23, 1528–1534 (2009)
2. Satish, L.: Short-time Fourier and wavelet transforms for fault detection in power transformers during impulse tests. In: Proceedings of the Institute of Electrical Engneering–Science, Measurement and Technology, vol. 145, pp. 77–84 (2002)
3. Jiang, X., Mahadevan, S.: Wavelet spectrum analysis approach to model validation of dynamic systems. Mechanical Systems and Signal Processing 25, 575–590 (2010)
4. Huang, N.E., Shen, Z., Long, S.R., Wu, M.L.C., Shih, H.H., Zheng, Q.N., Yen, N.C., Tung, C.C., Liu, H.H.: The empirical mode decomposition and the Hilbert spectrum for nonlinear and non-stationary time series analysis. Proceedings of the Royal Society of London Series A - Mathematical Physical and Engineering Sciences 454, 903–995 (1998)
5. Ricci, R., Pennacchi, P.: Diagnostics of gear faults based on EMD and automatic selection of intrinsic mode functions. Mechanical Systems and Signal Processing 25, 821–838 (2011)
6. Cheng, J., Yu, D., Tang, J., Yang, Y.: Application of frequency family separation method based upon EMD and local Hilbert energy spectrum method to gear fault diagnosis. Mechanism and Machine Theory 43, 712–723 (2008)

7. Lin, L., Hongbing, J.: Signal feature extraction based on an improved EMD method. Measurement 42, 796–803 (2009)
8. Wu, Z.H., Huang, N.E.: Ensemble Empirical Mode Decomposition: A Noise-Assisted Data Analysis Method. Advances in Adaptive Data Analysis 1, 1–41 (2009)
9. Lei, Y., He, Z., Zi, Y.: Application of the EEMD method to rotor fault diagnosis of rotating machinery. Mechanical Systems and Signal Processing 23, 1327–1338 (2009)
10. Zhang, J., Yan, R., Gao, R.X., Feng, Z.: Performance enhancement of ensemble empirical mode decomposition. Mechanical Systems and Signal Processing 24, 2104–2123 (2010)

# Regional Resource Allocation Algorithm and Simulation of the Adaptive Array Antenna

Cai Weihong, Liu Junhua, Lei Chaoyang, and Wen Jiebin

Changsha Telecommunication and Technology Vocational College, 410015
Changsha, Hunan, China
hncscwh@126.com, angchaolei317@sohu.com, wenjiebin99@163.com

**Abstract.** Array structure is adopted in adaptive array antenna of mobile base station, and the spatial cutting technique is used to divide the coverage area of base station into many differentiated small regions, improve utilizations of the system resources, reduce system interference and increase system capacity. This paper analyzes the space partitioning theory of adaptive array antenna signals, presents the traditional resource allocation algorithm of the adaptive array antenna area: resource fixed regional resource allocation algorithm and maximum and minimum regional resource allocation algorithm. Concerning problems existing in the traditional resource allocation algorithm of the adaptive array antenna area, the allocation algorithm for improvement of regional resources based on minimum standard deviation was put forward, and the principles of its regional resource allocation and calculation methods were introduced. Through simulation of call blocking rates for these three regional resource allocation algorithms, we can know that, the regional resource allocation algorithm of minimum standard deviation can reduce the call blocking rate compared with the traditional regional resource allocation algorithm, and the result in load balancing was best.

**Keywords:** adaptive array, antenna, resources, allocation, algorithm, simulation.

## 1 Introduction

Base station smart antenna can generally be divided into two kinds which are multiple-beam antenna and adaptive array antenna[1,2]. Multiple parallel beams are used to cover the entire user area for multiple-beam antenna, and the pointing of each beam is fixed, the width of beam is determined by the number of antenna elements. When the user moves in the area, the appropriate beam will be chosen from different beams in the base station to make the strongest received signal. Because the user signals do not necessarily lie in the beam center, when the user locates at the edge of the beam and the interference signal lies in the beam center, the reception is the worst, so the best reception of signal can not be achieved for multi-beam antenna, and it is generally used as a receiving antenna. But compared with the adaptive antenna arrays, multi-beam antenna has the merits of being simple in structure, needing not to determine the arrival direction of the user signal. 4 to 16 bay structures are generally adopted for adaptive array antenna, and the array elements spacing is half wavelength.

G. Lee (Ed.): Advances in Intelligent Systems, AISC 138, pp. 117–124.
springerlink.com          © Springer-Verlag Berlin Heidelberg 2012

Digital signal processing techniques are adopted to identify the arrival direction of user signal, and the antenna main beam is formed in this direction for adaptive antenna array system. Adaptive array antenna is the main type of smart antenna which can complete receiving and sending of user signals.

By using spatial cutting method to divide Omni-directional coverage area into many differentiated small regions for adaptive array antenna, system resource utilization can be improved, and system interference can be reduced which can improve the system capacity[3,4]. Adaptive array antenna can be widely used in 3G and 4G mobile systems.

## 2    Antenna Coverage of Adaptive Array Base Station

### 2.1    Principles of Adaptive Array Antenna Coverage

In mobile communication systems of adaptive array antennas, each cellular base station $C_i$ is the antenna array consists of a number of sensors[5], each sensor is only responsible for transceiving of signals of a small area within the $C_i$[6], the coverage area of all sensors can be divided into two categories: non-overlapping region and overlapping region. Here let us take the adaptive array antenna that has three sensors as an example to illustrate.

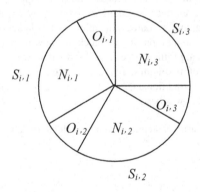

**Fig. 1.** Region of cellular base station $C_i$

In Figure 1, assume that the cellular base station $C_i$ contains $S_{i,1}$, $S_{i,2}$, $S_{i,3}$ three sensors, $S_{i,1}$ sensor signal can cover the range of { $O_{i,1}$, $N_{i,1}$, $O_{i,2}$}, $S_{i,2}$ sensor signal can cover the range of { $O_{i,2}$, $N_{i,2}$, $O_{i,3}$}, $S_{i,3}$ sensor signal can cover the range of { $O_{i,3}$, $N_{i,3}$, $O_{i,1}$}. {$N_{i,1}$, $N_{i,2}$, $N_{i,3}$} are the non-overlapping parts of signal coverage of the three sensors, { $O_{i,1}$, $O_{i,2}$, $O_{i,3}$} are the parts that can be overlapped for signal coverage of the three sensors.

Make N(X) represent the non-overlapping region that is adjacent to region X within the same base station, O(X) represent the overlap region that is adjacent to region X within the same base station, then in Figure 1 it has N $(O_{i,1})$ = { $N_{i,1}$, $N_{i,3}$}, N $(N_{i,1}UO_{i,1})$ = {$N_{i,3}$}, O $(N_{i,1})$ = { $O_{i,1}$, $O_{i,2}$} , O $(N_{i,1}UO_{i,1})$ = { $O_{i,2}$}.

Again make the $C_{i,j}$ represent the actual coverage of sensor $S_{i,j}$ in the base station $C_i$, that is the coverage of $C_{i,j}$ can only belong to the obtainable coverage $N_{i,j}UO$ $(N_{i,j})$ of the sensor $S_{i,j}$ signals in base station $C_i$, and assuming that there is no intersection between the ranges of service    $C_{i,j}$ and $C_{i,k}$ of any two different sensors $S_{i,j}$ and $S_{i,k}$, and all sensors within $C_i$ can cover the whole range of cellular base station.

**Definition 1.** A cellular base station $C_i$ composed by the range of services $C_{i,1},C_{i,2},\ldots,$ $C_{i\,m}$ provided by $S_{i,1}, S_{i,2},\ldots, S_{i,m}$,   amounting to m sensors must meet:

(1) $C_{i,j} \subseteq N_{i,j}UO$ $(N_{i,j})$,   ($N_{i,j}$ is non-overlapping region, $O(N_{i,j})$ is the overlapping region)

(2) Any two $C_{i,j}$和$C_{i,k}$ $(1 \leq j,\ k \leq m)$ , must meet the $C_{i,j} \cap C_{i,k} = \phi$

(3) $\overset{m}{\underset{j=1}{U}} C_{i,j} = C_i$ .

Still take Figure 1 for example, then $C_{i,1}= \{ N_{i,1}UO_{i,1} \}$ ，$C_{i,2}= \{ N_{i,2}UO_{i,2} \}$ ，$C_{i,3}= \{ N_{i,3}UO_{i,3} \}$ is a reasonable configuration that meets definition 1.

**Definition 2.** For a certain region $C_{i,j}$ of a cellular base station $C_i$ , the required call volume is defined as $\lambda \left( C_{i,j} \right) = t \times a$ . Of which is the average talk time of a telephone call in mobile station within $C_{i,j}$, a is the number of times of telephone calls of all mobile stations within $C_{i,j}$ in unit interval. n is the code number in bandwidth B $(S_{i,j})$. Where the code B $(S_{i,j})$ assigned to $C_{i,j}$ will affect the call interrupt rate of $C_{i,j}$. Call blocking rate of assessment region $C_{i,j}$ can be expressed as formula 1.

$$EB \left( n, \lambda \left( C_{i,j} \right) \right) = \frac{\dfrac{\lambda \left( C_{i,j} \right)^n}{n!}}{\displaystyle\sum_{k=0}^{n} \dfrac{\lambda \left( C_{i,j} \right)^k}{k!}} \tag{1}$$

## 2.2   Traditional Regional Resource Allocation Algorithm and Improvement of Adaptive Array Antenna

Currently, the traditional regional resource allocation algorithm of adaptive array antenna mainly contains: fixed regional resource allocation algorithm, maximum and minimum regional resource allocation algorithm.

1. Fixed Regional Resource Allocation Algorithm
Basic idea of fixed regional resource allocation algorithm is: the scope of responsibility $C_{i,j} = N_{i,j}UO_{i,j}$ of each sensor $S_{i,j}$ in cellular base station $C_i$ is fixed (j = 1,2,..., m) .

Then take Figure 1 for example, the range of $C_{i,1}$ is $\{N_{i,1}UO_{i,1}\}$, range of $C_{i,2}$ is $\{N_{i,2}UO_{i,2}\}$ and range of $C_{i,3}$ is $\{ N_{i,3}UO_{i,3}\}$.

2. Maximum and Minimum Regional Resource Allocation Algorithm.
Basic idea for maximum and minimum regional resource allocation algorithm is: principle of selection for the scope of responsibility $C_{i,j}$ of each sensor $S_{i,j}$ in cellular

base station $C_i$ is that one with the highest load in the overlapping area that has not been configured will be given priority to be configured to neighboring sensors with minimum load. $N_{i,j}$ in non-overlapping region will fixedly allocate to the $S_{i,j}$ sensors.

The specific allocation method is as follows:

Initialization: assigning the $N_{i,j}$ in the set of non-overlapping regions $\{N_{i,1}, N_{i,2}, \ldots , N_{i,m}\}$ of cellular base station $C_1$ to the corresponding $C_{i,j}$, $O(C_i)$ which forms a set of overlapping regions $\{O_{i,1}, O_{i,2}, \ldots , O_{i,m}\}$, where $j = 1,2,\ldots , m$, m is the number of sensors in the cellular base station.

Step 1: Determine whether the $O(C_i)$ is an empty set, if it is an empty set, it means that configuration has been completed in overlap region of cellular base station $C_1$, otherwise perform the following step 2.

Step 2: Find the region $O_{i,j}$ with maximum telephone traffic $\lambda (O_{i,j})$ in the set $O(C_i)$ of overlap regions in cellular base station $C_i$, then perform step 3.

Step 3: Identify the region $C_{i,k}$ with smaller traffic in set $N(Q_{i,j})$ of corresponding adjacent non-overlapping regions of the overlap region $O_{i,j}$ with the maximum telephone traffic $\lambda (O_{i,j})$, then perform Step 4.

Step 4: $O_{i,j}$ shall be managed by $C_{i,k}$, and remove $O_{i,j}$ from $O(C_i)$, re-execute step 1.

As coverage area of each sensor remains the same, it is impossible to dynamically follow the user traffic load variations of the coverage area of the base station to change by using fixed regional resource allocation algorithm. Though effects are greatly increased by using the maximum and minimum regional resource allocation algorithm compared with fixed regional resource allocation algorithm, distribution of traffic load of the whole base station is not integrated for this algorithm, which can not reach the best result of traffic balance. Thus, a new improved algorithm is proposed here, namely the regional resource allocation algorithm with minimum standard deviation.

3. Algorithm improvement ideas are as follows:

For a cellular base station $C_i$ consisting of m sensors, the whole base station coverage area can be divided into m overlapping regions $\{O_{i,1}, O_{i,2}, \ldots , O_{i,m}\}$ and m non-overlapping regions $\{N_{i,1}, N_{i,2}, \ldots , N_{i,m}\}$. Any non-overlapping region $N_{i,j}$ can only be covered by the sensor $S_{i,j}$, and coverage can be provided by two sensors for any overlap region $O_{i,j}$. For example: In Figure 1, sensors that can provide coverage for $O_{i,3}$ are $S_{i,2}$, and $S_{i,3}$. Therefore, for a base station that has m sensors, the configuration methods are 2m, so the set of the 2m configuration methods is the $A(C_i) = \{A_{i,1}, A_{i,2}, \ldots , A_{i,2m}\}$. Suppose $\lambda (C_i)$ is the total traffic of the $C_i$, then the average traffic $X_i$ that each sensor in $C_1$ required to bear is shown in formula 2, where m is the number of sensors in $C_i$.

$$X_i = \lambda (C_i) / m. \tag{2}$$

For example: In Figure 1, assuming the respective traffic of non-overlapping and overlapping regions are as shown in Table 1.

**Table 1.** The load of each region in subdistrict $C_i$ of cellular base station

| Coverage area | | Traffic  (*Erl*) |
|---|---|---|
| Non-overlapping region | $N_{i,1}$ | 2 |
| | $N_{i,2}$ | 3 |
| | $N_{i,3}$ | 4 |
| Overlapping region | $Q_{i,1}$ | 0.5 |
| | $Q_{i,2}$ | 0.4 |
| | $Q_{i,3}$ | 0.9 |

Then:

$$Xi = \lambda(Ci)/m = \ (2+3+4+0.5+0.4+0.9) \ /3=10.8/3=3.6 \ \ (Erl) \qquad (3)$$

As the number of sensors m = 3, therefore, there are 23 = 8 different coverage configuration methods altogether, all 8 configuration scenarios are shown in Table 2.

**Table 2.** Configuration of all subsets in A ($C_i$)

| | $S_{i,1}$ | $S_{i,2}$ | $S_{i,3}$ | $\sigma\left(A_{i,k}\right)$ |
|---|---|---|---|---|
| $A_{i,1}$ | $N_{i,1}UO_{i,1}$ | $N_{i,2}UO_{i,2}$ | $N_{i,3}UO_{i,3}$ | 0.9899 |
| $A_{i,2}$ | $N_{i,1}UO_{i,1}$ | $N_{i,2}UO_{i,2}UO_{i,3}$ | $N_{i,3}$ | 0.7874 |
| $A_{i,3}$ | $N_{i,1}$ | $N_{i,2}UO_{i,2}UO_{i,3}$ | $N_{i,3}UO_{i,1}$ | 1.1344 |
| $A_{i,4}$ | $N_{i,1}$ | $N_{i,2}UO_{i,2}$ | $N_{i,3}UO_{i,3}UO_{i,1}$ | 1.3953 |
| $A_{i,5}$ | $N_{i,1}UO_{i,1}UO_{i,2}$ | $N_{i,2}$ | $N_{i,3}UO_{i,3}$ | 0.9203 |
| $A_{i,6}$ | $N_{i,1}UO_{i,1}UO_{i,2}$ | $N_{i,2}UO_{i,3}$ | $N_{i,3}$ | 0.4970 |
| $A_{i,7}$ | $N_{i,1}UO_{i,2}$ | $N_{i,2}$ | $N_{i,3}UO_{i,3}UO_{i,1}$ | 1.1014 |
| $A_{i,8}$ | $N_{i,1}UO_{i,2}$ | $N_{i,2}UO_{i,3}$ | $N_{i,3}UO_{i,1}$ | 0.8832 |

For one configuration method $A_{i,k}$ in $C_1$, assuming $C^k_{i,1}, C^k_{i,2}, \ldots, C^k_{i,m}$ are the areas of responsibility of m sensors in $A_{i,k}$ configuration. The standard deviation $\sigma\left(A_{i,k}\right)$ of traffic that m sensors in configuration $A_{i,k}$ bear is shown as formula 4.

$$\sigma\left(A_{i,k}\right) = \sqrt{\frac{1}{m}\sum_{k=1}^{m}\left(\lambda\left(C_{i,k}\right)- X_i\right)^2} \qquad (4)$$

All of the configuration in $C_i$ are A ($C_i$) = {$A_{i,1}$, $A_{i,2}, \ldots, A_{i,2}{}^m$}, the standard deviation of configuration is $\sigma\left(A(Ci)\right) = \{\ \sigma\left(A_{i,1}\right), \sigma\left(A_{i,2}\right), \ldots, \sigma_{(A_{i,2}{}^m)}\ \}$.

The algorithm steps are as follows:

Step 1: Select a kind of resource configuration method $A_{i,k}$ from A ($C_i$) = {$A_{i,1}$, $A_{i,2}, \ldots, A_{i,2}{}^m$}, and its standard deviation $\sigma\left(A_{i,k}\right)$ is the minimum value of the set

$$\sigma\left(A(Ci)\right) = \{\ \sigma\left(A_{i,1}\right), \sigma\left(A_{i,2}\right), \ldots, \sigma_{(A_{i,2}{}^m)}\ \}$$

Step 2: $C_1$ assigns different regions to m sensors according to configuration method of $A_{i,k}$

Take Figure 1 for example, the respective traffic of overlapping regions and non-overlapping regions are shown in Table 2, calculate the standard deviations of 8 configuration methods $\sigma\left(A_{i,1}\right), \sigma\left(A_{i,2}\right),\dots, \sigma\left(A_{i,8}\right)$ according to the standard deviation algorithm as shown in Table 2. Perform step 1, that is selecting the configuration method $A_{i,6}$ with minimum standard deviation $\sigma\left(A_{i,6}\right)=0.4970$. In Step 2, assign to three sensors of $C_1$ according to configuration method of $A_{i,6}$ whose ranges of responsibility are $C_{i,1}=N_{i,1}\cup O_{i,1}\cup O_{i,2}$, $C_{i,2}=N_{i,2}\cup O_{i,3}$, $C_{i,3}=N_{i,3}$ respectively.

## 2.3   Simulation of Regional Resource Allocation Algorithm for Adaptive Array Antenna

1. Simulation parameter setting
To compare the difference in effects of coverage for the above-mentioned three algorithms, simulation can be conducted for verification. The basic parameters of analog simulation are shown in Table 3.

**Table 3.** The basic parameters table of simulation

| Parameters | Numeric values | | |
|---|---|---|---|
| The number of cellular base stations | 3 × 3 cellular base stations, each cellular base station contains six sensors | | |
| Yardage of each cellular base station | 60 | | |
| Interference range | Two small regions | | |
| The total number of mobile users | 10000 people | | |
| The average time per call | 4 minutes | | |
| The average frequency of calls per hour | 0.0 ~ 2.5 times | | |
| Population distribution | Proportion of people in overlapping areas: 10%, 20% | | |
| | 5 high load area | | |
| | The number of people in high-load area accounts for 10% | | |

2. Simulation comparison
Through simulation, simulation comparisons of simulated blocking rates of three different regional resource allocation algorithms are shown in Table 4.

**Table 4.** Simulated blocking rates of the three regional resource allocation algorithms

| Proportion of the number of people in overlapping areas | Allocation algorithm | The average call blocking rate (%) | | | | | |
|---|---|---|---|---|---|---|---|
| | | The average frequency of calls of mobile users | | | | | |
| | | 0 | 0.5 | 1.0 | 1.5 | 2.0 | 2.5 |
| 10% | Fixed algorithm | 0 | 0.1 | 1.2 | 3.2 | 5.5 | 6.3 |
| | Maximum and minimum n algorithm | 0 | 0.05 | 0.2 | 1.1 | 3.0 | 3.5 |
| | Minimum standard deviation algorithm | 0 | 0.03 | 0.18 | 1.0 | 2.9 | 3.4 |
| 20% | Fixed algorithm | 0 | 0.09 | 1.1 | 3.1 | 5.3 | 6.2 |
| | Maximum and Minimum n algorithm | 0 | 0.04 | 0.1 | 0.4 | 1.8 | 2.8 |
| | Minimum standard deviation algorithm | 0 | 0.02 | 0.09 | 0.3 | 1.7 | 2.7 |

It can be seen from Table 4:

(1) When the number of people in overlapping regions accounts for 10%: the call blocking rate of fixed regional resource allocation algorithm is 0 to 6.3%, the call blocking rate of the maximum and minimum regional resource allocation algorithm is 0 to 3.5%, and the call blocking rate of regional resource allocation algorithm with minimum standard deviation is 0 to 3.4%;

(2) When the number of people in overlapping regions accounts for 20%: the call blocking rate of fixed regional resource allocation algorithm is 0 to 6.2%, the call blocking rate of the maximum and minimum regional resource allocation algorithm is 0 to 2.8%, and the call blocking rate of regional resource allocation algorithm with minimum standard deviation is 0 to 2.7%;

It can be seen from this that: under the same condition, the regional resource allocation algorithm with minimum standard deviation has the minimum traffic blocking rate for coverage area of base station.

## 3    Conclusions

Carrying capacities of different sensors are not mean for fixed regional resource allocation algorithm, therefore, it is certain that load balancing effect of it is the worst compared with the latter two; greater effect can be achieved by using the regional resource allocation algorithm with minimum standard deviation compared with the maximum and minimum regional resource allocation algorithm for all of the other two configuration methods are taken into account for this algorithm, and then a set of configuration with least difference of carrying capacities among sensors shall be selected from all configuration results, in order to achieve load balancing. Therefore, the best balancing effect of loads covered by regions can be achieved for the regional resource allocation algorithm with minimum standard deviation.

# References

1. Zhang, X., Li, H.: Applications of Minimum Mean Square Error Criterion in Radar Digital Beam Forming. Silicon Valley (13), 173–174 (2010)
2. Ma, Y., Hu, Y.: Performance Improvement of the CDMA System by Smart Antennas. Electronic Technology (02), 74–76 (2008)
3. Zhang, H., Qiu, X., Meng, L., Gao, Z., Zhang, X.: Independent Load Balancing Management Methods of TD-SCDMA Radio Access Network. Journal on Communications (01), 9–19 (2011)
4. Wu, Q., Xiao, Q.: Multi-antenna Technique Based on Wireless Channel Capacity Solution. Ship Science and Technology (04), 73–75 (2010)
5. Kuang, P., Wang, J.: A Kind of Improved Beamformer for Linear Constraints of Feature Space. Electronic Countermeasures (02), 38–41 (2009)
6. Kim, S., Varshney, P.K.: Adpative Fault Tolerant Bandwidth Management Framework for Multimedia Cellular Networks. IEEE Proceeding-Communications (10), 932–938 (2005)

# Asymptotic Stabilization of Nonlinear DAE Subsystems Using Artificial Neural Networks with Application to Power Systems

Zang Qiang[1] and Zhou Ying[2]

[1] School of Information and Control Engineering, Nanjing University of Information Science & Technology, Nanjing 210044, China
[2] College of Automation, Nanjing University of Posts and Telecommunications, Nanjing 210003, China
autozang@163.com, zhouying@njupt.edu.cn

**Abstract.** The problem of robust stabilization for such class of uncertain nonlinear Differential-Algebraic Eqyatuion subsystems is considered in this paper. The robust stabilization controller is proposed based on backstepping approach using two-layer Artificial Neural Networks (ANN) whose weights are updated on-line. A new adaptive algorithm is proposed to update the weights of ANN such that all signals in the closed-loop systems are bounded and the states are convergent asymptotically to the equilibrium through the proposed controller. Finally, using the design scheme proposed in this paper, a governor controller is designed for one synchronous generator in a multi-machine power systems. The simulation results demonstrate its effectiveness.

**Keywords:** DAE Systems, Subsystems, Artificial Neural Networks, Backstepping, Power Systems.

## 1 Introduction

Many physical systems are described by differential-algebraic equation (DAE) systems[1]. Some great progress has been made recently for nonlinear DAE systems. The sufficient conditions are presented for the stability of nonlinear DAE systems in [2]. The exact linearization of nonlinear DAE systems is researched in [3]. The output stabilizaiton for nonlinear DAE systems is considered in [4].

However, the controlled systems in many practical engineering applications are often nonlinear DAE subsystems within a large-scale system[5]. There exists constraint between the controlled DAE subsystems and the rest of the large-scale systems, which constraint arises naturally from the point of physics. The controlled DAE subsystems are influenced by the rest of the large-scale systems. A so-called power systems component structural model formulated in [6] just falls into this category. As far as the writer knows, the research for nonlinear DAE subsystems has seldom been found.

Owing to its excellent ability to approximate a nonlinear function with satisfactory accuracy, Artificial Neural Network (ANN) has been applied to system identification or identification-based control[7]. The goal of this paper is to investigate the robust

G. Lee (Ed.): Advances in Intelligent Systems, AISC 138, pp. 125–134.
springerlink.com      © Springer-Verlag Berlin Heidelberg 2012

stabilization problem for uncertain nonlinear DAE subsystems. Firstly an equivalent system is achieved through a local diffeomorphism and a feedback. Then the robust stabilization controller is proposed for the equivalent system based on backstepping approach using ANN whose weights are tuned on-line with a new adaptive algorithm. At last a governor controller is designed for one synchronous generator in multi-machine power systems to show the effectiveness of the proposed scheme in this paper.

## 2    System Description and Problem Formulation

We consider the following uncertain nonlinear DAE subsystems within large-scale systems [5]:

$$
\begin{aligned}
\dot{\xi} &= f_1(\xi, \gamma, \overline{v}, \theta) + g(\xi, \gamma, \overline{v}, \theta)u \\
0 &= f_2(\xi, \gamma, \overline{v}, \theta) \\
y &= h(\xi, \gamma, \overline{v}, \theta)
\end{aligned}
\tag{1}
$$

where $\xi \in R^n$ is the vector of differential variables, $\gamma \in R^m$ is the vector of algebraic variables, $\overline{v} \in R^s$ is the interconnection input acted on (1) by the rest of the large-scale systems, $u \in R$ is the control input, $y \in R$ is the control output, $\theta$ denotes both parametric and nonparametric uncertainties, $f_1, g, f_2, h$ are all sufficiently smooth vector fields. The interconnection input $\overline{v}$ are local measurable and bounded. The origin is supposed to be the isolated equilibrium of (1).

The following basic assumptions are made for systems (1).

*Assumption 1.* The Jacobian matrix $\dfrac{\partial f_2}{\partial \gamma}$ of $f_2$ with respect to $\gamma$ is nonsingular.

*Assumption 2.*    The DAE subsystems (1) has uniform relative degree $n$ at the origin [5].

*Assumption 3.* Nonlinear DAE subsystems (1) can be transformed equivalently into following form

$$
\begin{aligned}
\dot{x}_1 &= F_1(x, \overline{v}, \theta) + G_1(x_1, \overline{v})x_2 \\
&\vdots \\
\dot{x}_n &= F_n(x, \overline{v}, \theta) + G_n(x_1, x_2, \cdots, x_n, \overline{v})u \\
0 &= f_2(\xi, \gamma, \overline{v}, \theta) \\
y &= x_1
\end{aligned}
\tag{2}
$$

where $G_i$'s are known and nonsingular.

The refence [5] presented the sufficient conditions for above transformation.

*Assumption 4.* For each $F_i, i = 1, \cdots, n$ in system (1), the following conditions hold:

$$F_i(\underbrace{0, \cdots, 0}_{i}, x_{i+1}, \cdots, x_n, \overline{v}, \theta) = 0, i = 1, \cdots, n \tag{3}$$

The objective in present article is to design a smooth state-feedback control law such that the equilibrium of the closed-loop system (1) is asymptotically stable.

## 3    Main Results

Now we will give the design procedure of the controller based on Backstepping method. The uncertain functions $F_i's$ are accounted for by a two-layer ANN as shown as following.

*Step 1.* Consider the first subsystems

$$\dot{x}_1 = F_1(x, \overline{v}, \theta) + G_1(x_1, \overline{v})x_2 \tag{4}$$

and define the error variables $z_1 = x_1, z_2 = x_2 - \alpha_1$, where $\alpha_1$ is the first virtual controller. Define

$$\overline{G}_1(z_1, \overline{v}) = G_1(x_1, \overline{v}), \phi_1(z_1, x_2, x_3, \cdots, x_n, \overline{v}, \theta) \triangleq F_1(z_1, x_2, x_3, \cdots, x_n, \overline{v}, \theta) \tag{5}$$

By virtue of Assumption 3, it can be seen that $\phi_1$ is a smooth vector function and the condition $\phi_1(0, x_2, x_3, \cdots, x_n, \overline{v}, \theta) = 0$ is satisfied. From the smoothness of $\phi_1$, there must exist a smooth matrix function $\phi_1^* = \phi_{1,1}^*(z_1, x_2, x_3, \cdots, x_n, \overline{v}) \in R^{r \times r}$ such that

$$\phi_1(z_1, x_2, x_3, \cdots, x_n, \overline{v}, \theta) = \phi_{1,1}^*(z_1, x_2, x_3, \cdots, x_n, \overline{v}, \theta)z_1 \tag{6}$$

where $\phi_{1,1}^*(z_1, x_2, x_3, \cdots, x_n, \overline{v}, \theta) = \int_0^1 \left. \frac{\partial \phi_1(\xi, x_2, \cdots, x_n, \overline{v}, \theta)}{\partial \xi} \right|_{\xi = \sigma z_1} d\sigma.$

The following assumption is made here for system (1) that will be used in the next steps:

*Assumption 5.*

$$\left| \lambda_{\max} \left( \phi_{i,j}^*(z_1, \cdots, z_i, x_{i+1}, \cdots, x_n, \overline{v}, \theta) \right) \right| \le B_i(z_1, \cdots, z_i), j = 1, \cdots, i, i = 1, \cdots, n \tag{7}$$

where $B_i's$ are unknown bounding functions and $z_j's$ are error variables defined in the next steps.

*Assumption 6.* $B_i$ can be represented by $n$ two-layer ANN for some constant ideal weights $W_i$, i.e.,

$$B_i = W_i^T \psi_i + \varepsilon_i, |\varepsilon_i| < c = cons, i = 1, 2, \cdots, n \tag{8}$$

where $\psi_i$ are the radial basis functions for the $ith$ ANN. Define ANN functional estimate of $B_i$ by

$$\hat{B}_i(x) = \hat{W}_i^T \psi_i(x), i = 1, 2, \cdots, n \tag{9}$$

with $\hat{W}_i$ the current ANN weight estimate provided by tuning algorithm to be designed.

Then the first Lyapunov function candidate for subsystem (4) can be chosen as $V_1 = \frac{1}{2} z_1^T z_1$. From Assumption 4 to Assumption 6, the time derivative of $V_1$ along subsystem (4) is

$$\dot{V}_1 = z_1^T (\phi_1 + \bar{G}_1(z_1, \bar{v}) x_2) \le z_1^T \bar{G}_1(z_1, \bar{v}) z_2 + z_1^T B_1 z_1 + z_1^T \bar{G}_1(z_1, \bar{v}) \alpha_1 \tag{10}$$

Obviously the first virtual controller can be chosen as

$$\alpha_1(z_1) = \bar{G}_1(z_1, \bar{v})^{-1} \left( -c_1 z_1 - (n-1) z_1 - \hat{B}_1(z_1) z_1 \right) \tag{11}$$

where $c_1 > 0$ is the design parameter. Substitute (11)into(10) we get

$$\dot{V}_1 \le -c_1 z_1^T z_1 - (n-1) z_1^T z_1 + z_1^T \bar{G}_1 \left( z_1, \bar{v} \right) z_2 + \tilde{B}_1 z_1^T z_1, \tilde{B}_1 = B_1 - \hat{B}_1 \tag{12}$$

*Step* $k(k = 2, 3, \cdots, n-1)$. Similar to step 1, until to step $(k-1)th$, we have defined error variables

$$z_1 = y - y_r, \cdots, z_k = x_k - \alpha_{k-1} \tag{13}$$

where

$$\alpha_{k-1} = \bar{G}_{k-1}^{-1}(-c_{k-1} z_{k-1} - (n-k+1) z_{k-1} - \bar{G}_{k-2} z_{k-2}) + \bar{G}_{k-1}^{-1}(-(k-2)\hat{B}_{k-1}^2 z_{k-1} - \hat{B}_{k-1} z_{k-1}) \tag{14}$$

and $c_{k-1} > 0$ is the design parameter. Then define the $(k+1)th$ error variable $z_{k+1} = x_{k+1} - \alpha_k$ where $\alpha_k$ is the $kth$ virtual controller to be designed. From the definition of $z_k$ and $\alpha_{k-1}$ we have

$$\dot{z}_k = F_k + \bar{G}_k \alpha_k + \bar{G}_k z_{k+1} - \sum_{i=1}^{k-1} \frac{\partial \alpha_{k-1}}{\partial z_i^T} \dot{z}_i \tag{15}$$

Then $\alpha_k$ can be chosen as

$$\alpha_k = \bar{G}_k^{-1} \left( -c_k z_k - (n-k) z_k - \bar{G}_{k-1} z_{k-1} + \right) + \bar{G}_k^{-1} \left( -(k-1)\hat{B}_k^2 z_k - \hat{B}_k z_k \right) \tag{16}$$

where $\bar{G}_k(z_1, z_2, \cdots, z_k, \bar{v}) = G_k(z_1, z_2 + \alpha_1 \cdots, z_k + \alpha_{k-1}, \bar{v})$.

We get the $kth$ subsystems as following

$$\dot{z}_1 = -c_1 z_1 - (n-1)z_1 - \hat{B}_1 z_1 + F_1 + \bar{G}_1 z_2$$
$$\vdots \tag{17}$$
$$\dot{z}_k = -\bar{G}_{k-1} z_{k-1} - c_k z_k - (k-1)\hat{B}_k^2 z_k - \hat{B}_k z_k + F_k - \sum_{i=1}^{k-1} \frac{\partial \alpha_{k-1}}{\partial z_i^T} \dot{z}_i + \bar{G}_k z_{k+1}$$

Define

$$\phi_k(z_1, \cdots, z_k, x_{k+1}, \cdots, x_n, \bar{v}, \theta) \triangleq F_k(z_1, \cdots, z_k + \alpha_{k-1}, x_{k+1}, \cdots, x_n, \bar{v}, \theta) \tag{18}$$

With Assumption 4, we have $\phi_k(0, \cdots, 0, x_{k+1}, \cdots, x_n, \bar{v}, \theta) = 0$. Due to the smoothness of $\phi_k$, there exists a function $\phi_k^* = [\phi_{k,1}^*, \phi_{k,2}^*, \cdots, \phi_{k,k}^*] \in R^{r \times kr}$, $\phi_{k,i}^* \in R^{r \times r}$, $i = 1, 2, \cdots, k$ such that

$$\phi_k = \phi_k^* \begin{bmatrix} z_1 \\ \vdots \\ z_k \end{bmatrix} = \phi_{k,1}^* z_1 + \cdots + \phi_{k,k}^* z_k \tag{19}$$

where $\phi_k^*(z_1, \cdots, z_k, x_{k+1}, \cdots, x_n) = \int_0^1 \frac{\partial \phi_k(\varsigma, x_{k+1}, \cdots, x_n)}{\partial \varsigma} \bigg|_{\varsigma=\sigma \begin{bmatrix} z_1 \\ \vdots \\ z_k \end{bmatrix}} d\sigma$.

The Lyapunov function $V_k$ is picked as $V_k = \frac{1}{2} \sum_{i=1}^{k} z_i^T z_i$, then the time derivative of $V_k$ along (17) is

$$\dot{V}_k \leq -\sum_{i=1}^{k-1} c_i z_i^T z_i - (n-k+1)\sum_{i=1}^{k-1} z_i^T z_i + \sum_{i=1}^{k-1} \sum_{j=1}^{i} \tilde{B}_i \left| z_i^T z_j \right|$$
$$+ B_k \left| z_1^T z_k \right| + \cdots + B_k \left| z_{k-1}^T z_k \right| + B_k z_k^T z_k + z_k^T (\bar{G}_k \alpha_k + \bar{G}_{k-1} z_{k-1}) + z_k^T \bar{G}_k z_{k+1} \tag{20}$$

It can be easily verified that the following inequalities hold $-z_i^T z_i + \hat{B}_k \left| z_i^T z_k \right| - \hat{B}_k^2 z_k^T z_k \leq 0$, so we have

$$\dot{V}_k \leq -\sum_{i=1}^{k} c_i z_i^T z_i - (n-k)\sum_{i=1}^{k} z_i^T z_i + z_k^T \bar{G}_k z_{k+1} + \sum_{i=1}^{k} \sum_{j=1}^{i} \tilde{B}_i \left| z_i^T z_j \right| \tag{21}$$

*Step $n$.* At the last step the actual controller $u$ is proposed. Define $z_n = x_n - \alpha_{n-1}$, then

$$\dot{z}_n = F_n + G_n u - \sum_{i=1}^{n-1} \frac{\partial \alpha_{n-1}}{\partial z_i^T} \dot{z}_i \qquad (22)$$

The actual controller $u$ is chosen as

$$u = \bar{G}_n^{-1} \left( -c_n z_n - \bar{G}_{n-1} z_{n-1} - (n-1)\hat{B}_n^2 z_n - \hat{B}_n z_n \right) \qquad (23)$$

where $\qquad \bar{G}_n(z_1, \cdots, z_n, \bar{v}) \triangleq G_n(z_1, \cdots, z_n + \alpha_{n-1}, \bar{v}) \qquad$ .Define

$\phi_n(z_1, \cdots, z_n) \triangleq F_n - \sum_{i=1}^{n-1} \frac{\partial \alpha_{n-1}}{\partial z_i^T} \dot{z}_i$ , similar to above discussion, there is function

$\phi_n^*$ such that $\phi_n = \phi_n^* \begin{bmatrix} z_1 & \cdots & z_n \end{bmatrix}^T$ ,where $\phi_n^* = \int_0^1 \frac{\partial \phi_n(\varsigma)}{\partial \varsigma} \Big|_{\varsigma=\sigma\begin{bmatrix} z_1 \\ \vdots \\ z_n \end{bmatrix}} d\sigma$ .

For whole closed-loop error systems, Lyapunov function is chosen as
$V_n = \frac{1}{2} \sum_{i=1}^{n} z_i^T z_i$ , then we have

$$\dot{V}_n \leq -\sum_{i=1}^{n} c_i z_i^T z_i + \sum_{i=1}^{n} \sum_{j=1}^{i} \tilde{B}_i \left| z_i^T z_j \right| \qquad (24)$$

Now we will give the main Theorem of this note.

*Theorem*    Take the control input (24) with ANN weight tuning be provided by

$$\dot{\hat{W}}_i = \sum_{j=1}^{i} \Gamma_i \phi_i \left| z_i^T z_j \right|, i = 1, 2, \cdots, n \qquad (25)$$

with constant matrices $\Gamma_i = \Gamma_i^T > 0, i = 1, \cdots, n$ . Then all signals in the closed-loop systems (17)(with $k = n$ ) and (25) are bounded and the states are convergent asymptotically to the equilibrium.

*Proof*: Consider the following Lyapunov function candidate for the whole he whole error system (17) (with $k = n$ ) and (25) is

$$V = \frac{1}{2} \sum_{i=1}^{n} z_i^2 + \frac{1}{2} tr(\tilde{W}^T \Gamma \tilde{W}) \qquad (26)$$

where $\tilde{W} = diag\{\tilde{W}_1, \cdots, \tilde{W}_n\}, \Gamma = diag\{\Gamma_1, \cdots, \Gamma_n\}$ . It can be verified that the time derivative of $V$ is

$$\dot{V} \leq -\sum_{i=1}^{n} c_i z_i^T z_i + cn \sum_{i=1}^{n} z_i^T z_i \qquad (27)$$

where $c > 0$ some constant. So as long as choose $c_i$'s satisfy $c_i - cn = \bar{c}_i > 0$, $i = 1, 2, \cdots, n$, we have

$$\dot{V} < -\sum_{i=1}^{n} \overline{c}_i z_i^T z_i \leq 0 \qquad (28)$$

So all the signals in the whole closed-loop error system are stable, i.e., $z_i \in L_\infty, i = 1, 2, \cdots, n$. From (28) it can be obtained that $z_i \in L_2, i = 1, 2, \cdots, n$. By (24) we have $\dot{z}_i \in L_\infty, i = 1, 2, \cdots, n$. By the well-known *Barbalat Lemma*, we have $z_i \to 0(t \to \infty), i = 1, 2, \cdots, n$. This completes the proof of the theorem.

## 5    Application to Power Systems

The speed governor control of turbo-generator set G1 (DAE subsystems) in the two-area four-machine large power systems is studied. The controlled systems are described as following:

$$\dot{\delta} = \omega - \omega_0 + \Delta(t)$$

$$\dot{\omega} = \frac{\omega_0}{H} P_H - \frac{D}{H}(\omega - \omega_0) - \frac{\omega_0 E_q' V_s}{H x_{d\Sigma}'} \sin\delta$$

$$- \frac{\omega_0 V_s^2}{2H} \left( \frac{x_d' - x_q}{x_{d\Sigma}' x_{q\Sigma}} \right) \sin 2\delta + \Delta(t)$$

$$\dot{P}_H = -\frac{P_H}{T_{H\Sigma}} + \frac{P_{H0}}{T_{H\Sigma}} + \frac{C}{T_{H\Sigma}} U_C + \Delta(t)$$

$$0 = f_2^1(\cdot) = \frac{I_t(Q_t + x_q I_t^2)}{\sqrt{(Q_t + x_q I_t^2)^2 + (P_t + r_a I_t^2)^2}},$$

$$0 = f_2^2(\cdot) = I_q - \frac{I_t(P_t + r_a I_t^2)}{\sqrt{(Q_t + x_q I_t^2)^2 + (P_t + r_a I_t^2)^2}},$$

$$0 = f_2^3(\cdot) = E_q' - \frac{(P_t + r_a I_t^2)^2 + (Q_t + x_d' I_t^2)(Q_t + x_q I_t^2)}{I_t \sqrt{(Q_t + x_q I_t^2)^2 + (P_t + r_a I_t^2)^2}},$$

$$0 = f_2^4(\cdot) = \theta_U - \delta - \text{arc } ctg \frac{x_q I_q - r_a I_d}{E_q' - x_d' I_d - r_a I_q} \qquad (29)$$

where differential variables $(\delta, \omega, P_H)^T$ are relative power angle between G1 and G4, rotate speed deviation of G1 and the high pressure mechanical power respectively, algebraic variables $(I_d, I_q, Q_t, \theta_U)^T$ are the d-axis current, the q-axis current, the reactive power and the angle of voltage respectively, and the interconnection input $(I_t, P_t)$ are the generator stator current and active power. The governor position $U_c$ is the control input, and the output of G1 is chosen as $y = \delta$. The disturbance noise is supposed to be $\Delta(t) = \sin t$.

Based on MATLAB, simulation is conducted under the following operating status: at the beginning the systems operate with double lines in a stable state; at 0.5s a three-phase symmetrical short-circuit to ground on one of the lines occurs at point $k = 0.1$; at 0.65s the fault is cleared, and the systems operate in two-line mode. The designed parameters is chosen as

$$c_1 = c_2 = c_3 = 10, \Gamma_1 = 10, \Gamma_2 = 20, \Gamma_3 = 30 \tag{30}$$

For the space limited, we only give following simulation results as shown from Fig.1 to Fig.3.

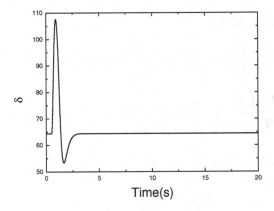

**Fig. 1.** Power angle $\delta$

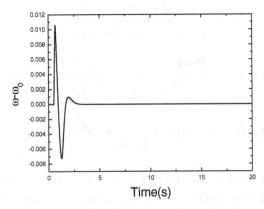

**Fig. 2.** Rotate speed deviation $\omega - \omega_0$

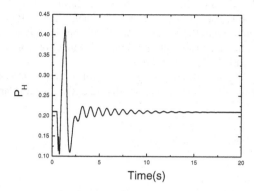

**Fig. 3.** The high pressure mechanical power $P_H$

Apparently, all the states of the power systems asymptotically converge to the non-zero equilibrium with a good performance of convergence.

## 6    Conclusions

For a class of uncertain nonlinear DAE subsystems, the robust stabilization controller design scheme is proposed in this paper based on backstepping approach using ANN whose weights are updated on-line. All the signals in the closed-loop systems are bounded and the equilibrium of the closed-loop system is asymptotically stable.

**Acknowledgements.** This work is supported by National Natural Science Foundation of China (61004001); Nanjing University of Information Science & Technology Science Foundation (S8110046001); Nanjing University of Posts and Telecommunications Climbing Project(NY 210013).

## References

1. Campbell, S.L., Nichols, N., Terrell, W.J.: Duality. Observability and Controllability for Linear Time-Varying Descriptor Systems. Circuits, Systems, Signal Process. 10(3), 455–470 (1991)
2. Hill, D.J., Mareels, I.M.Y.: Stability Theory for Differential/Algebraic Systems with Application to Power Systems. IEEE Transactions on Circuits and Systems 37(11), 1416–1423 (1990)
3. Jie, W., Chen, C.: Nonlinear Control of Differential Algebraic Model in Power Systems. Proceedings of the CSEE 21(8), 15–18 (2001)
4. Zang, Q., Dai, X.: Output Feedback Stabilization Control for Nonlinear Differential-Algebraic Equations Systems. ACTA Automatica Sinica 35(9), 1244–1248 (2009)

5. Zang, Q., Dai, X., Zhang, K.: Backstepping Control for a Class of Nonlinear Differential-Algebraic Equations Subsystems with Application to Power Systems. In: 7th World Congress on Intelligent Control and Automation, pp. 4663–4667 (2008)
6. Dai, X., Zhang, K.: Interface Concept and Structural Model of Complex Power Systems. Proceedings of the CSEE 27(7), 7–12 (2007)
7. Kwan, C., Lewis, L.F.L.: Robust Backstepping Control of Nonlinear Systems Using Neural Networks. IEEE Transactions on Systems, Man and Cybernetics-Part A: Systems and Humans 30(6), 753–766 (2000)

# Evaluation on the Efficiency for the Allocation of Science and Technology Resources in China Based on DEA Model

Wang Bei[1,2], Liu Wei-dong[1], and Zhang Jian-bo[3]

[1] Key Laboratory of Regional Sustainable Development Modeling, CAS, Beijing 100101, China
Institute of Geographical Sciences and Natural Resources Research, CAS, Beijing 100101, China
[2] Graduate University of Chinese Academy of Sciences, Beijing 100049, China
[3] School of Resource and Environmental Science, Wuhan University, China
awangb.09b@igsnrr.ac.cn

**Abstract.** With the advent of knowledge economy age, science and technology resources, as one of the most significant strategic resources, will play an important role in the future competition. In this circumstance, it is necessary to pay attention to the problems relating to the allocation of S&T resources. According to the design of input-output index system of S&T resources, this paper evaluates the allocation efficiency of S&T resources in China with the CCR model of DEA method, points out the reasons for the allocation inefficiency and suggests the ways to improve the allocation potential. In this basis, it summarizes the following conclusion:1) The spatial pattern of allocation efficiency of S&T resources in China presents clustered distribution, and the efficiency level appears a descending trend from the east to the west. 2) For most of the provinces in China, the lower utilization rate of S&T resources resulting in the output insufficiency of economic benefit is the most important reason affecting the allocation efficiency. Therefore, we should enhance the utilization rate of S&T resources and boost the transformation of S&T achievements as the main ways to improve the allocation efficiency of S&T resources in China.

**Keywords:** allocation efficiency, S&T resources, evaluation, DEA, China.

## 1 Introduction

With the advent of knowledge economy age, knowledge industry is becoming the mainstream of industries in the world. Meanwhile, science and technology(S&T hereafter) resources will play an important role in this process. Differing from other traditional resources, S&T resources possess lots of characteristics, for example, they could be used long-term and repeatedly, and often obtain an increased output than input. As we all know, our country has a special history, and also ignored the development and utilization of S&T resources in a long time, which results in the situation that most of the current problems are related to the unreasonable allocation of S&T resources[1,2]. Under this background, this problem concerning how to regard the current distribution of S&T resources, how to evaluate the allocation efficiency of S&T resources, and how to make the allocation pattern more reasonable, is becoming a hot issue in making decisions by government.

G. Lee (Ed.): Advances in Intelligent Systems, AISC 138, pp. 135–140.
springerlink.com          © Springer-Verlag Berlin Heidelberg 2012

Since the end of last century, the research on the allocation of S&T resources has attracted academics' attention, which main point is on the field of allocation pattern or allocation efficiency[3,4]. According to the traditional view, the research on resource allocation is mainly affected by management or economics, so it is rarely studied on the field of spatial disparity, efficiency and other issues from the perspective of geography. Based on this, the paper takes full account of the allocation system of S&T resources being a typical input-output system, so it uses the entropy method and DEA model to analyze the spatial disparity and efficiency for the allocation of S&T resources. According to the design of input-output index system of S&T resources, this paper takes the 31 province-level administrative units in our country for example(due to the unavailable data, Hong Kong, Macao and Taiwan are excluded). With the CCR model of DEA method, we could calculate the relative efficiency of S&T resources allocation for every principle unit, in order to explore the reason of low efficiency and the way to improve it.

## 2   Data and Methods Adopted

S&T resources consist of human resources, financial resources, material resources and information resources. In a narrow sense, only the human resources and financial resources are included in the definition of S&T resources. In this paper, we select the most commonly used international indicators, the total investment of R&D expenditure and full-time equivalent of R&D personnel, as the input part of the allocation system of S&T resources[5]. The output of allocation system of S&T resources can be measured in many ways, not only including the scientific papers and books on behalf of the direct achievements in scientific research, but also the inventive patents representing technological innovative achievements, as well as the output value of high-tech products standing for the transformation capability of S&T achievements into economic benefits. Hence, the output indicators are composed of direct knowledge output and indirect economic benefit respectively.

In addition, due to an important rule needed obey, the number of decision making unit should exceed twice the sum of input-output indicators. Otherwise, the distinction of efficiency evaluation with DEA model will decline. Therefore, when the number of decision making unit is fixed, we should try the best to reduce the variables of input-output index, in order to improve the distinction capability of efficiency evaluation[6]. Table 1 shows the input-output index system of S&T resources. Besides, considering the time lag in input-output system[7], we choose 2-year as the lag period. And all the data in this paper are based on the Survey regarding the S&T Activities by National Bureau Statistics, an annual survey conducted for all S&T phenomenon. So the input data is coming from Chinese S&T Statistical Yearbook(2008), and the output is from Chinese S&T Statistical Yearbook(2010).

As table 1 shows, the input variables are the S&T human and financial resources, and the knowledge output and economic benefit resulting from the investment are the output variables. Suppose we have a set of n peer DMUs, {DMUj: j=1, 2,..., n}, which produce multiple outputs $y_{rj}$ (r=1, 2,..., s),by utilizing multiple inputs $x_{ij}$ (i=1, 2,..., m). Let the inputs and outputs for DMUj be $X_j=(x1_j, x2_j,..., x_{mj})_t$ and $Y_j=(y1_j, y2_j, ..., y_{sj})_t$, respectively, so the CCR model is as follows:

**Table 1.** Input-output index system of S&T resources

| Target | Indicator | Principle |
|---|---|---|
| Input of S&T resources | Gross amount of R&D expenditure | Investment of S&T financial resources |
| | Full-time equivalent of R&D personnel | Investment of S&T human resources |
| Output of S&T resources | Number of scientific papers Number of invention patent granted | Direct knowledge output of scientific research |
| | Value of high-tech product Amount of transaction contract in technical market | Indirect economic benefit of scientific innovation |

$$min\theta$$
$$s.t.$$
$$\begin{cases} \sum_{j=1}^{n} X_j\lambda_j + S^- = \theta X_0 \\ \sum_{j=1}^{n} Y_j\lambda_j - S^+ = Y_0 \\ \lambda_j \geq 0, S^+ \geq 0, S^- \geq 0 \end{cases}$$

In the model, $\theta$ $(0 < \theta \leq 1)$ is the efficiency value, and $\lambda_j$ is the weight variable. Besides, $S^-$ and $S^+$ are respectively slack variable and surplus variable. According the meaning of DEA model, if the value of $\theta$ is close to 1 very much, it accounts for the high ratio of input-output, which represents the high efficiency of resources allocation. When the value of $\theta$ is equal to 1, the input-output level in this DMU is located in the optimal production frontier, and the allocation efficiency reaches the highest. Meanwhile, when the value of $\theta$ is less than 1, $S^+$ and $S^-$ must be positive. And though the value of $S^+$ and $S^-$ we could recognize what the main affecting feature is and how much space the efficiency could be improved.

## 3    Evaluation on the Allocation Efficiency of S&T Resources in China

**Spatial pattern of the allocation efficiency of S&T resources in China.** According to the calculation on the input-output index system of S&T resources with CCR model of DEA method by the software of MaxDEA, we could get the result, which is called "$\theta$", of the allocation efficiency of S&T resources in 31 provinces. And then, we divided this "$\theta$" into five levels representing the five types of efficiency zones, the highest as the first type and the lowest as the fifth. Meanwhile, we put it into the statistic clustering method in ArcGIS, where the result of spatial disparity pattern for the allocation efficiency of S&T resources could be visual.

Based on Fig.1, the characteristics of the allocation efficiency in China could be summarized in four aspects. 1) Generally speaking, the spatial pattern of allocation efficiency of S&T resources in China presents clustered distribution, and the efficiency level appears a descending trend from east to west. 2) From a regional perspective, on the one hand, the high-level zones of allocation efficiency consist of the Jing-jin region, the Pearl River Delta region, the Hubei-Hunan region, the Shanxi-Gansu region and the Northeast region. On the other hand, the lower-level zones mainly include most of the West regions bordering on the other countries, as well as Henan-Shanxi region in Middle area. 3) Judging from the type area, the number of provinces in different type area is almost equivalent, that is to say, the level of the allocation efficiency of S&T resources appears the hierarchical and progressive distribution. 4) For the composition of province number in every type area, the East region accounted for 86% and the West region accounted for 14% in the whole 7 provinces of the first type area; in the 5 provinces of the second type area, the East and West regions accounted for 40% respectively, and the Middle area accounted for 20%; in the 6 provinces of the third type area, the East region accounted for 50%, the Middle area accounted for 17%, and the West region accounted for 33%; in the 7 provinces of the fourth type area, the East region accounted for 29%, the Middle area accounted for 14%, and the West region accounted for 57%; in the 6 provinces of the fifth type area, the Middle area accounted for 33% and the West region accounted for 67%. As mentioned above, the allocation efficiency in East area is superior to that in Middle area, which in Middle area is better than in West area.

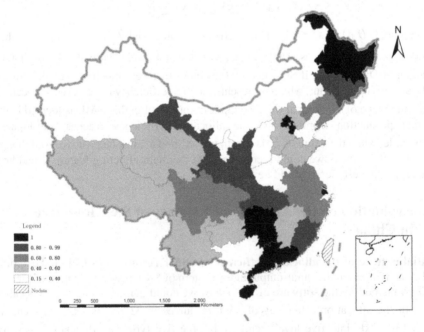

**Fig. 1.** Spatial disparity pattern of the allocation efficiency of S&T resources in China

**Reasons leading to the allocation inefficiency of S&T resources.** According to the analysis of $S^+$ and $S^-$, there are different reasons in different provinces. The provinces influenced greatly by the input redundancy of human and financial resources include Chongqing, Liaoning, Guizhou, Yunnan, Shandon, Qinghai, Tibet, Hebei, Jiangxi, Henan, Guangxi, Shanxi, Xinjiang, Inner Mongolia and Ningxia, and the rate of input redundancy in which is in excess of 60%. What's more, the rate in the other provinces except Chongqing, Liaoning, Guizhou and Yunnan, exceeds 100%. It is clear that although the scarcity of S&T resources has been recognized by the common people, the waste of them is still serious in most parts of China. Especially for the central and western regions, S&T resources are more limited than that in the eastern region. In the circumstances, failed to make full use of S&T resources is the main reason for allocation inefficiency. Besides, the provinces mainly influenced by the output insufficiency of high-tech product value of indirect economic benefit include Gansu, Yunnan, Qinghai, Guangxi, Xinjiang and Inner Mongolia, and the rate of the output insufficiency is between 60%-96%. All the provinces are distributed in the western region of China, so less contact with other regions is the main reason for lacking innovation leading to low value of high-tech production. And the provinces mainly influenced by the output insufficiency of the amount of transaction contract in technical market include Hubei, Shanxi, Fujian, Jilin, Jiangsu, Sichuan, Zhejiang, Anhui, Guizhou, Shandong, Tibet, Hebei, Guangxi, Shanxi and Xinjiang, their rate of the output insufficiency is between 60%-100%. As we can see, these provinces have a large number and distributed widely. To some extent, it reflects a fact that the transformation capability of S&T achievements should be improved in the whole country. In addition, the provinces mainly influenced by the insufficiency of the scientific paper number of direct knowledge output include Tibet and Qinghai, and influenced by the insufficiency of the granted invention patent number include Gansu and Fujian. Except the four provinces, the rate of the direct knowledge output in the other provinces is below 60%, which show the direct knowledge output is much higher, and there is limit space to raise the insufficient rate.

As mentioned above, the reasons for the allocation inefficiency of S&T resources in China can be summarized as the four following aspects.1) For the provinces that allocation efficiency less than 100%, the input redundancy of S&T resources is the key reason for the allocation inefficiency; 2) For the provinces that allocation efficiency less than 50%, the input redundancy is the most important reason for their inefficiency. These provinces are mainly distributed in the central and western regions, and have a great improvement potential due to their input redundancy of scientific human and financial resources exceeding 100%. 3) Compared with the direct knowledge production, the output deficiency of indirect economic benefit is the main problem for the provinces of the allocation efficiency less than 100%. The provinces accounting for 92% have a great space for improvement in the indicator of amount of transaction contact in technical market, and 75% provinces could be enhanced in the value of high-tech product. 4) For the whole 31 provinces across the country, the situation of direct knowledge output operates well.

# 4   Conclusion and Discussion

Nowadays, as an important strategic resource, the potential value of S&T resources is difficult to measure. And the characteristic of rareness orders the realization of reasonable allocation in order to serve the development of society well. This paper designed the input-output index system of S&T resources, built the evaluation model of allocation efficiency, and finished the evaluation of spatial allocation efficiency of S&T resources in China. Meanwhile, this paper pointed out the reasons for the allocation inefficiency and suggested the way to improve the allocation potential. In this basis of the current research, we summarizes the following conclusion.1) The spatial pattern of allocation efficiency of S&T resources in China presents clustered distribution, and the efficiency level appears a descending trend from the east to the west. 2) Although the total factors leading to the allocation inefficiency in China are the input redundancy and the output insufficiency, the different province affected by the focus is diverse. For most of the provinces in China, the lower utilization rate of S&T resources resulting in the output insufficiency of economic benefit is the most important reason affecting the allocation efficiency. Therefore, we should enhance the utilization rate of S&T resources and boost the transformation of S&T achievements as the main way to improve the allocation efficiency of S&T resources in China.

In this paper, there are still some limitations and shortcomings of the research on the allocation efficiency of S&T resources in China. Firstly, due to the inaccessible data, it is difficult to analyze the whole allocation efficiency in the micro-scale view based on the municipal-level administrative unit. Secondly, this paper selected only one time point in the research, which reflected the current situation of allocation efficiency of S&T resources in China. So it is needed we select a few more time points to study on the variation process of allocation efficiency of S&T resources. Overall, the selection to the typical municipal-level administrative unit for the scale, or the more time points reflecting the process of efficiency variation, will be the direction for the further research.

# References

1. Ding, H.D.: The new challenges on S&T resource allocation and strategies analysis. Studies in Science of Science 23(4), 474–480 (2005)
2. Liu, H.F.: Research on the Problems of Disposition of Sci-Tech Resources in China. World Sci.-Tech. R&D 21(1), 95–98 (1999)
3. Wei, S.H., Wu, G.S.: Research on the efficiency of regional science and technology (S&T) resource allocation. Studies in Science of Science 23(4), 467–473 (2005)
4. Li, H.X., Li, W.S.: Research on efficiency rating of science and technology resources deployment and space variance in our country. Scientific Management Research 28(4), 35–40 (2010)
5. The Yellow Book on Science and Technology, 2002. China Science and Technology Indicators (2002)
6. Liu, Y., Huang, J.Y.: The assessment of regional vulnerability to natural disasters in China based of DEA model. Geographical Research 29(7), 1153–1162 (2010)
7. Hu, Z.H., Liu, D.C.: Performance Evaluation of Scientific Research Input upon Promoting Regional Economic Growth in China. China Soft Science (8), 94–100 (2009)

# Research on Reform in the Teaching of Software Engineering Course

Jianjun Li[1] and Xiaorong Wang[2]

[1] Lushan College, GuangXi University of Technology, LiuZhou, 545006, China
[2] Department of Computer Engineering, GuangXi University of Technology,
LiuZhou, 545006, China
31063@qq.com, ccnuxnxs@163.com

**Abstract.** From the teaching work in these years, we deeply felt that the software engineering course is so important for students in major of computer science and technology. How to turn this course from a difficult and boring course to an interesting course is becoming urgent issue. In this paper, we analyse the problems during the teaching of software engineering course and proposed some reform methods. We wish these methods can motivate students'enthusiasm, thrust their learning and thus increase teaching effect greatly.

**Keywords:** software engineering, reform, education.

## 1 Introduction

Continues growth of software industry has increased the need for software engineering professionals, and consequently, the need for effective software engineering educational programs. Accordingly, many universities have included several courses covering software engineering concepts in their computer science/engineering curriculum. Software engineering is an integral part of the computer science and engineering undergraduate curriculum. One main goal of software engineering curriculum is to teach students how to develop and manage large-scale, long-lived software systems in a costeffective manner. Software engineering is the disciplined application of engineering principles to the creation of complex software systems. It is an amalgam of people, process, and technology [1].Students need to learn both technical knowledge, such as requirements elicitation, architectural design, implementation, testing, etc., but also non-technical aspects, such as project management and development of know-how for a particular application domain.

*Software engineering* (SE) is a multidimensional field that involves activities in various areas and disciplines, such as computer science, project management, system architecture, human factors, and technological evolution. Several efforts have been made to map the different dimensions of SE and to design a proper curriculum that addresses them all [2,3]. Accordingly, this domain diversity is often reflected in computer science (CS) and SE curricula, in which separate courses address the different topics.

G. Lee (Ed.): Advances in Intelligent Systems, AISC 138, pp. 141–144.
springerlink.com          © Springer-Verlag Berlin Heidelberg 2012

This paper will discuss how to make students study this course more easily. And how to carry out teaching method more effectively according to students on various levels will be discussed.

## 2    Character of Software Engineering Course

Like any other engineer, the software engineer must master:

- the theoretical foundations ofthe discipline;
- the design methods ofthe discipline;
- the technology and tools ofthe discipline.
  In addition, he/she has to be able to:
- keep his/her knowledge current with respect to the new approaches and technologies;
- interact with other people (often not from the same culture);
- understand, model, formalize, analyze a new problem;
- recognize a recurring problem, and reuse or adapt known solutions;
- manage a process and to coordinate the work ofdifferent people;

From character[4] of software engineering, we can find oftware Engineering I (SE361) is the initial course offering in our department's program. The typical student will take SE361 in the second quarter of their sophomore year. Prior to that they will have completed a sequence of introductory computer science (CS) courses. The CS courses focus on developing programming skills (using Java and Ctt) and address traditional CS topics - data structures, algorithms, object-oriented concepts and some exposure to concurrent and network programming techniques. Students individually complete lab and project assignments with little opportunity, or need, for collaboration with fellow classmates. The CS course sequence introduces some software engineering concepts, but at a cursory level. Far example, students are able to identify the major activities associated with the waterfall life cycle of development. Students arrive to SE361 with accomplished individual programming skills and finegrained problem solving techniques.

The course is composed of a lecture section and project lab section. Teams of five to six students form during the first lab session and work together for the remainder of the quarter. The lab section meets formally for two hours each week. The scope of project activities normally requires teams to meet outside of regular class and lab tines.

## 3    The Main Problem in SE Educate

What is even worse they may get an impression that there aren't any problem seem to be well defined, intuitively modeled by known techniques and methods, soluble in a semester [5]. Students often consider subjects usually taught on software engineering courses as being obvious and/or expressed in fuzzy ways. However real systems rarely have neat solutions and the requirements for those systems are influenced by many technical and non-technical factors. Aspects like management, organization, leadership, planning, communication, cooperation to large extent are dominant factors of success in software engineering projects [6].

In software engineering education, another of the main problems is the definition of an industrial strength software project, i.e., a project that cannot be solved by one or two students on their own, a project that incorporates existing and evolving third party software without any documentation, a project with changing requirements, and a project that has not got a clear outcome right from the beginning. Moreover, the project should be interesting for the students and challenging by allowing the use of state-of-the-art technology. On the other hand, the project should not be too large, such that a team of students will be able to finish it within a university term. In software engineering research, one of the main problems is recruiting new students for research projects. These students should be sufficiently skilled in software engineering, should have some experience in the respective research area and application domain, and should be highly motivated .

The existing discrepancy between university education and industry needs is at least partly responsible for the criticism made of many university courses by employers that the courses do not equip students for real work.

## 4    Proposed Reform Method

**Transforming Research Project to A Student Project.** Given the importance of project-based work in introductory software engineering courses, we are often inclined to believe that real-life projects, i.e. those which come from an actual user community, such as a company or public organization, would be preferable to synthetic projects, developed expressly for the course. Our current research in the field of electronic timetable or database application provided a suitable candidate.

As is well known by software engineering instructors however, real-life projects offer several difficulties which make them less advantageous than they seem [7]. Some of these are:

(A) The organization owning the project may not be able to designate an official to work with the academic project who can devote sufficient time and effort to it. (B) The project may be too complex or too routine to be used in the course. (C) The project may not be sufficiently defined as yet by the owner organization for students to start work on it immediately. (D) The hardware and software requirements of the project may not be available in the university. (E) Meetings may be difficult to arrange, which will hinder project's progress.

**Integration of Software Tools in Software Engineering Education.** There are significant benefits to be gained from promoting extensive use of software tools and environments in software engineering education, providing that they are educationally appropriate. Using a "purpose-built" teaching support environment specifically designed to emphasise the systematic nature of the processes and tools involved, support for the teaching of a range of programming paradigms and software prototyping via the use of (executable) formal specifications. It also enables the production, subject to rigorous set of constraints, of software systemswhich exhibit powerful behaviour at an early stage.

**Leisure Education in Professional Background.** The leisure education with the professional background is a higher-level education model which against too much emphasis on vocational training of traditional higher education, it requires the relaxation of non-vocational training as an important part of education. In the education process, the Leisure Education brings up people's appreciation, interests, skills and the ability to create recreational opportunities, people could organize their leisure time in a useful way, accordingly implement the process of "be human"[8].

## 5   Summary

From the teaching work in these years, we deeply felt that the software engineering course is so important for students in major of computer science and technology. How to turn this course from a difficult and boring course to an interesting course is becoming urgent issue. In this paper, we analyse the problems during the teaching of software engineering course and proposed some reform methods. We wish these methods can motivate students'enthusiasm, thrust their learning and thus increase teaching effect greatly.

## References

1. Tilley, S.: Software Engineering 2, http://www.cs.fit.edu/~stilley
2. Software Engineering 2004 (SE 2004): Curriculum guidelines for undergraduate degree programs in software engineering (2008), http://sites.computer.org/ccse/
3. Software engineering body of knowledge (2008), http://www.swebok.org/index.html
4. Ghezzi, C., Mandrioli, D.: The challenges of software engineering education. ICSE 2005, 637 - 638
5. Klint, P., Nawrocki, J.R.: In: Proc. Software Engineering Education Symposium SEES 1998. Scientific Publishers OWN, Poznan (1998)
6. Lethbridge, T.C., LeBlanc, R., Sobel, A.: SE 2004: Recommendations for Undergraduate Software Engineering Curricula. IEEE Software, 19–25 (November 2006)
7. Lethbridge, T.C.: What Knowledge is Important to a Software Engineer? IEEE Computer 12(5), 44–50 (2000)
8. Dai, S.: Leisure Education: Students Ideological and Political Work of the new task. Chinese Higher Education 24, 20 (2008)

# A Novel Charge Recycling Scheme in Power Gating Design

Huang Ping, Xing Zuocheng, Yang Xianju, Yan Peixiang, and Jia Xiaomin

School of Computer, National University of Defense Technology, Changsha, 410073, China
huangping@nudt.edu.cn

**Abstract.** Power gating has become a popular technique to reduce the ever-increasing leakage power for commercial microprocessors or SoCs. The reactivation energy and delay cost weaken its performance. This paper firstly proposes a novel charge recycling scheme to reduce the transition energy and delay, and then gives its equivalent model. The experiment results show that, comparing to the traditional power gating implementation, it can achieve 19.66% reactivation energy reduction, 9.28% peak leakage reduction, and 23.36% wakeup delay reduction, at 25°C, at the cost of 2.75% area increasing. At the same time, the circuit reliability is improved since the ground bounce reduced.

**Keywords:** power gating, charge recycling, low power design, VLSI.

## 1   Introduction

With the CMOS technology scaling down, the supply voltage is reduced accordingly to avoid device failure due to high electric fields in the gate oxide and the conducting channel under the gate. This quadratically reduces the dynamic component of circuit power dissipation, but, unfortunately, also decreases the switching speed of transistors. To compensate for the performance loss, the transistor threshold voltages are reduced, which in turn causes an exponential increase in the subthreshold leakage current. With the process scaling down to 65nm and below, leakage power has now become a dominant component of the total power dissipation in the CMOS circuits [1-5].

Power-gating, also called MTCMOS (Multi-threshold CMOS), is one of the most effective standby leakage reduction techniques recently developed [2-5]. It uses a high Vth PMOS/NMOS to connect the virtual power (VVDD) / virtual ground (VGND) and the real power / ground supply of a circuit, and turns off the circuit during the standby periods, as shown in Fig. 1. Accordingly, the PMOS switch is called header, while the NMOS switch is called footer [4].

Previous researches show that power gating can reduce the leakage power by 10-1000x [2]. However, power gating is not a cost-free technique. Adding a switch will results in performance loss, the area cost, the reliability, mode transition cost and so on. Since the target is to make sure the benefit of leakage power reduction from power gating overwhelms the introduced power and delay penalties, we will focus on the mode transition energy minimization in this paper.

G. Lee (Ed.): Advances in Intelligent Systems, AISC 138, pp. 145–153.
springerlink.com          © Springer-Verlag Berlin Heidelberg 2012

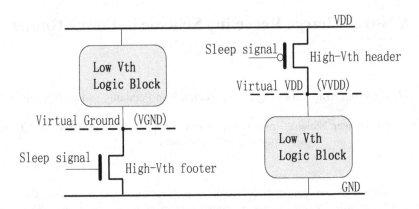

**Fig. 1.** Power gating structure

Let's firstly check where the mode transition energy comes from. Here we take a footer implementation as example. Firstly we assume that there is a long enough idle time so th at the VGND is fully charged up to a voltage (Vvg) near to VDD (V), by leakage current of the circuit. And we assume that the Cg is the gate capacitance of the footer, and the Cvg is the capacitance in VGND. In the sleep mode, the nSLEEP is zero to cut off the footer, while the VGND is gradually charged up to $V_{vg}$ by the leakage current. In the active mode, the nSLEEP is charged to V, which turns on the footer and makes the VGND discharge to a voltage near to zero soon. Easy to find, during the sleep-to-active transition, power supply should consume an energy of $C_gV^2$ to turn on the footer, while the energy stored in VGND is wasted. If we can make use of the wasted energy, we may save more power supply energy. Charge-recycling is a good way to implement this idea.

The remainder of this paper is organized as follows. In section 2, we review previous work in the area of charge-recycling power gating design. Section 3 introduces the concept of our improved charge-recycling scheme. We give the equivalent model in section 4, and the experiment results in section 5. Section 6 concludes.

## 2    Previous Work

Charge-recycling is a good way to save power during the mode transition. Pakbaznia et al used a transmission gate to connect the virtual power supply and the virtual ground supply between two different modules, where one is implemented by footer and another is implemented by header [8]. They turn on the transmission gate just before and after they turn on/off the sleep transistor, to implement the charge recycling during the mode transition, and got 50% mode transition energy reduction. However, to ensure the signal integration of the power supply, only one power switch is used in sub-90nm design, either footer or header. This limits the usage of this kind implementation of charge-recycling.

Tada et al proposed a state reservation power gating scheme by making use of the charge recycling between virtual ground and the gate [9]. Henzler et al [10] took the same way. They used a small PMOS transistor to connect the VVDD and the gate of the header, and got 25% energy reduction. However, two sleep enable signals, like SEL1 and SEL2 here, are needed to turn on or off the switch in Henzler's work, as shown in Fig. 2. Their generation is not an easy thing, and their routing cost will doubled accordingly, which will result in routeability in the modern distributed sleep transistor network implementation.

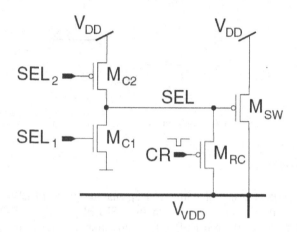

**Fig. 2.** Charge-recycling between the gate and the virtual power supply

This paper will show how to implement the similar charge recycling in the traditional implementation where only on sleep enable signal is needed, and get a considerable gain.

## 3    Our Proposed Design

As shown in Fig. 3, a charge-recycling NMOS (CR-NMOS) is added in the traditional power gating circuit. Then using the original controlling logic, without adding any sleep enable signal, we can get charge sharing between the VGND and the nSLEEP (the gate of the sleep transistor).

One thing needs to be pointed out. Different from Pakbaznia's work, the phase of the CR signal has to change. In Pakbaznia's work, they turned on/off the CR signal just before the sleep signal during the wakeup transition. Here, however, the CR signal should be just after the transition of the SLEEP signal. If we turn on the CR signal just before turning on the footer, the CR-NMOS and the NMOS transistor MN will turn on simultaneously, and the charge stored by the node VGND will discharge through the CR-NMOS and the NMOS MN, which make the charge-recycling lose some possible gain.

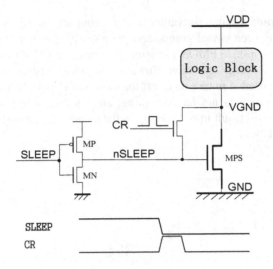

**Fig. 3.** Our proposed charge-recycling circuit

# 4    Model Analysis

When the CR-NMOS is turned on with the sleep enable signal SLEEP simultaneously, VGND will share the stored charge with the nSLEEP through CR-NMOS, while power supply will charge the nSLEEP at the same time, as shown in Fig. 4. Here we assume the pulse will end just when the voltage of VGND is equal to that of the nSLEEP.

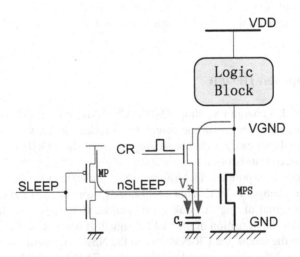

**Fig. 4.** Charging process during wakeup transition

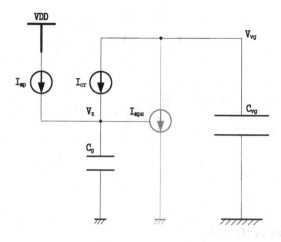

**Fig. 5.** Equivalent model of the proposed circuit

At the beginning of the pulse, the VGND is at voltage $V_{vg}$, the nSLEEP is zero. At the end of the pulse, the VGND is st ill above or equal to the nSLEEP. We can easily get the following results.

1) MP is working at velocity saturation state
2) MCR is working at saturation state
3) MPS works in sub-threshold leakage when $V_x$ is smaller than $V_{tn}$, and then works in velocity saturation state when $V_x$ is larger than $V_{tn}$.

Based on these results, we get the following equivalent model during the wakeup transition, as shown Fig. 5. The $I_{mp}$ is the charge current through MP, the $I_{cr}$ is the sharing charge through CR-NMOS transistor, the Imps is the current flows through the footer MPS, $C_{vg}$, Cg are the capacitance of the VGND, and the gate capacitance of the footer. $V_{vg}$ and $V_x$ are the voltage of VGND and nSLEEP respectively.

When $V_x$ is smaller than the threshold voltage of the MPS, the $I_{mps}$ is the sub-threshold leakage. This model can be expressed by the following equations.

$$I_{cr} + I_{mps} = C_{vg}\frac{dV_{vg}}{dt} \tag{1}$$

$$I_{cr} + I_{mp} = C_g\frac{dV_x}{dt} \tag{2}$$

$$I_{mps} = I_S\, e^{\frac{V_x}{nkT/q}}\left(1 - e^{-\frac{V_{vg}}{kT/q}}\right) \tag{3}$$

$$I_{cr} = \frac{1}{2}\mu_n C_{ox_n}\left(\frac{W_{cr}}{L}\right)(V - V_x - V_{tn})^2 \tag{4}$$

$$I_{mp} = v_{sat_p} C_{ox_p} W_{mp} \left( V - V_{tp} - \frac{V_{DSAT_p}}{2} \right)$$
(5)

In equation (3), $I_s$ and n are empirical parameters, kT/q is the thermal voltage. $V_{DSATp}$ is the velocity saturation voltage for PMOS.

When the $V_x$ is charged over the $V_{tn}$, MPS will work in velocity saturation mode. The equation (3) should be changed to equation (6).

$$I_{mps} = v_{sat_n} C_{ox_n} W_{mps} \left( V_x - V_{tn} - \frac{V_{DSAT_n}}{2} \right)$$
(6)

## 5    Experiments Results

In our experiments, a commercial 65nm process is used, and the standard power supply is 1V. We use a CR-NMOS to recycle the charge in a traditional power gating circuit, in which two cascaded inverter chain are gated by a footer. The inverter chain is a low threshold circuit with the threshold $V_{th\_pmos}$=0.196V, $V_{th\_nmos}$=0.130V. The high threshold we used here is $V_{th\_pmos}$=0.424V, $V_{th\_nmos}$=0.458V.

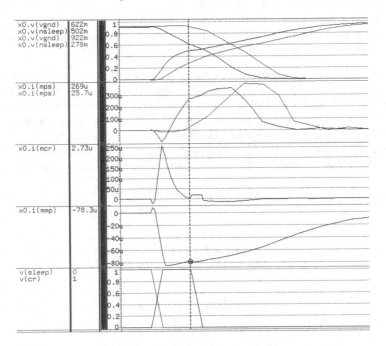

**Fig. 6.** Current during wakeup transition

Fig. 6 shows the simulation results for the current of different transistors during the pulse. The results verifies the equivalent model. During the pulse, the voltage of nSLEEP, $V_x$, with CR-NMOS is apparently larger than that of without it. This excessive comes from the recycling charge through CR-NMOS. As described in the equivalent model, $I_{mcr}$ is quadratically reduced and $I_{mp}$ is monotonously reduced with the increasing of $V_x$. The $I_{mps}$ keep increasing with the $V_x$.

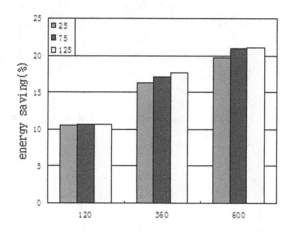

**Fig. 7.** (a) Size and temp effects on wakup energy saving

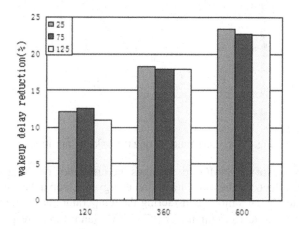

**Fig. 7.** (b) Size and temp effects on wakup delay reduction

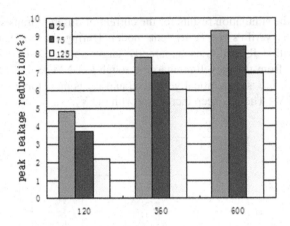

**Fig. 7.** (c) Size and temp effects on peak leakage reduction

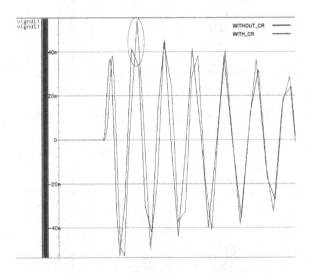

**Fig. 8.** Ground bounce during the wakeup transition

Since the resistor of CR-NMOS determines the efficiency of charge recycling, we checked the size effects on the mode transition energy, wakeup delay and peak leakage through the footer. What's more, considering the temperature has greatly effect on the leakage. We test our test circuit under 25℃, 75℃ and 125℃ respectively. Fig. 7 compares the size and temperature effect on the leakage current, peak leakage, wakeup energy and wakeup delay. Fig. 7 (a) and (b) tell us the wider CR-NMOS can save more wakeup energy and delay. Compared to the larger difference between different size, the temperature affect little on the wakeup energy and delay. However, it has a significant influence on the peak leakage flow through the footer because of the exponential dependent on the temperature, as shown in Fig. 7 (c). The simulation shows a 600nm CR-NMOS can achieve 19.66% reactivation energyreduction, 9.28% peak leakage reduction, and 23.36% wakeup delay reduction at 25℃.

The significant reduction in peak leakage can reduce the ground bounce. For electrical parameters of the package interface, we used L=2nH, R=70mΩ and C=0.2pF for the TQFP package [11]. Fig. 8 compares the ground bounce between charge recycling using the CR-NMOS and traditional power gating. The ground bounce reduces significantly, especially at the second positive peak, where the ground bounce reduces from 54.6mV to 41.3mV. This help to increase the reliability.

# 6    Conclusion

Power gating will be a basic techniques to cut the leakage power in future MPSoc era. This paper proposes a charge-recycling scheme to reduce the transition energy and delay. It added an NMOS transistor to connect the gate of the switch and the virtual ground to realize the charge sharing. Authors also analysis the charge recycling circuit and give the equivalent model. Experiment results verify the equivalent model, and show larger CR-NMOS works better, and the scheme can achieve 19.66% reactivation energy reduction, 9.28% peak leakage reduction, and 23.36% wakeup delay reduction, at the cost of 2.75% area increasing, at   25°C. The reduction in peak current reduces the ground bounce greatly.

**Acknowledgement.** This work is supported by the National Natural Science Foundation of China (60873016).

# References

1. De, V., et al.: Technology and design challenges for low power and high performance. In: ISLPED, pp. 163–168 (1999)
2. Roy, K., Mukhopadhyay, S., Mahmoodi-Meimand, H.: Leakage Current Mechanisms and Leakage Reduction Techniques in Deep-Submicrometer CMOS Circuits. Proceedings of the IEEE 91(2) (February 2003)
3. Kumar, R., Hinton, G.: A family of 45nm IA processors. In: ISSCC 2009 Digest of Technical Papers, pp. 58–59 (February 2009)
4. Shi, K., Howard, D.: Challenges in Sleep Transistor Design and Implementation in Low-Power Designs. In: DAC, pp. 97–102 (2006)
5. Hu, Z., et al.: Microarchitectural techniques for power gating of execution units. In: Proc. ISLPED 2004, pp. 32–37 (2004)
6. Kim, S., et al.: Understanding and minimizing ground bounce during mode transition of power-gating structures. In: ISLPED (August 2003)
7. Pakbaznia, E., Fallah, F., Pedram, M.: Charge Recycling in MTCMOS Circuits: Concept and Analysis. In: DAC (2006)
8. Tada, A., Notani, H., Numa, M.: A novel power gating scheme with charge recycling. IEICE Electronic Express 3(12), 281–286 (2006)
9. Henzler, S., Nirschl, T., Skiathitis, S., Berthold, J., Fischer, J., Teichmann, P., Bauer, F., Georgakos, G., Schmitt-Landsiedel, D.: Sleep Transistor Circuits for Fine-Grained Power Switch-Off with Short Power-Down Times. In: ISSCC (2005)
10. Kao, J., Chandrakasan, A., Antoniadis, D.: Transistor sizing issues and tool for multi-threshold CMOS technology. In: DAC (1997)
11. Heydari, P., Pedram, M.: Ground bounce in digital VLSI circuits. IEEE Trans. VLSI Systems 11(2), 180–193 (2003)

# Research of Communication Platform of Intelligent Public Transportation System Based on GPRS

Dong Yongfeng, Guo Zhitao, Liu Peijun, and He Min

Hebei University of Technology, Tianjin 300401, China
dongyf@hebut.edu.cn, mrnow@jsmail.hebut.edu.cn

**Abstract.** To solve the increasingly serious urban transportation problem, the research and application of intelligent transportation system becomes the inevitable choice for all countries. The paper introduces design and implementation of a communication platform for intelligent transportation system. It focuses on the process of communication, the design of communication protocol and the implementation of communication software. The platform achieved real-time communication and it can monitor, track and locate buses. A great deal of test and experiment show that the platform is feasible. This platform has been successfully applied to specific projects and achieved good effect.

**Keywords:** communication platform, wireless communication technology, GPRS, data frame.

## 1   Introduction

With the development of economy and the acceleration of living rhythm, people's demand for public transportation is bigger and bigger. But due to various reasons, the public transportation system in our country is not optimistic. The situation that vehicles increase, routes extends and bus trips grow in big cities makes the traffic heavier. No matter in development countries or in developing countries, this question is more and more serious. City traffic problem can't be solved completely by increasing roads.

Intelligent transportation system includes GPS, GPRS, and GIS technology. It shows as electronic map. It realizes real-time surveillance and dispatch of buses with collecting, transporting and handling the information about bus locations, conditions, passengers and so on so. By using this system, buses can be adjusted rapidly and reasonably. The system improves the efficiency of the bus operation, makes the best use and allocate of existing resources for public transport enterprise.

## 2   System Design

**System Composition.** Intelligent transportation system[1] mainly includes monitoring dispatching center, passenger traffic acquisition system, car terminal

G. Lee (Ed.): Advances in Intelligent Systems, AISC 138, pp. 155–161.

system, communication network and electronic stop system. Monitoring dispatching center receives and handles data, monitor and dispatch buses. Passenger traffic acquisition system is placed on the bus to count the number of people who get on and off at every bus stop. Car terminal system consists of ARM embedded motherboard, GPS satellite receiving module and GPRS wireless communication module. Communication network mainly uses the GPRS network to realize data transmission system so that it can reduce data transmission cost and enhance the delivery time and veracity. Electronic stop system with GPRS module and screen display bus location information.

**Data Transfer Process.** Data transfer process[2] refers to the process that data transfer through GPRS network. The relationship and transformation of communication states show in figure 1. Full lines represent the transformation in ideal situation. But in actual circumstances, we find it that data packet may have inspection error or lost situation. It will cause state overtime. In these circumstances, the states will be represented by dotted lines.

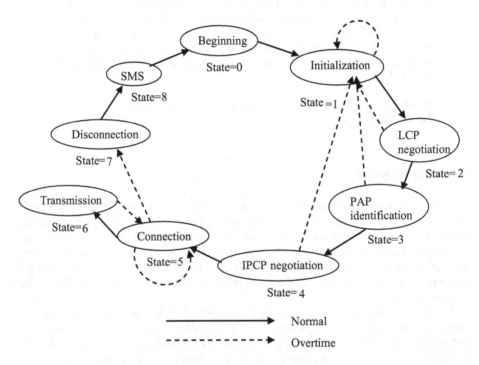

**Fig. 1.** Transformation of communication state

Firstly, GPRS module consult with GPRS gateway on communication links after dial-up. It means to negotiate point-to-point configuration parameters. Consultation process keeps LCP, PAP, IPCP and other protocols. LCP protocol is used to establish, build and test links. PAP protocol is used for processing verifying password. IPCP protocol is used for setting network protocol environment, and assigns IP addresses. Then, it transmits IP packet according to the standard of consultation. According to the different application, the IP packet may carry TCP message, also can be the UDP or ICMP messages. This paper uses TCP message to transfer data information.

# 3    Frame Type Definition and Instructions

**Communication Protocol.** Data exchange between the communication server and the car terminal system is based on communication protocol and the data will transmit by frame. Communication protocol defines 3 special characters: '$' - upload begins; '@' – downlink begins; '#' – ends. All the data are in hexadecimal format. If these three special characters appear in a place where it is not the stating prefix or the end suffix, they will be translated to double byte. $=0x5e+0x01, #=0x5e+0x02, @=0x5e+0x03, 0x5E=0x5e+0x04. If a data is a multibyte data, rank the high byte in front and the low byte behind.

**Communication Protocol Frame Format.** The format of upload datagram which is transmitted from car terminal system to communication server is shown in Table 1.

**Table 1.** Format of Upload Datagram

| start prefix | command word | priority | data block length | data block | calibration bytes | end suffix |
|---|---|---|---|---|---|---|
| 1byte | 1byte | 1byte | 2bytes | 0~255bytes | 1byte(XOR) | 1byte |

The format of download datagram which is transmitted from communication to car terminal system is shown in Table 2.

**Table 2.** Format of Download Datagram

| start prefix | command word | priority | data block length | data block | calibration byte | end suffix |
|---|---|---|---|---|---|---|
| 1 byte | 1 byte | 1 byte | 2 bytes | 0~255bytes | 1byte(XOR) | 1 byte |

Upload datagram and download datagram is different in start prefix. The former is start with '$' and the latter start with '@'. Command word is a symbol to differentiate the data frames. Data block length (before translated) shown in bytes is the length of parameters related to command word or the data a frame has. The effective length is 0 ~ 65535 bytes. The low byte is in front, the high byte is behind. Data block is the parameters and data. Calibration byte is the result of XOR of all data between start prefix and calibration byte (before translated).

**Key Data Block Frame.** Equipment number: 5 bytes, the first 2 bytes are GPRS line number; the last 3 bytes are bus number. Driver number: 2 bytes. On-board machine SIM card serial number: 3 bytes. Longitude and latitude definition: both are hexadecimal data of 4 bytes. The first byte is degree; the second byte is point; the third and forth bytes are decimal fraction of the point. The third byte is the high part and the forth byte is the low part. These two bytes make up of a word which is a hexadecimal number. The point rounded to four decimals. Time tag definition: 6 bytes. Vehicle running logo: 1 byte. The meaning of every bit is shown in Table 3.

**Table 3.** Format Of Vehicle Operation Estate Datagram

| state | D7 | D6 | D5 | D4 | D3 | D2 | D1 | D0 |
|-------|----|----|----|----|----|----|----|----|
| 0 | out | not speeding | new event | middle door open | back door open | front door open | up | not service |
| 1 | in | speeding | ole event | middle door closed | back door closed | frontdoor closed | down | service |

**Establishing Connection Datagram.** When the car terminal power open, error reconnection or disconnected reconnection successful, car terminal system send establishing connection datagram to communication server. The datagram need response from communication server. As shown in Table 4.

**Table 4.** Formate of Establishing Connection Datagram

| Equipment number | Driver number | On-board machine SIM card serial number | Connection time | On-board machine line version number |
|------------------|---------------|------------------------------------------|-----------------|--------------------------------------|
| 5bytes | 2 bytes | 3 bytes | 6 bytes | 10 bytes |

**GPS Datagram.** GPS module on bus will send GPS information to communication server every 10 seconds for the convenience of monitoring and dispatching buses. GPS information don't need response from communication server. As shown in Table 5.

**Table 5.** Format of GPS Coordinate Information Datagram

| Equipment NO | Event time | North latitude | East longitude | speed | Azimuth | Height | Vehicles running | Next stop | GPRS signal |
|--------------|------------|----------------|----------------|-------|---------|--------|------------------|-----------|-------------|
| 5bytes | 6 bytes | 4 bytes | 4 bytes | 1 byte | 1 byte | 1 byte | 1 byte | 1 byte | 1 byte |

**Receiving Response Datagram.** When car terminal system or communication server receive information that need response, they will send receiving response datagram. This datagram has two types: one is for manual, the other is for automatic. It doesn't need response from receiver. As shown in Table 6.

**Table 6.** Format of Receiving Response Datagram

| Equipment number | Response time | North latitude | East longitude | response command number | command result |
|---|---|---|---|---|---|
| 5bytes | 6 bytes | 4 bytes | 4 bytes | 4 bytes | 4 bytes |

Download Task Information Datagram.This datagram is used to dispatch buses for monitoring dispatching center. As shown in Table 7.

**Table 7.** Format of Download Task Information Datagram

| Event time | Command number | Command length | Command introduction | Manual response |
|---|---|---|---|---|
| 6bytes | 1byte | the length of command introduction | no definition | 1byte |

## 4    Design of GPRS Communication Server

As a way that data transferred and a link that information exchange, GPRS communication server [3-5] play an important role in intelligent transportation system. Its main function is receiving car terminal system connection and communication with monitoring dispatching center.

We use socket network programming technology to realize data transmission and multithreading technology to achieve communication between a server and multiple terminals. Sending and receiving data between the client and the server adopts nonblocking methods. The nonblocking communication mechanism is realized by classes in packet of java.nio.

## 5    System Testing and Analysis of the Results

**Car Terminal System Connection Test.** By writing a test tools, setting up different number of car terminal to connect the server, we can test the reliability and real-time of the server. After lots of tests, we get an average value of terminal connection number. The result is shown in Table 8.

**Table 8.** Testing Results

| Connecting thread count expected | Has connected thread count | Not connected thread count | CPU usage |
|---|---|---|---|
| 50 | 50 | 0 | 0%-3% |
| 100 | 100 | 0 | 0%-3% |
| 200 | 200 | 0 | 0%-3% |
| 300 | 300 | 0 | 0%-3% |

**Table 8.** (*continued*)

| 500 | 500 | 0 | 0%-3% |
|---|---|---|---|
| 800 | 800 | 0 | 1%-3% |
| 1000 | 1000 | 0 | 1%-4% |

**Data Receiving Test.** In order to make the test results more reliable and to make the test for communication server more comprehensive, data receiving test for the server is needed. We test the situation that a server receiving 50 commands contemporarily, then increase the number of sending command. The result is shown in Table 9.

**Table 9.** Testing Results

| sending command count | received command count | not received command count | CPU usage |
|---|---|---|---|
| 50 | 50 | 0 | 0%-3% |
| 100 | 100 | 0 | 0%-3% |
| 150 | 149 | 1 | 0%-3% |
| 200 | 198 | 2 | 1%-4% |
| 300 | 295 | 5 | 1%-4% |

When increasing car terminal connections, the CPU usage of communication server changes little. Furthermore, no connection is missed. On the aspect of data receiving test, with the sending data increasing, memory usage and CPU usage of communication server has no apparent change. When sending data over a certain amount, CPU usage will increase, and small amounts of data will be missed. However, in actual situation, the amount of sending data won't be so much. And we can use higher configuration computer in practical production as a server to reduce server load pressure.

## 6    Conclusion

This paper studies the basic function of communication platform in intelligent transportation system, gives detail introduction about communication protocol between the platform and car terminal, proposes communications platform design scheme and realizes it. For nonblocking and blocking mixed mode communication server, we should have further research to make it start as little thread as many to process more client communication problem.

**Acknowledgment.** This research is supported by scientists and engineers service businesses (2009GJA10020), application base and cutting-edge technology research plan of Tianjin (10JCYBJC00200).

# References

1. Liang, S., Liang, Y., Chen, J.-N.: Implementation of Communication Platform for Intelligent Public Transportation System based on GPRS. Communications Technology 40(10), 25–40 (2007)
2. Zhou, G.: Research and Realization of the Transmission Technology in Advanced Public Traffic System, pp. 28–31. Central China Normal University (2008)
3. Chen, B., Cao, W.: An Efficient Algorithm Vehicle License Plate Location. In: IEEE ICAL 2008 Conference, pp. 221–225 (2008)
4. Zheng, B., Tian, B.: Automation Detection of Technique of Preceding Lane and vehicle. In: IEEE ICAL 2008 Conference, pp. 201–205 (2008)
5. Li, N., Xu, Z.-X.: The application of GPRS technique to vehicle monitoring system. Applied Science and Technology 32(6), 34–36 (2005)

# The Research and Analysis on Digital Animation Courses Setup

Peng Shengze, Wen Yongge, and Liu Zhibang

School of Mathematics and Computer Science, Mianyang Normal University,
Mianyang, China
Department of Science and Technology, Mianyang Normal University, Mianyang, China
Beijing Institute of Technology, School of Computer Science & Technology, Beijing, China
shengze_p@yahoo.com.cn, wenyongge@sohu.com

**Abstract.** Vocational education in schools should focus on the consistency of learning and practical work, the paper production process for animation projects and animation specialty in the state of curriculum, specialized courses on how to apply to the production of animation projects conducted in-depth the key points study. Animation of a vocational school class how the professional approach taken and the implementation of the project should pay attention to practical problems.

**Keywords:** production process, curriculum, project teaching, teaching, teaching problem.

## 1    Introduction

Teaching in vocational institutions will be introduced to the teaching work of the project is relatively common form of teaching. Its real purpose is to guide students to apply theoretical knowledge and skills to solve practical problems in the work. The curriculum for the animation class animation projects and enterprises is the combination of professional animation to one of the important issues, but also an important part of artistic practice course. In teaching practice, we should pay attention to practical work in schools to learn and consistency, thus highlighting the openness of the process of teaching, professional and student initiative. However, courses teaching how to bind to the production of animation projects, is worthy of our in-depth analysis of the key issues[1].

## 2    The Characteristics of the Animation Production Pipeline Project

Animation in the field of artistic creation is a major industry, animation production is a complex law has a certain production and processing. In the animation of the entire process of production, whether it is any kind of animation, they are much the same basic processes, all with stage, continuity, integrity, technical characteristics. To the

G. Lee (Ed.): Advances in Intelligent Systems, AISC 138, pp. 163–168.
springerlink.com        © Springer-Verlag Berlin Heidelberg 2012

process of animation is concerned, can be roughly divided into: general design, design, concrete production and editing four stages. By the overall time can be divided into early, middle and late stages. Concrete analysis: two dimensional animation production work flow in general, including: overall planning - the story script - a shooting platform of the - of the original painting design - painting - Check - description line - color - animation - special effects production - editing - voice - synthetic output . Simply put, two-dimensional animation production process from script - storyboard - original painting - animation - color - Synthesis - Voice composition. Three-dimensional animation of the production process generally includes: The Story of the script - a shooting - the original painting design - modeling - Material - Animation DESIGN - Design - Rendering - voice - Edit synthesis - output. Most of the current animation production work applications (especially three-dimensional animation) computer to complete. Therefore, some of the animation production components will be more reasonable or update the presentation of the production means. But no matter how the production process, in each of the sessions are covered by a large number of technical requirements. From which also reflects the characteristics of a large number of professional and technical personnel assigned amounts. This is what we want to study through its animation features the basic curriculum system[2].

## 3    The Animation Specialty in the State of Curriculum

In most vocational schools in the animation profession, teaching is still using more traditional single-disciplinary approach to the closure. In this way the learning process in the sub-disciplines in the continuity of the lack of technical knowledge, systematic and practical. Student learning objectives, learning environment and learning organization as a whole there numerous educational disadvantages. Curriculum in the animation program, generally are open are: drawing, color, sketch, original painting design, scene design, animation, basic, Flash animation, Photoshop, 3Ds max, Maya, video effects, video post, photography, video, etc.. Although the course covers most of the animation produced most of the knowledge and skills. However, in actual teaching often do not see clear teaching, is nothing more than a certain number of training courses out excellent students. This phenomenon in secondary vocational schools is particularly acute in the animation majors. Most vocational schools in the teaching model to 2 +1 to the development of teaching programs. Which, in the first school year and a large number of basic cultural lessons (such as language, mathematics, foreign languages, ethics, basic computer applications, etc.). Can only be opened in the curriculum plan sketch, color, or a small amount of animation based on theory. The rest will have to set up the second year in a large number of professional courses. However, only one year of professional study courses, facing third-year internship Shiyou. In this way, all professional courses are under pressure in the second year. Because the animation is difficult courses, closely linked with each other, and no sufficient time for graduation. If you simply open the door a few specialized courses so-called good jobs, but also worried about the careers of students face too narrow, and thus also affect the students future career development in the future of animation. Proved, but simply classes by subject, single subject graduation, students completing their studies for their professional development is basically a very slim prospect. The feedback information indicates that there are many, many students still unable to figure out before practice exactly how to make a cartoon out, but it is in part what kind of animation technicians. So they appear to learn very Kulei, closed it hastily, practice is very helpless, very disappointing corporate chain reaction. Some of the animation industry is still interested in re-learning students to choose, re-training, as well as some students have chosen a different career, had given up favorite professional[3-7].

## 4    The Professional Curriculum and Animation Projects the Organic Integration of Production Processes

In the animation teaching, vocational education for the current situation of serious study and teaching model to adapt to the road to success is education. How to curriculum reform and teaching practice to enhance teaching effectiveness, enhance their comprehensive technical capabilities and application of skills to enhance student interest in learning and confidence in employment is the main problem we have to consider.

In teaching practice, must find the process with enterprise project consistent animation teaching animation talent to adapt to today's market demand.

Through research and analysis, our actual animation project as the subject of the second grade curriculum and teaching objectives. To make students more focused and mobilize the enthusiasm of students and participation. We "successfully embark on career paths for students," as the starting point, the animation instruction in the use of practical projects identified in the core of the studio-style teaching model to second-year production cycle for the project throughout the semester. Students learning by doing, students completing the course of the project continuously improve skills, learning and work to achieve real integration. Specific implementation is as follows:

**Animation workflow in accordance with and in accordance with the form of cartoons.** (Such as three-dimensional animation projects produced by the group: Original stage - modeling stage - material stage - stage of the animation - rendering stage - the late stages) at each stage of the task with the appropriate professional courses. (Example: creation of animation based on original stage theory, folk songs research, write the script, character design, scene design, storyboard and so on. Modeling stage set up, 3Ds max technology, Maya techniques, role modeling, scenario modeling. Rendering stage to open a lighting technology, special effects technology.) relatively fixed by a professional teacher is responsible for teaching a single stage.

**All teachers involved in the project is responsible for the course to a collective need for a project under the animation of specific teaching program.** And in the process of teaching the system to take collective focus on lesson planning, decentralized responsibility, the collective focus on teaching evaluation system and other measures.

**Several groups of students according to the distribution of students in each group elect a good job as a management and technical producer.**

**Student achievement by stages to sub-project appraisal and completion of final assessment by two comprehensive assessment.**

This program previously distributed by the professional animation production process into the project module. Students that clear objectives, and identify the learning direction, but also enhance the learning interest and motivation in the whole production process in the form of task-driven learning to lead the completion of their mission, professional teachers will also be completed animation projects for a common goal, develop and teach students in the production process all the necessary theoretical knowledge and technical points, integrated services for the professional improvement of the quality of teachers has also brought more touches.

Professional courses through the combination of production and animation projects, technically breaking the disciplinary education system. Dilute the discipline, focus on process; dilute the results, pay attention to the task. Major construction work and study animation teaching model, the formation of the field of animation tasks and learning the corresponding system. Application of task-driven, project-oriented curriculum into teaching in the animation. Truly professional education from the animation class status and the relationship between demand for animation industry professionals to start, is conducive to the study of real work situations tasks, work processes and learning between the docking area.

## 5    Note the Issue and Reflect the Teaching

**In the course of the arrangements.** To co-ordinate the entire production cycle of a reasonable amount of tasks, take full account of the basic stages of production time, including the new knowledge and new technology point of a reasonable explanation and analysis of actual production. Otherwise easily lead to a certain extend or shorten the cycle, giving a cycle after the impact.

**Topics of the project tasks to fully consider the amount of students.** The students do the actual level of basic skills and prediction. It should also form in the animation style to the current contents of most of the animation company's items for reference, script ideas based on the taste of the animation, so that students can always learn in a fun, the easy to the successful completion of tasks.

**Formation and management of the project.** Based on self-control in the level and quality of students in the project independently or less were used for the benefit of cooperative projects to avoid some of the students in the production of the various links of interdependence plot. Can be used in the production of special studio-style teaching methods, to adopt a group-oriented management.

**Reflected the problems of teaching.** Teaching in the project will reflect the shortage of technical teachers. Animation requires a strong comprehensive ability of teachers, teachers teaching time, should participate in the business of animation production, to ensure that teachers are not divorced from social production practice. Younger students in vocational, quality is poor, usually early in the production of excited, medium fatigue, numbness late. In teaching in different stages of motivation to encourage and guide them tide over the crisis. Continued use of scenarios is an ideal teaching method.

Example, students participate in role-play, so they really melt in the role of medium, through simulation exercises to enable students to appreciate and understand the concepts quickly, receive knowledge and skills to further enhance their interest in learning.

The teaching process in the implementation of the project, the teacher is no longer dominant, but the guide, explore, collaborators and managers. Strong autonomy of students, large degree of freedom, increased teacher preparation, management, and more things. Meanwhile, the quality of student work evaluation will be diversified. Analysis of teaching management how to adapt to this change, would help promote the teaching reform.

Fifth, the professional curriculum of traditional animation projects produced little steering feel

The first is its necessity in order to fully promote employment-oriented, innovative teaching model is teaching animation classes necessary road, according to the characteristics of corporate jobs, combined with the dual resources of businesses and schools, develop a practical people, to be in production-oriented, and the project is teaching classes from the current animation industry, the production mode, and, in the project or teaching is the student, give full play to the student's initiative, but also fully embodies the " people-oriented "scientific point of view.

Second is the feasibility of the development of vocational education is the focus of national education, projects and programs will combine corporate culture consistent with the teaching mode, only the animation is introduced into the classroom teaching, can be shifted to simulate the real classroom curriculum design design, students can design from the previous work into the project design. After teaching practice has been shown to project works as the core of the teaching model of education, a goal that everyone can do it, everyone has the direction, everyone has the achievements, the practice has been to see the students learning ability, communication skills, ability to cooperate the project has been significant production increase. When students see their hard work to complete the animation works, for their efforts and proud of it, but also for their professional development path into the future to add more courage and confidence.

To sum up, taking courses in vocational education programs teaching is to cultivate a professional skilled personnel and effective teaching methods. Teachers and students to explore, the guiding students to apply research skills and knowledge to solve problems. Teaching, student-centered teaching, teaching the classroom curriculum into the project as the central practice center; students not only self-learning, and reflect a high degree of passion for learning and strong interest in learning how to better and faster the attention of a student learning task. Practice shows that the introduction of the professional skills courses in the production process to the project is to improve the quality of the animation class teaching effective way.

**Acknowledgment.** The work is supported by the Natural Science project of Sichuan province Education Department of China under Grant Nos. 10ZC029 and Animation Research Centre project in Sichuan province under Grant Nos. 201010030.

# References

1. Li, S.: Enterprise project teaching courses in film and animation practice teaching. Western Science and Education Forum (October 2009)
2. Yan, H.G.: The practice of vocational education teaching and research project. Education Forum (October 2008)
3. Introduction to digital media technology. In: The 21st Century Planning Materials in Digital Media Colleges
4. Jia, X.: The status and problems of modern distance education technology development. China Science & Technology Education Innovation (14), 317–318 (July 2008)
5. Peng, Y.: Learning support services based on problem-solved in rural distance education. Journal of Yunmeng (3), 135–136 (March 2009)
6. Wang, Z.: Discussion of distance education platform based on cooperative work technology. Science & Technology Information (2), 15–17 (February 2008)
7. Li, H.: Issues concerning the support-service system for distance education. Journal of Guangdong Radio & TV University (1), 1–3 (January 2003)

# The Digitize of Traditional Chart Handles a Research

Yang Weijian

Institute of Computer, Sichuan University of Science & Engineering, Zigong, China

**Abstract.** This paper aims at traditional chart to obtain data method one and obtains a data result, the error is partial to great bad situation and put forward the irregular curve chart progress number quantize transaction that the traditional hand work draw, the exploitation gets a homologous digitize information in identifying the foundation of curve chart progress storage. This paper put forward the mold mass transaction way of thinking for turning on the foundation of comprehensive analysis chart information data extract principle and combined an Image Processing technique and carried out curve chart of follow identify and the accurate extraction of chart data.

**Keywords:** Traditional Chart, Digitize, Curve Identification, Image Processing.

## 1 Introduction

Most of the traditional storage chart diagram of direct store documents, a small amount of direct storage of the chart document scans. Utilization of the former is not high, look for time-consuming, difficult to horizontal comparison; which takes a lot of storage space. Both are relying on access to chart data by hand, time-consuming, error is too large. Using computer technology, only the feature points stored in the chart. When users need to view the chart, the chart features a computer fitted according to a specific chart. This would not only save storage space and significantly improve the utilization of the chart. Traditional paper charts of this situation, through the transformation curve diagram recognition, combined with the specific characteristics of the image, to accurately read the information on the purpose of the chart [1].

## 2 The Overall System Analysis

System processing line: Image pre-processing and storage of image recognition → → → Coordinate Calculation of extraction results. From this we can see that the image pre-processing and curve identification are the two most critical parts, as can coordinate calculation formula by selecting the appropriate received.

Image pre-processing, including image gray, thinning, contour extraction, etc. [2]. Mainly for color image grayscale, the grayscale intensity of the key formulas (system optional grayscale formula: Gray = 0.299R + 0.587G + 0.114B) choice. Reasonable formula would make the image grayscale grayscale processing more natural, but also for subsequent processing provides a good material. Thinning, the thinning algorithm design is more important, the low efficiency of the refinement algorithm will not only

G. Lee (Ed.): Advances in Intelligent Systems, AISC 138, pp. 169–174.

slow down the processing speed and image distortion will make it but will increase the burden on the system, reducing the recognition rate. Image contour extraction is mainly effective starting point for regional and image scanning to determine accurately the curve of the contour extraction is the prerequisite and basis for identification. That is the image transformation curve identification is based on contour extraction based on the data obtained.

Chart image recognition refers to recognition of hand-drawn curve. Irregular curve in this section recognition, the full rate and recognition algorithms time complexity of the algorithm is to choose the algorithm standard. Identification of the curve, the algorithm must ensure the full recognition will be a curve, a curve when there is fault, not the full recognition, the scalability curve can be used, the use of technology will be broken curve Crochet completion. Meanwhile, the curve identification is based on the pixel, for a bitmap it is often a lot of pixels, so that required running time not too long, must be completed in a shorter time curve recognition.

Curve identification results in the curve of the distribution of markers in the pointer variable, specifically, is to curve the coordinates of pixels on the record in preparation for the data extraction. When the user complete chart information, based on the actual relationship between the coordinates and device coordinates to calculate the coordinates of user needs.

## 3    The Key Module Handles

Grayscale image processing and image contour extraction, in accordance with the literature [1] [2] introduced the method can be. Image refinement introduced here, to identify and coordinate the curve read three modules.

**Image refinement.** Image processing, image thinning is a very important preprocessing step, the refinement efficiency of a direct impact on recognition. Although detailed and there are feature extraction, training and recognition.

The core module of the system curve for the identification chart, and if there are more images in a redundant pixels, but rather falling tone recognition rate, interfere with recognition of the results. Refinement of the image curve removed redundant pixels, retaining only the backbone curve of pixels, such treatment does not affect the curve of the recognition, but more prominent contours of curve, reduce the computational complexity [3]. After the refinement is the so-called layers of peeling, removing the figure from the original number of points, but still maintain the original shape, until you get the image of the skeleton. Skeleton, can be understood as the central axis of the image, for example, the skeleton of a rectangle is its length direction of the axis; the square of the skeleton is its center point; round the skeleton is its center, the skeleton is its own line, isolated skeleton point is itself.

Thinning algorithm must be met: convergence; guarantee after thinning thread of connectivity, to maintain the basic shape of the original, reducing the distortion of the intersection of strokes, refine the results of the center line of the original image, the rapid and detailed fewer iterations.

So how to determine whether a point to remove it? Find the skeleton point? According to its eight neighboring points to determine the situation, as shown in Figure 1 (a) are a few examples.

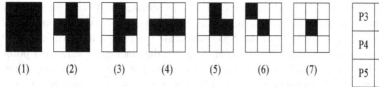

**Fig. 1.** Regional distribution and image thinning

In Figure 1 (a), according to a certain point the case of the eight adjacent points to determine whether to delete the point. ① can not be deleted because it is an internal point, we ask that framework, if even the internal points are deleted, the framework will also be hollowed out; ② can not be deleted, and ① is the same reason; ③ can be deleted, so that the point is not a skeleton; ④ can not be deleted, because the delete, the original part of the disconnected connected; ⑤ can be deleted, so the point is not a skeleton; ⑥ can not be deleted because it is the endpoint of a straight line, if such a point deleted, then the final the entire line was deleted; ⑦ can not be deleted, because the skeleton is an isolated point of its own.

In conclusion: (1) Internal points can not be deleted; (2) isolated point can not be deleted; (3) Linear end point can not be deleted; (4) If P is a boundary point, remove the P, if the connected component does not increase, then P can be removed .

Thinning Algorithm: p1 neighborhood eight points in Figure 1 (B) below: (where p1 is the black point, if the following four conditions are satisfied, then delete p1, main role is p1 = 0) refine the deleted condition:

A, $2 \leq NZ (p1) \leq 6$ / / $NZ (p1) = p2 + p3 + p4 + p5 + p6 + p7 + p8 + p9$, used to exclude isolated points and internal points

B, $Z0 (p1) = 1$ / / $Z0 (p1)$ p1 point for the branch number, branch number is 1 illustrates a border point, connected component does not increase after deletion

C, $p2 * p4 * p8 = 0$ or $Zo (p2)! = 1$ / / down to remove, and avoid being interrupted black line

D, $p2 * p4 * p6 = 0$ or $Zo (p4)! = 1$ / / to the right to delete, and to avoid the black line is interrupted

**Contour extraction and scanning to determine the initial point.** When the image pre-processing after the end of the chart has become a piece of material to be identified. In the curve of the chart to identify and track, we also need to determine the effective area of the chart, the chart means that the so-called effective area of the curve where the function curve, simply, that is, the coordinates of the chart range to be set up not out. Because a chart can be divided into two regions, effective regional and invalid region, and if we can identify the function of the coordinates of the chart before the regional division, then, when the system is doing the chart identify the function curve, it can not Detection of invalid region, this program will reduce the computation, but also improve the recognition rate function curve [3].

The chart coordinate system with the basic composition is the function curve, as long as the program can identify the coordinates of curve, then the basic framework to coordinate curve draw a square map, then the square figure is within the effective area of the chart. System as a whole chart N * M (assuming a width of the chart height N M) of the coordinate system to function curve to the coordinates of Y-axis for example (Note: The chart has two coordinates, one is the function chart curve XY coordinate system, the other one is the icon of the NM coordinate system), is a coordinate chart coordinates N = N1 on a vertical line, after binarization, the black pixel value of 1 point, whereas the white point pixel is 255, then a vertical line means that the coordinate system in the chart, XY coordinate system is the chart Y-axis coordinate system axis perpendicular to the N line up of black pixels, the same way we can get XY X-axis coordinate system is the chart coordinate system axis perpendicular to the M line up of black pixels. By this theory, the system can easily find the function curve of the XY coordinate system. That is, the coordinate system of charts and all the straight lines perpendicular to the horizontal axis the number of pixels in the black, then sort, then the pixel is the line most function curve of the XY coordinates of the Y axis, the same way the system can coordinate statistical charts lines perpendicular to the longitudinal coordinates of all the black lines in the number of pixels, and then sort, then the pixel is the line most function curve of the XY coordinates of the X-axis, have the function of the XY axis coordinate system after the inferior race as Square of the framework of painting to get the chart is the effective area.

Be effective in the region on the basis of the chart and began scanning the starting point to determine the so-called scanning function curve that is the starting point of beginning or end, when the system determines the starting point after scanning, then the function along the curve to determine the trace point. System through scanning starting point to determine in this way: the effective area in the chart, from left to right scanning, until it encounters a black pixel to stop, then the point on the curve as a function of the scanning starting point. Treatment effect of the module shown in Figure 2:

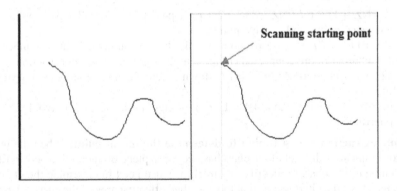

**Fig. 2.** Determine scanning starting point

**Image scanning tracking and identification.** Digital processing of the chart was intended to identify the chart's horizontal and vertical coordinates of a point, if you want to be able to identify the coordinates, the system must do is to identify the function curves in the chart, that is from the scanning starting point tracking function curve and record the track its trajectory, the characteristics of the function curve (ie, pixel X coordinates and Y coordinates) recorded in the array. When users enter a coordinate value, based on previous user input coordinate system is perfect information, find the function curve on the chart coordinate values and pixel coordinates corresponding to the formula, a conversion chart based on this formula the corresponding pixel value, the pixel value to record the function curves of the array to search, when the search point to the corresponding pixel to be marked on the map. Pixel location according to user needs to calculate and display the coordinates [4].

Chart tracking function curve is the core of the system identification module, the starting point in the chart have been identified in scanning, tracking, identification module first thing to do is to start the scan basis to identify the entire function curve. Before the implementation of the system, there are two bills for the system selection.

The first option is to use the traditional way, the effective area of the chart in the beginning to the end of the scanning pixel, as the chart of the binary processing, so it is easy to search for a black pixel to record all non-coordinate Department of black pixels. Finally, according to record data to calculate the function curve of the track. This method has the advantage of searching out all the black pixels, that pixel will not scan any less the case, but this method also has great defects exist, first of all, one by one the effective area of the chart scanning, processing speed will become very slow. Second, the scan is complete, how to calculate the trajectory of the function curve is the most complex, especially in how to identify effective points and interference points become very complicated.

The second is the system used in tracking algorithm. Starting point in determining the scan, based on the continuity of the function curve, it can be anchored to scan a starting point to search for the direction of the surrounding eight pixels. Figure 3 (1) below:

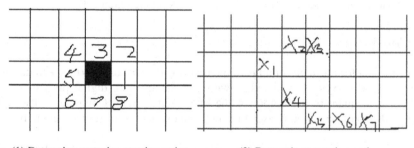

(1) Determine scanning starting point          (2) Determine scanning path

**Fig. 3.** Determine scanning starting point

In order to avoid duplication of work, the point must be searched for and marked, the system using a dynamic two-dimensional array to complete the search before all the array values set to 0, when the search for the black pixel is the value of the array set 2, if the white is set to 1. That is a pixel in the search process before the array will check whether the value corresponds to 0, that is the point for the search had no access to the search.

Search end condition is the starting point of all the eight directions after the search, it will end the search. Before the search process is only refundable, meaning that when the search process from the black pixels into 1 pixel in the direction of point 2 and the black pixels as a search basis point 2, if pixel 2 are not met, programs will not fall back to a test pixel in the direction of other points, and enter the search, but the end of this search. If this situation occurs in Figure 3 (2) below, the error will appear: the current search point is X1, X1, when in the search for the black pixels found in X2, so the X2 as a search point, into the X2, and then follow this line of thought into the X3, then X3 is not a suitable surrounding pixels, the program will end the search. However, we can see that in fact the main contours of the function curve is X4, X5, X6, X7 such extension direction. This error condition in the early more common when the algorithm. To solve this search errors, the system must increase the fallback mechanism, that is about the search is completed when the discovery X2, not the end of X2 to continue the search but fall back to the search, the search is complete when the discovery X2 X1 to search then fall back Search this idea into the X4 which can also function curve can be done a complete identification.

The image quality is not very clear is that you can modify the step step, while in each condition is satisfied, to the corresponding point in the corresponding array value is set 2, that is, the curve in front of Crochet way to connect.

# 4    Chart Data and Analysis of Test Results

Data extraction for the shortcomings of the traditional chart, using VC + +, design a hand-drawn to the irregular curve of the chart for the collection, digital processing and digital data to quantify the output of the integrated system [5]. When testing different shapes selected pieces of the chart for testing. Vice different by more than chart processing, can be seen from the results, the system identification of the function curve has a certain accuracy. This in turn proved the availability of the divergent method. Recognition in terms of coordinates, but also very precise, to the previous design requirements.

# 5    Conclusions

Data extraction for the shortcomings of the traditional chart, using VC + + to design a set of irregular curves of the hand-drawn charts acquisition, digital processing and digital data to quantify the output of the integrated system. Computer software to complete the chart scanning, recognition and data extraction functions accurately.

**Acknowledgment.** The work is supported by Fund: 863 National Key Foundation (2008AA11A134), Sichuan Province, scientific and technological projects (07GG004-011 re-Points), Sichuan Key Laboratory of Artificial Intelligence (2010RY002) funding.

# References

1. Guo, L.: Engineering Drawings Recognition and vector. Xinjiang University (2003)
2. Long, Z., Liu, M.: Binarization threshold value when the adaptive selection method and Visual C + + implementation. Harbin Railway Science and Technology 1, 8–10 (2006)
3. Zhang, H., Cai, R.: Visual C + + Digital Image Pattern Recognition Technology and Engineering Practice, vol. 2, pp. 23–57. People's Posts and Telecommunications Press, Beijing (2003)
4. Dai, F., Zhang, X., Xu, D.: Curve matching method identified the realization of the deformation. Science Technology and Engineering 7(3), 406–408 (2007)
5. Yang, S.: Image pattern recognition technology-VC + +, pp. 135–158. Tsinghua University Press, Beijing (2005)

# The Application of Network Simulation Software NS-2 Based on SVM

Wu Renjie

Hebei North University, 075000, Zhangjiakou, China

**Abstract.** Network simulation is an effective means of studying network technology. Simulation methods include the use of special software or preparation of the corresponding simulation procedures. As a powerful network simulation software, NS-2 is favored by researchers, more and more people begin to study and use NS-2. This article mainly introduces the ways of studying and using simulation software NS-2 and illustrates the basic principles of it .Moreover, the article also achieves a simulation example of TCP and UDP protocols, offers the simulation program, explains how to put the simulation procedure into the file and displays the analytic results by tools.

**Keywords:** NS-2, network simulation, simulation program, TCP, UDP.

## 1 Introduction

NS (Network Simulator) is the literal translation of the network simulator, also known as the network simulator is a network technology for open source, free software simulation platform. Is the NS-2 Network Simulator version 2, or Network Simulator version 2, developed with the UC Berkeley researchers can easily use the network technologies [1].

NS is generally believed to have originated in 1989 by the UC Berkeley network simulator developed by REAL (REAL network simulator). In fact, REAL network simulator developed at Columbia University Network Testbed NEST (Network Simulation Testbed) improved on the basis of. REAL network simulator is mainly used for simulation of a variety of IP networks. The software was originally developed for UNIX systems based on network design and simulation carried out. 1995, NS Development of the U.S. military DARPA VINT (Virtual InterNetwork Testbed) project funded by the USC/ISI, Xerox PARC, LBNL, and UC Berkeley to develop [2]. Currently, NS-2's development has been DARPA SAMAN (Simulation Augmented by Measurement and Analysis for Network) project and the U.S. National Science Foundation-funded project [3].

## 2 Principle of NS-2 Introduction

**NS-2's general architecture.** NS-2 contains the Tcl/Tk, Otcl, NS, Tclcl. Which is an open Tcl scripting language used to program the NS-2; Tk is the Tcl graphical interface development tool that helps the user in a graphical environment to develop graphical

G. Lee (Ed.): Advances in Intelligent Systems, AISC 138, pp. 175–181.
springerlink.com     © Springer-Verlag Berlin Heidelberg 2012

interfaces; Otcl is based on Tcl/Tk for object-oriented extensions, there are Its class hierarchy; NS-based core package is an object-oriented simulator, written with C++ to Otcl interpreter as a front end; Tclcl NS and Otcl provides the interface to make objects and variables appear in both Language [4]. For visual observation and analysis of simulation results, NS-2 also offers an optional Xgraphy and Nam. NS-2 structure shown in Figure 1.

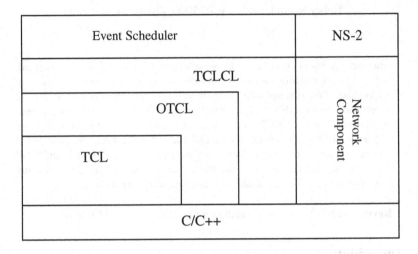

**Fig. 1.** NS-2 structure

**NS-2 function modules.** NS-2 simulator package a number of functional modules, the most basic is the node, link, agent, packet formats, the following were to introduce each module.

Event Scheduler: NS is a discrete event driven network simulator. It uses the Event Scheduler hope to complete all components of the work and plans of the working time occurred in the list and maintenance. NS-2 currently offers four different scheduler data structures, namely the linked list, heap, calendar and real-time scheduler.

Node (node): Objects by TclObject composite components, the NS-2 can be expressed in the end nodes and routers.

Link (link): Composite made by a number of components used to connect to the network nodes. All the links are in the queue to manage the group in the form of the arrival, departure and discarded.

Agent (agent): Responsible for the network layer packet generation and reception can also be used in all levels of protocol implementations. Each agent is connected to a network node from the node to assign it a port number.

Packet (packet): The NS-2, a "package" consists of a header (Header), Stack (Stack) and an optional data spaces. In general, packet header only, no data part. This is because, in a simulated environment, the actual data transfer is meaningless.

# 3    Development of Language and Code of NS-2

NS-2 using the C++ and Otcl as a development language. NS can be said Otcl script interpreter, which contains the simulation event scheduler and network component object libraries, and network component model libraries.

On the one hand, the specific agreement of the detailed simulation and implementation requires a systems programming language, it can efficiently operate byte header and other data structures, to achieve an appropriate algorithm in large-scale operation on the data collection, and such algorithms often is the need to constantly repeat. For such tasks, the running speed is very important, and turnaround time (run simulation, find bug, fix bug, recompile, re-run) becomes less important [5].

On the other hand, many research networks around the network components are the specific parameters and the environment, the need for frequent re-set and modify the simulation scenario, the need for a shorter time to develop a large number of scenarios. In these cases, the turnaround time (change the model is run again) is even more important [5]. As simulation scenario configuration need only be executed once at the beginning of the simulation, so this part of the task running time is not important.

In order to meet the needs of these two different tasks, NS use two languages, C++ and Otcl. C++ program runs faster, is to compel the type of language (for strict data type checking), easy to implement complex data types, easy to achieve precise, sophisticated algorithms, but to modify, debug and recompile the time it takes to be longer Therefore, C++ is suitable for the specific protocol implementation [6]. Otcl run slower, but can be easily (and interactive) to amend, not compiled, but not mandatory Otcl type, not prone to error, so it is suitable configuration used for simulation [6].

# 4    Tcl and Otcl

**Tcl command language extended.** Tcl is the acronym for Tool Command Language, Tcl consists of two parts: a scripting language and the corresponding interpreter [6].

Tcl only supports one data structure: string (String). All commands and parameters, the results of the command, all the variables are strings. Tcl command of the basic syntax is:

Command arg1 arg2 arg3 ...

One built-in command Command name or on behalf of Tcl procedure, arg1, arg2, etc. is the order parameter, with the line or a semicolon to the end of a command [7].

The following example uses a simple command to achieve the classic "Hello World!" Routine.

puts stdout "Hello World!"
Hello World!
The second line "=>" indicates the output.

# And use Tcl; # to comment, comment a whole line of them can use the symbolic beginning of a line # (before the # sign can also have blank spaces or tabs and other characters), and; # can be used in command behind, making later in a comment. For example:

# This is a remark.
set a -3; # a =- 3

**Otcl ---- object-oriented Tcl.** Otcl is a Tcl object-oriented language extension introduced in Otcl the concept of classes and objects, object-oriented mechanisms for the accession of the original Tcl become more powerful and more convenient to use. Otcl the object and class concepts and C++ and other object-oriented programming language, the concept of the same. Although Otcl and C++, object-oriented languages the same concept, but the concrete implementation and syntax, or there is a big difference [7].

**NS main code base.** There are 6 NS base class:

(1) Tcl categories: C++ code and Tcl code to bridge between;

(2) TclObject categories: base class for all simulation objects;

(3) TclClass class: defines the class to explain the class hierarchy, and allows the user to instantiate TclObject, and TclObject one correspondence;

(4) TclCommand class: encapsulates the C++ code and Tcl code to invoke the command each method;

(5) EmbeddedTcl class: encapsulates the higher level of built-in command loading method;

(6) InstVar categories: access to C++ member variables, such as Otcl variable methods.

**Tcl class.** Tcl class six the most important functions are: to gain access to the entrance of Tcl instance; through the interpreter calls Otcl process; outgoing or incoming from the results of the interpreter; report an error condition and exit in a uniform manner; store, find TclObject class object ; taking the handle of the interpreter (if the above method is still not enough, then we must get a handle to the interpreter, to write our own function).

**TclObject class.** Class TclObject is in the interpretation and compilation of all levels of the base class. Were the subject of each class TclObject user created in the interpreter, an equivalent level of the shadow of the object while also being compiled to create these two objects close together. Class contains the implementation of this mapping TclClass mechanism.

The following example is the configuration of a SRM agent (class Agent/SRM/Adaptive):

```
set srm [new Agent/SRM/Adaptive]
$ Srm set packetSize_1024
$ Srm traffic-source $ s0
```

Ns in accordance with the conversion, the class Agent/SRM/Adaptive is the class Agent/SRM sub-class, and Agent/SRM is a subclass of Agent, and Agent is TclObject subclass. Compile the corresponding relationship between the classes derived: ASRMAgent derived from the SRMAgent, SRMAgent derived by the Agent, and ultimately by TclObject derived [7]. The first line in the example above is a subclass of class TclObject the creation (or revocation); the next line configures a bound variable; the last line of interpretation of the object to activate a C++ method as an instance of the process.

**TclClass class.** Compile classes TclClass is a pure virtual class. Derived from this base class to class provides two functions: to establish and build class image interpretation class and provide examples of the TclObject method. Each derived class and a class hierarchy specific compiler compile class association, and can instantiate new objects associated class. For example, the class RenoTcpClass. It is derived from the class TclClass, and class RenoTcpAgent association. It will instantiate a new object class RenoTcpAgent. RenoTcpAgent compiled class hierarchy is: RenoTcpAgent derived from the TcpAgent, TcpAgent derived by the Agent, the Agent is derived from the TclObject [7]. RenoTcpAgent defined as follows:

```
static class RenoTcpClass: public TclClass
static class RenoTcpClass:public TclClass
{
public:
RenoTcpClass():TclClass("Agent/TCP/Reno"){ }
TclObject*create(int argc,const char*const*argv)
{
return(new RenoTcpAgent());
}
}class_reno;
```

**TclCommand class.** Ns class TclCommand is to provide a globally executed by the interpreter within a simple command mechanism. There are two functions, they are defined in ~ ns/misc.cc in.

Defines a command class   ns-version."VersionCommand   It does not require parameters and returns a string that the current version of ns. Such as:

% Ns-version; # get the current version number

Defines a command class ns-random."RandomCommand   ns-random does not need argument and returns an integer. If a parameter is given, it will put this argument as a seed. If the seed is 0, the command will use a custom seed value; Otherwise, it will set a specific random number generator seed to obtain a specific value [7]. For example:

% Ns-random; # returns a random number

3174857054

% Ns-random 0; # custom set seed

858,190,129

% Ns-random 23786; # set a specific value of the seeds are given

**EmbeddedTcl class.** ns allows compiled code or interpreted code, extensions, extended code will be executed in the initialization. For example, in the script ~ tclcl/tcl-object.tcl or ~ ns/tcl/lib in the script, the extensions to load and assignment is done through the class EmbeddedTcl .

EmbeddedTcl code where there are three things to note: First, if an error is encountered during execution, ns will stop running. Second, the user can display any code in the overloaded script. In particular, they can make their own amendments redefine the entire source file library script. Third, if you add a new script or change the original script, the user must recompile in order to be effective [7].

**InstVar class.** InstVar class defines the shadow object will compile a C++ member variables to explain to bind to the corresponding object instance variables in Otcl the methods and mechanisms. This binding enables either variable at any time from the interpreter and can compile code from within the set and get.

Class has a member variable tracedvar_ InstVar point class TracedVar objects, classes TracedVar object encapsulates information about the compiled object and method, defined as (~/ns/tclcl/tracedvar.h):

```
class TracedVar{
public:
TracedVar();
virtual char*value(char*buf,int buflen)=0;
inline const char*name(){return(name_);}
inline void name(const char*name){name_=name;}
inline TclObject*owner(){return owner_;}
inline void owner(TclObject*o){owner_=o;}
inline TclObject*tracer(){return tracer_;}
inline void tracer(TclObject*o){tracer_=o;}
protected:
TracedVar(const char*name);
const char*name_;
TclObject*owner_;
TclObject*tracer_;
public:
TracedVar*next_;
};
```

InstVar base were derived from five sub-categories: class InstVarReal, class InstVarTime, class InstVarBandwidth, class InstVarInt, and class InstVarBool. Corresponding binding real, time, bandwidth, integer and boolean types of variables [7].

# 5     Simulation of NS-2 to TCP and UDP protocols

**NS-2 script Introduction.** NS-2 network simulation script generally follows the structure:

```
# Generate a simulated object
set ns [new Simulator]
# Define an end of the program
Proc finish {} {
    Exit 0
}
# Below you can add some, such as network infrastructure and application code
set
# Begin simulation
$ Ns run
# Activate the finish at the right time to end the simulation program, following the
first 5 seconds that the end of the simulation
# Ns at 5 .0 "finish"
```

**TCP and UDP analog implementation steps**

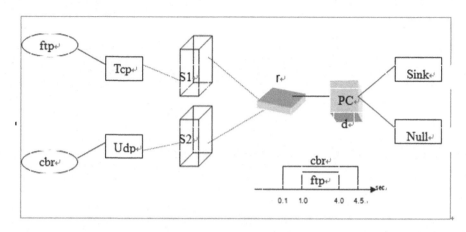

**Fig. 2.** Network topology

# 6    Conclusions

NS-2 is a powerful network simulation tools, the paper by NS-2 TCP and UDP protocol analog and awk programming language used on the network packet loss rate is analyzed. Packet loss rate is only one parameter of network performance, as well as end to end delay, jitter and throughput and other performance parameters of their analysis needs further research.

# References

1. China IT laboratories. Network Simulator NS-2 simulation software environment Introduction. China Machine Press, Beijing (2007)
2. Bin, Y., et al.: NS-2 with network simulation. People's Posts and Telecommunications Press, Beijing (2007)
3. Hui, W.: NS-2 network simulator and application of the principle. Northwest Industry University Press, Xi'an (2008)
4. Fang, L., et al.: NS-2 network simulation fundamentals and applications. National Defence Industry Press, Beijing (2008)
5. Liu, B.: NS-2 Simulator and educational applications. Computer Knowledge and Technology (Natural Science) (10) (2007)
6. Zhu, X., et al.: NS Teaching in the network. Shandong TV University (1) (2007)
7. Wang, H.: NS-2's MFlood protocol extension. University of Texas (2) (2007)

# Discussion from the Contest to the Innovation Education

Xionghua Guo, Huixian Han, and Bosen Zhang

Hunan Mechanical & Electrical Polytechnic, Changsha, China

**Abstract.** The paper analyses the proposition and contest results of the spring CNC Vocational Skills Competition Hunan Province in 2010, points out the problems of the Hunan Vocational College CNC teaching and contest, and proposes a few suggestions on innovation in the teaching of CNC occupational skills.

**Keywords:** CNC skill, contest, personnel training, innovative design.

## 1   And Ideas Based on Proposition

According to the National Vocational Skills Competition of the technical documentation requirements, NC Vocational Skills Competition 2010 theme is: complex part modeling, multi-axis programming and machining. Which contains seven assessment points, ① 3D modeling; ② part programming and machining; ③ preparation processing machine parts; ④ innovative design; ⑤ machine breakdown maintenance; ⑥ civilization safety; ⑦ Plans for teachers. According to the requirements of the NC Skills Competition in 2010 than in previous years the race had a larger increase. First, in the programming process on the axis from the axis into the second, the way competition from students to teachers and students also participate in individual competition contest; the third, the first time the innovative design competition requirements; Fourth, in the contest content with the machine maintenance projects. These requirements also pointed out the urgent need to strengthen vocational education in the area. Not just for the competition to competition and race, but want to get into the contest to promote the reform of vocational education, leading the direction of education reform.

The requirements of the National Competition CNC control, feel a bit more difficult proposition. Difficult in the area? Difficult in the "innovation" word. People engaged in mechanical work, most of the eyes to see the parts square, hole, shaft with the way the structure, in the mechanical field, there are many smart ideas for innovation, but a handful of the part in order to achieve limited functionality, functional innovations have greater difficulty, creative time is limited, pursue innovation, and the direction of where? What is that? On reflection, the final design of the rotating vase, turn fountains, turning the torch, turn the four themes of the impeller. Figure 1-4. Some of these topics in the life of the items, some inspired by the Olympic torch, which seem to point machining stay further apart, but in the daily lives Bai students. Select the number of the competition theme, is trying to decorate these items as a subject, leaving his players to play, so not only to the players that the creative direction, also left a relatively large player creative. The gist of the main players in the assessment of basic quality and 3D modeling capabilities.

G. Lee (Ed.): Advances in Intelligent Systems, AISC 138, pp. 183–190.

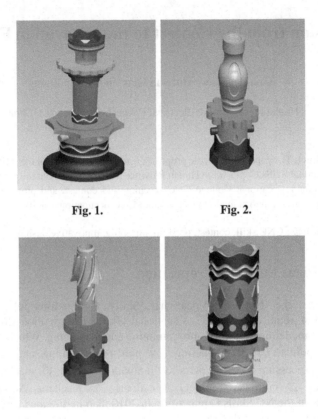

Fig. 1.                    Fig. 2.

Fig. 3.                    Fig. 4.

## 2     Competition Analysis

**Test Analysis.** Spring 2010 CNC Hunan Vocational Skills Competition in April 27 in Zhuzhou Hunan Railway Professional Technology College kicked off, after 3 days 3 nights of competition, to the successful conclusion of April 30. Throughout the race tight, orderly, and no error occurred and major accidents. To achieve the desired objectives

The competed in a total of 32 race teams, race team, the actual entry of 28 students and 84 teachers of 27 players plus players, the actual total number of 111 entries. Proposition for this competition and game content, in the contest, teachers and students with some players were a sub-item questionnaire. Poor adjustment included machine breakdown maintenance, innovative design, process documentation, modeling, and other complex parts put on. Statistical analysis shown in Table 1.

From the survey results, the difficulty of this race more appropriate questions, contests the content more comprehensive, and high recognition of the participants questions. Zhuzhou testing unit of this competition a few old technician for the Association tournament title this evaluation is: There is a high technology capability and machining difficult, to measure a player's skill level.

**Competition Analysis.** Looking at the results from the competition, the team has made some very good results, showing excellent processing skills and the overall technical level, some teams although the results were not satisfactory, an increase of knowledge through competitions, see the gap between the efforts to find direction, is also a harvest.

**Achievements.** Looking at the results from the competition, with the results of the competition in previous years, compared with 3D modeling in the preparation process, the players have made great progress, all players in the game also reflects the good professional quality.

The first is the team spirit is good, the race question is to ask the three players together to complete the processing of a set of components, including turning, milling complex process arrangements, innovative design and 3D modeling of the match, with the preparation of technical documents here must be arranged, division of labor, the process of tacit understanding, to get the fastest speed and good results.

The second arrangement is subject to the good quality not afraid of hardship. This time the race conditions, some teams have to arrange for the evening, even in the morning to play. Should be said that competition in this period, in spirit, energy should be affected on, but the players were willing to accept the arrangement without any complaints.

**Table 1.** Summary of questionnaire survey

| Survey Project | Content | | Recognition ratio (%) | Survey Project | Content | | Recognition ratio (%) | Survey Project | Content | | Recognition ratio (%) |
|---|---|---|---|---|---|---|---|---|---|---|---|
| Troubles hooting tools | Troubleshooting time | Too Tight | 67 | Innovative Design | Difficulty | Large | 17 | Process documentation | If reflect the level | Can | 89 |
| | | oderate | 33 | | | Appropriate | 83 | | | Cann't | 11 |
| | | long | 0 | | | Easier | 0 | | | | |
| | Troubleshooting more difficult | Large | 23 | | content | More | 46 | | Form Design | Reasonable | 89 |
| | | Difficult | 77 | | | Appropriate | 54 | | | Unreasonable | 11 |
| | | Easy | 0 | | | Too little | 0 | | | | |
| Modeling and assembly of complex parts | Quantity | Appropriate | 70 | Parts processing and assembly | The difficulty of processing | Large | 55 | plans designed | need to do | Necessary | 78 |
| | | Too much | 30 | | | Appropriate | 45 | | | Not necessary | 22 |
| | | Too little | 0 | | | Easier | 0 | | | | |
| | Difficulty | Large | 33 | | The difficulty of 3D modeling | Large | 22 | | Preparation time | More | 11 |
| | | Appropriate | 63 | | | Appropriate | 67 | | | Appropriate | 67 |
| | | Easy | 6 | | | Easier | 11 | | | Too little | 22 |

The third is a brave Pinbo not give up the tough fight. Regardless of race or difficulties in the smooth, the players were able to adhere to the last minute, not players drop out, this spirit is commendable.

Scoring rate from the sub-project point of view, in the process planning and 3D modeling, the players made more than 60% of the scoring rate, reflecting the process of teaching in vocational schools in our province has made great progress. Remember the higher the spring of 2008 NC game, but also the contents of a process planning, preparation of technical documents was not by conventional standards to sub-points. And this can achieve 63% of the scoring rate is indeed progress is not small. This just shows the competition of the teaching guide.

**Problems.** Some of this competition the team to achieve good results, but there are some shortcomings. First results from the analysis of the distribution of games start to explore the question of which reflect. See chart below:

From the chart reflects the following questions:

1. The polarization results, the highest score 79.025, the lowest point is only 8.7, a difference of 9 times.

2. Non-normal distribution curve shape, more low-scoring team in the region, in much the team score in the region.

The case, highlighting the province integrated unbalanced development of knowledge and skills training. Also reflect the higher skill level of the province to be further strengthened.

# 3    Analysis

1. Comprehensive basic knowledge of uneven development. The skills competition is not just a simple game, the basic outlines of six assessment points in all aspects of CNC machining knowledge, to get high scores, only the knowledge have been balanced development. Scores from the sub-project situation, not good enough. Maintenance and innovative design tools have two "0" points were 15 teams with 14 teams participating teams accounted for 53.6% and 50%. So many teams have "0" seriously affect the overall results. And the safety of civilization will be more of a loss of points, there are 15 teams in the drop points in this, and some teams lose points also are more up to -17 points, 10 points above there are several teams that ERA is a pity.

2. Parts are not familiar with the process. Structure more complex, longer part machining process must have a priority arrangement, a car, milling processes must first arrange car parts general process, roughing and finishing to separate arrangements. In this race, some runners use directly on the milling of rough, leading to subsequent processes can not, indicating that the player does not process the concept, lack of basic processing and common sense.

3. The series co-ordinating the production schedule is not strong. In actual production, a product must be made up of many component parts, these parts of the production sequence must have a co-ordination arrangements, the general process for production long before the parts, otherwise it will cause the product to delay progress, parts, accessories and semi-finished inventory increase increased production costs. Although the topic of this game is only 4-5 parts, but there are problems in this regard.

Some players choose the most simple pin before processing, the processing of complex parts to the delay. This reflects the thinking of players and the overall concept of the process is not strong.

4. 3D modeling capabilities to be further strengthened. 3D modeling is the core part of the "features" method of analysis and forming two. The main topics are not very complicated shape, in the shape of the decorative use only two methods: surface, "offset" and "variable geometry scan" in the shape of the sine function. Some players feel that this is not enough processing time, a considerable part of the reason is the shape of more time delay, resulting in no time after processing. A team processing center has been parked, only to come in the last 30 minutes when started, that is basically all the time is spent in style on the. This 3D modeling software that will further in-depth training.

5. Processing machine operator training to be strengthened. There are many schools that are not familiar with the machine tool system, which reflects several problems: ①, in explaining the numerical control programming, the principle may explain the relative neglect of NC, NC instructions may not have the internal links to make it clear that students in the practical application can not comprehend by analogy. ②, do not take full advantage of some commonly used numerical simulation software for numerical control system the necessary training. ③, four-axis machine tool less, there is no effective training. Some participating teams leading to the unsatisfactory results of processing.

6. The competition is not thoroughly understand the topic, the form of competition problems do not adapt. The topics to give players a creative theme, a bit out of the purely mechanical part of the frame, with a number of process modeling ideas. Players may be less than usual this talent like. In fact, the basic shape of these questions are relatively simple. For example, the "spendthrift" the basic shape of the subject, only two outside the park and an inner hole (see Figure 5), modeling as long as 5-10 minutes to complete, no NC machining can be completed very soon. Shape of the outer surface of the decorative use only two methods: surface, "offset" shape and the sine function of the "variable geometry scan" shape. Here is a little more difficult and the sine curve of convex shape, the so-called "difficult" is not technically anything special, but students may not have noticed this function normally, in fact, can be seen at any time in life, such as curves on the bottle of mineral water . In line with the general accuracy is only one match inside and outside the park. So this topic is not difficult. However, the average score of the subject is not high.

**Fig. 5.**

7. Strengthen the teaching innovation. "Innovation" is the first time the competition a new topic, undergraduates each year a national "innovative design" competition, mainly functional innovation. Teaching in vocational skills with innovative teaching should be a new topic. The game scoring rate of innovation is only 17.39%, expected. However, teaching the skills for the future put forward higher requirements.

# 4    Several Suggestions to NC Teaching

**Further improve the teaching and practice facilities and expand competition surface.** In terms of the number of entries from more than 100 people, but face competition in the schools, not very wide, some schools have set up CNC professional but did not have a race, the main reason may not be four-axis machining centers, usually not made This training shows that the school is not perfect on the device configuration.

**Expand the depth and breadth of numerical control technology teaching.** Emphasis on numerical principle of teaching, full use of simulation system to strengthen the operation of different systems training machine. NC principle is an important part of teaching, to make it clear in the CNC system can be completed in the production which features the same function in each system may have different orders of expression, the number of functions are also different. As long as the essence of programming to understand, to see each numerical control system is like the difference between understanding, coupled with the training simulation software, operating various CNC machine tools will be easier.

For multi-axis CNC programming, but also the production of an essential part of the operation with the popularity of three-axis machine, the multi-axis CNC machining operation must be the subject of teaching should be improved. In the multi-axis machining, five-axis is the foundation, use more in production, so the teaching process should focus on multi-axis into five-axis machine tool programming.

**Enhance innovative teaching.** Innovation is an eternal theme of teaching, specifically to improve the way through what the students ability to innovate'll be a topic. First of all, what is considered "innovative"? We provide more than the basic quality of students, there are many aspects of basic quality, ethical aspects of human qualities, there is the cultural quality of knowledge, skills qualities, there are physical qualities, musical quality, there are aesthetic qualities and so on. Schools offering a variety of courses, designed by various practical teaching and skills training, all students in order to develop or enhance these qualities. Then, "Creative Education" belongs to what category? Innovation is a higher level of quality is the result of a number of basic qualities reflected after shaping a manifestation of wisdom. Usually referred to a facile imagination, new ideas, daily in piecemeal reform, is a reflection of innovation. Functional innovations on innovation, but also the appearance of innovation. Expanded its use of a range of items, or to improve its use, but also an innovative; it looks more beautiful, more like people, this is an innovation. Here we must be careful not to "innovation" overstating the "invention" of the hierarchy, and own the box, set up obstacles.

How to carry out innovative teaching? In strengthening the basic quality of training should also increase the teaching or practice the following elements:

① the aesthetic ability to increase the industrial model of training. Each industry are the "United States" as a distillation of the industry standard up as a reflection of innovation. Take clothing, food, live it, which is not the "United States" as a goal the industry to pursue it. "Clothing" is the basic function of insulation to cover them, and developed a wide range, contests, popular clothes, do not for a "beauty" it? "Food" to feed their families the basic function is to supplement the calorie nutrition, developed to have the color, smell, taste, and even carved phoenix Diao Cooking; "housing", housing, basic function is to shelter, to develop the pursuit of a harmonious and beautiful the architecture. Take us in the machinery industry, from the dead machine, to the human form of industrial design major, is the "beauty" as the promotion criteria. Therefore, the aesthetic quality of education as an aspect of innovation is important. A strong aesthetic sense who, in pursuit of beauty, when he was doing everything not Mama Huhu; to the pursuit of beauty, he would go to great pains to discover the inner things perfect harmony and beauty and appearance ; the pursuit of beauty is a kind of process is the process of thinking and discussion, but also a creative thinking process.

② develop self-learning abilities and habits. People say "Math" is the thinking of gymnastics, is a good way to develop independent thinking. I think you can play the same self-efficacy, but also to get instant success. People who like to be like the self-brains, long, long time will be able to improve people's understanding, develop good habits of thinking ground. For example, many professional software to learn, is a good a good way to exercise self-learning ability.

③ broaden the areas of life, more hands-on, take part in a variety of practical activities, to society, to the plant to absorb nutrients, so people could live. Issued by the accumulation of ideas will naturally fight sparks innovation to bear fruit.

## 5    Strengthen the Teaching and Practice of Craft Ideas

Process is to have learned, and scattered in various professional courses in the embodiment of comprehensive application of knowledge. Parts manufacturing process focused on the whole process of consideration, which the subjects used a single study, a single deal with the problem of learning technology, is a turning point is an increase. Explain why teachers in the knowledge of all the subjects, should deliberately put this knowledge in part the role of the manufacturing process and the intrinsic link between them make it clear, and examples of actual production process planning to do more training, to gradually improve their ability to deal with process issues.

**Acknowledgment.** The work is supported by Educational Commission of Hunan Province of China (No.10CO156).

# References

1. Wang, X.-M., Wang, D., Liu, H.-F.: Higher Mathematical Modeling Study and Practice of the organization. Henan Mechanical College 09 (2007)
2. Sun, H.Q.: To contest as an opportunity to guide students to independent learning, training students the spirit of innovation. Entrepreneurial World 07 (2008)
3. Lun, Z.J.: NC Skills Competition Technology. Manufacturing Technology & Machine 01 (2005)
4. Dan, X.: To improve vocational skills, the effect of teaching numerical control. Nantong Shipping College 03 (2007)
5. Asia, X.Y.: To carry out innovative design competition for students to explore creative ability and practice. Equipment Maintenance Technology 02 (2004)

# Application of Interactive Interface Design as Teaching Materials in Chinese for Children

Lin Chien-Yu[1,2], Lin Chien-Chi[3], Lai Yichuan[2,4], and Du Jie-Ru[2,5]

[1] Department of Special Education,
National University of Tainan, 33, Sec. 2, Shu-Lin St. Tainan, Taiwan
linchienyu@mail.nutn.edu.tw
[2] Graduate Institute of Assistive Technology,
National University of Tainan, 33, Sec. 2, Shu-Lin St. Tainan, Taiwan
[3] Institute of Ocean Technology and Marine Affairs,
National Cheng Kung University, 1, University Rd, Tainan, Taiwan
[4] Kaohsiung Municipal Chiu-Cheng Primary School, 47,
Liantan Rd., Zuoying District, Kaohsiung, Taiwan
[5] Hop-Shine Elementary School, 91, Hexing Rd.,
Zhongpu Township, Chiayi County 606, Taiwan

**Abstract.** This study focuses on how to design teaching materials of words with Chinese for children. The participants are children from kindergarten and elementary school, parts of whom are plagued with learning disabilities. This study is divided into training and testing steps. The training step allows researchers to develop a Chinese unit course. In the testing step, children use the Chinese teaching materials as an assistive technology. Different needs from children with learning disabilities are important, so in this study it design customer-made teaching materials. The purpose of the study is to explicate why the design worked well and how it can be improved on different case study. This study applies interface design of Chinese as assistive technology for children, which enables to improve the interactivity of teaching aid for children. The design of Chinese teaching materials offers both teachers and students a novel learning method.

**Keywords:** Chinese, user-friendly, interface, assistive technology, teaching materials.

## 1 Introduction

Today Chinese is the first or second language of some 1.2 billion people from all regions of the Chinese-speaking world[1].Words are generally regarded as the basic meaningful unit of language[2].User-friendly is the first condition of interface design. The study utilizes equipments of assistive technology and applications of flash software so that teachers teaching for normal or resource classes in kindergarten and elementary school enable to improve students' learning interest , especially for students who have difficulty following the scheduled learning progress[3]. Computer-based instruction is widely used in special education. Special education teachers may sometimes have

G. Lee (Ed.): Advances in Intelligent Systems, AISC 138, pp. 191–198.
springerlink.com     © Springer-Verlag Berlin Heidelberg 2012

trouble teaching their students to use different computer tools[4],and advances in computer technology allow translation of traditional paper questionnaires into novel display versions[5].

Through the assistance of assistive technology and the demonstration of multimedia design teachers have enough ability to produce the learning materials of custom-made design in order to support learning disabled students to absorb knowledge[6][7], user-friendly design is defined as the structural design of an interface that presents the features [8] [9],and teaching interaction procedure is a systematic form of teaching where the teacher describes the behavior[10].

This study is suitable for students with learning disabilities in resource class and children in kindergarten. The research  takes pupils as the main body and divides into two part—the design of teaching materials and the demonstration of students' works. Teachers teaching for resource classes in elementary schools can not only participate in this study directly but also perform the outcome of the study to students right away[11].The study proceeds from case study and develops into operating interface and assistive equipments. Under the practical exercises and operating processes of teachers and students from kindergarten and elementary school, the research can observe the obvious differences between the traditional and the custom-made design of spoken Chinese teaching material when displaying students' works. Students also very enjoy custom-made spoken Chinese teaching materials. Therefore, according to the outcome of the study, the study is successful. Information and communication technology is a powerful tool for learning, which helping teachers explain difficult concepts, giving access to a huge range of examples and resources, and engaging pupils easily[12].

HCI exists at the junction of  computing sciences,   design arts, and social sciences. Human-computer interaction is a discipline concerned with the design, evaluation and implementation of interactive computing systems for human use[13]. The goal of interaction design subfield within HCI is to improve the experience for learning disabled students of direct interaction with the computer.

Computer-mediated communication facilitates understanding of communication patterns, forms, functions and subtexts, which can in turn engender an understanding of how to derive meanings within such contexts[14], patterns or figures are increasingly being used not just in education but also in many other areas such as software engineering, engineering and business management, and are also frequently being advocated for teaching human–computer interaction   principles[15].

The interfacial design of assistive teaching materials makes use of a monitor to demonstrate Chinese characters. With one mouse click corresponding sound appears, which enables to help students learn according to their own preference.

This assistive technology of interactive design for spoken Chinese describes a new design for teaching materials developed in the frame of a research project supported by information interface tools[16]. It is important that Introduction of basic education or special education-based computer aided tools in the routine development process of education.

## 2   Case Study

Because children in kindergarten always need repeat training, students with learning disability have difficulty concentrating on teaching materials; this study therefore attempts to introduce the concept of custom-made learning for spoken Chinese teaching materials. The learning materials are customized for students themselves so as to improve their learning interest. The range of custom-made application is considerably extensive, including attire, architecture, medical care, rehabilitative instrument, etc. The research reports on a case series of 4 teachers who work for resource class at elementary school or kindergarten.This study begins from computer assistive teaching, which provides diverse learning methods in the design of teaching materials in spoken Chinese. Owing to the introduction of custom-made concept, children can see the Chinese leading words show on the screen. Using touch screen or mouse can link to corresponding information, which enables to increase the attraction and intimacy of teaching materials. Here are three cases and explanations of this study.

**Case 1.** Case 1 is a collaborative teaching plan to develop the exercise of Chinese characteristics and pronunciations from four elementary school teachers. Members in the research are all teachers teaching in various elementary schools. There are teachers teaching for resource classes. The research discusses a serial class for spoken Chinese class for children with learning disabilities.

Firstly, in system design the research classifies Chinese words into 8 categories: fruit, animal, furniture, kitchenware, stationery, transportation, location and recreational sport. There are 9 Chinese common words in each category to perform Chinese voice recordings and graphic designs. In animal category, for instance, there are 9 common Chinese words were designed with concerning animals. With the assistance of Flash software design, users enable to hear corresponding Chinese pronunciations and images when learning Chinese characters.

Participants in the study amount to twenty students who belong to various disability categories respectively. Students felt interested when operating Chinese teaching materials which are illustrated by the design of informational interface. The subject in Figure 1 is a first grade student in elementary school named Ke who is plagued with speech and language disorders. He has difficulty connecting Chinese characters and pronunciations during the course of Chinese acquisition.

Developing applications on interactive teaching materials for students in special education programs is of priority concern.

**Fig. 1.** Invisible button design (To protect the child, the figure used blurred effect)

Because children with learning disabilities are worse than normal students, it is difficult for them to learn Chinese pronunciations with Chinese characters as soon as normal children, which is also the central problem that teachers attempt to solve.A teacher firstly demonstrated the operative method of mouse to student Ke. Subsequently, the student had an exercise to utilize the mouse in person so as to promote his familiarity with the relationship between the mouse and the cursor. This implementation process can assist student Ke to understand how to agree with sounds and images correctly. According to his current level of literacy, the presentation mode of sounds and images by means of a mouse click toward Chinese character is really appropriate for his learning requirement.

Communication is an arduous challenge for children with learning disabilities. Chinese pronunciation is one of the important and essential process for communication. Accordingly, the teacher group takes advantage of sound recording by means of the assistance of flash software. This case in the study invents invisible buttons which contribute to Chinese pronunciation units on the screen when a user can learn Chinese pronunciations through a user-friendly interface. Diverse explanatory sounds appear while the user presses each Chinese character with an invisible button so that the interference of Chinese pronunciations is therefore solved. Afterward teachers participating in this study make students personally operate teaching materials. Thanks to repeatedly practical operations slow learners have more opportunities to recognize Chinese characters and pronunciations. Teachers are also able to revise the basis of teaching materials according to distinct subjects in Chinese courses. In the scope of applications it is a great case with good extension and expansion.

**Case 2.** Tr. La is a teacher teaching for special educational class in Kaohsiung, who thought the tone of Chinese spoken education is an important issue for students in special educational class.

In Taiwan, one of the vital parts in Chinese lessons is focus on pronunciation, especially on tone marks, which usually carries out on grade 1. In Mandarin Chinese, a word can have different meanings depending on tonal contrasts signaled by modulations in pitch during articulation; when pronounced in a high and level pitch (Tone 1), the word ma means 'mother', but when produced with a pitch that rises from mid-range to high (Tone 2), ma means 'hemp' (hereafter, pitch distinctions like these are indicated using numerical notation; e.g., ma1, ma2)[17]. For example, phonetic symbols are divided into five accents. It is boring for children to remember each accent in direct and simple way. Most of teachers will teach phonetic symbols with simple physical actions to impress students.

On the course design of phonetic symbols, Tr. La separates the pronunciation of Chinese characters into five parts and students have to shoot relative films in each part. It is easier to draw students' attentions when they watch the films that they shot. Students are familiar to the characters in the films whom are their classmates, which is the key element to increase the attraction of teaching materials. Tr. Chen based on each step to shoot a film and applied flash software to generate a multimedia displayed by SWF archive. With the single switch of assistive technology it is unnecessary for students to learn how to use keypads and mouse. Only press the big buttons can students gradually understand relative information. Figure 4 is the research operating demonstration.

**Fig. 2.** Teaching material of Chinese tone in spoken lesson

This single switch allows an individual to control all mouse functions using a combination of five single switches or any multiple switches. The reliable, durable, colorful single switch is activated by pressing anywhere on the top surface. These Buttons are suitable for people who need a large switch with auditory and tactile feedback.As for students who fall behind the scheduled learning progress, repetitive practices indeed have a great effort to support them.

**Case 3.** Case 3 is an exercise of Chinese characters and pronunciations about figures and directions.

Developing applications on touch screen for students in special education programs is of priority concern.In the aspect of curricular arrangement in Taiwan, the recognition of figures and pointing directions is a vital unit for top class children in kindergarten. This case therefore demonstrates figures and directions through Chinese characters in coordination with speech sounds as well as animated cartoons. It might be difficult for children at a young age to operate mouse, because children taking part in the case are all under 7-year-old. Considering this reason, we determine to take advantage of touch screen to execute the case. The responses of children are all positive during the process of operation. Children will give immediate reactions by repeating what they hear. Children at a young age are full of interest about the demonstration of dynamical images; therefore, children will continuously press the screen and anticipate for the appearance of voices and images.

**Fig. 3.** Course of "up and down" in kindergarten

Accordingly, the reachers makes use of sound recording and animation about the learning unit   in order to support Chinese words with the as sistance of flash software and touch screen. This case in the study invents invisible buttons which contribute to the whole learning unit on the screen when a user looks at the screen. Diverse explanatory sounds appear while the user presses the places that     appear Chinese word with a invisible button   so that the interference of spoken Chinese is therefore solved. In addition, the display of touch screen is also able to distinctly convey the concept of the process of Chinese learning in order to decrease the communicational obstruction of young students.

## 3     Conclusion and Recommendation

Teachers have already constructed the basic section of flash model. When designing different levels of courses about Chinese pronunciations, teachers just need to perform sound-recordings and the substitutions of Chinese characters. As for teachers, in the design of teaching materials, this digitized learning process is certainly friendly. Therefore, it is easy to popularize digitized teaching materials to other teachers so as to accomplish teaching materials together. Underachievers will have an easier way to learn Chinese pronunciations.

Interacting with computers reflects the movement within the human computer interaction community. Some of the topics in designed teaching units make interactive interface design of assistive technology as their learning goals. Technology-based learning focuses on content learning to explore more user-centred and collaborative approaches to learning. The importance of information and communication skills of interface design for the future has been asserted to support the development of these skills and tools in schools. In this study, the application of technological innovations relies upon user-interface design and the concept of custom-made, which facilitates users' control ability and interaction with an innovation, to convert their technical capabilities into a usable and friendly teaching material. The research anticipates more teachers teaching spoken Chinese at kindergarten and elementary school to participate in this study so as to enhance the variety and abundance of teaching materials. Consequently, not only normal children but also children with learning disabilities are able to receive more resources when learning so as to gain interesting learning and lives.During the process of Chinese learning, slow-learners need more repeated practices than other children in order to improve their impressions of characters and pronunciations. Furthermore, some of the students with learning disabilities have difficulty realizing contents when they are reading. It is beneficial for this type of students to be guided with voices when reading the text.

The case study in the research, for children with reading disorders, begins from the exercise of Chinese phrases. The guidance of multimedia can prevent them from the pressure of asking for others' help. Pupils also enable to learn the connections between characters and pronunciations by means of the constant exercises.Furthermore, thanks to application information interface, students are inclined to absorb knowledge actively and aggressively. They can learn independently with no need of other's assistance. Interactive media interface for children with learning disabilities has become an important and helpful computer-aided design of teaching materials. The interface arrangement also gives students assistance when they attempt to learn different Chinese units, such as Chinese characters. In addition to the stroke order of Chinese characters, the demonstration of pictures and videos also can improve the ability for children to remember Chinese characters. There is a close link between assistive technology, special education and communication design research as well as studies that examine how interactive design of teaching materials can influence learning. Human–computer interaction is a valuable issue for learning disabled students in the future.

**Acknowledgment.** This work was partially supported by the National Science Council, Taiwan, under the Grant No. 98-2410-H-024-018- and No.97-2410-H-024-018-).

# References

1. Li, C.W.C.: Conflicting notions of language purity: the interplay of archaising, ethnographic, reformist, elitist and xenophobic purism in the perception of Standard Chinese. Language & Communication 24, 97–133 (2004)
2. Li, X., Rayner, K., Cave, K.R.: On the segmentation of Chinese words during reading. Cognitive Psychology 58, 525–552 (2009)
3. Klatt, B.A., Goyal, N., Austin, M.S., Hozack, W.J.: Custom-fit total knee arthroplasty results in malalignment. The Journal of Arthroplasty 23, 26–29 (2008)
4. Shimizu, H., McDonough, C.S.: Programmed Instruction to teach pointing with a computer mouse in preschoolers with developmental disabilities. Research in Developmental Disabilities 27, 175–189 (2006)
5. Goodhart, I.M., Ibbotson, V., Doane, A., Roberts, B., Campbell, M.J., Ross, R.J.M.: Hypopituitary patients prefer a touch-screen to paper quality of life questionnaire. Growth Hormone & IGF Research 15, 384–387 (2005)
6. Kawate, K., Ohneda, Y., Ohmura, T., Yajima, H., Sugimoto, K., Takakura, Y.: Computed tomography–based custom-made stem for dysplastic hips in japanese patients. The Journal of Arthroplasty 24, 65–70 (2009)
7. Lin, C.Y., Lin, C.C., Chen, T.H., Hung, M.L., Liu, Y.L.: Application infrared emitter as interactive interface on teaching material design for children. Advanced Materials Research (accepted, 2011)
8. Cho, V., Cheng, T.C.E., Lai, W.M.J.: The role of perceived user-interface design in continued usage intention of self-paced e-learning tools. Computers & Education 53, 216–227 (2009)
9. Kim, Y.J.: The effects of task complexity on learner–learner interaction. System 37, 254–268 (2009)
10. Leaf, J.B., Dotson, W.H., Oppeneheim, M.L., Sheldon, J.B., Sherman, J.A.: The effectiveness of a group teaching interaction procedure for teaching social skills to young children with a pervasive developmental disorder. Research in Autism Spectrum Disorders 4, 186–198 (2010)
11. Lin, C.Y., Lin, H.H., Jen, Y.H., Wang, L.C., Chang, L.W.: Interactive technology application program of experience learning for children with developmental disabilities. Key Engineering Materials (accepted, 2011)
12. Waite, S.J., Wheeler, S., Bromfield, C.: Our flexible friend: The implications of individual differences for information technology teaching. Computers & Education 48, 80–99 (2007)

13. Rosinski, P., Squire, M.: Strange bedfellows: human-computer interaction, interface design, and composition pedagogy. Computers and Composition 26, 149–163 (2009)
14. Bower, M., Hedberg, J.G.: A quantitative multimodal discourse analysis of teaching and learning in a web-conferencing environment–the efficacy of student-centred learning designs. Computers & Education 54, 462–478 (2010)
15. Kotzé, P., Renaud, K., Van Biljon, J.: Don't do this – Pitfalls in using anti-patterns in teaching human–computer interaction principles. Computers & Education 50, 979–1008 (2008)
16. Lin, C.-Y., Wu, F.-G., Chen, T.-H., Wu, Y.-J., Huang, K., Liu, C.-P., Chou, S.-Y.: Using Interface Design with Low-Cost Interactive Whiteboard Technology to Enhance Learning for Children. In: Stephanidis, C. (ed.) HCII 2011 and UAHCI 2011, Part IV. LNCS, vol. 6768, pp. 558–566. Springer, Heidelberg (2011)
17. Malins, J.G., Joanisse, M.F.: The roles of tonal and segmental information in Mandarin spoken word recognition: An eyetracking study. Journal of Memory and Language 62, 407–420 (2010)

# Application of Virtual Interface of Interactive Teaching Materials for Children with Developmental Disabilities

Lin Chien-Yu[1,2], Wang Li-Chih[2], Lin Ho-Hsiu[2,3], Jen Yen-Huai[2,4],
Te-Hsiung Chen[1], and Chang Ling-Wei[2]

[1] Graduate Institute of Assistive Technology,
National University of Tainan, Tainan, Taiwan
linchienyu@mail.nutn.edu.tw
[2] Department of Special Education,
National University of Tainan, Tainan, Taiwan
[3] Tainan Municipal Shengli Elementary School, Tainan, Taiwan
[4] Department of Early Childhood, TransWorld University, Yunlin, Taiwan
[5] National Tainan School for the Hearing Impaired, Tainan, Taiwan

**Abstract.** This research is focus on application of virtual interface of interactive teaching materials for children with developmental disabilities. Based on the base technology of interactive technology, the infrared emitter is the important tool; children control the tool via micro switch, thereby bringing the children the experience of the interactive technology application. There are two demonstrations on this research. The equipments are actually applied on children with developmental disabilities, the research focus on using low-cost equipment, then the relative activities will be easy follow for children. Case 1 and 2 are children with developmental disabilities took an infrared emitter to control the interactive teaching materials. All the interactive teaching materials design by flash software. The participants took infrared emitter as a mouse via wii remote. In this research, the devices relied upon user-friendly design, reducing the working load. The research applies wii remote as an interactive technology, technically, the concept is derived from the infrared receiver on the wii remote, the main purpose is introduce one kind of teaching skill in the resource classes.

**Keywords:** virtual interface, interaction, interface design, infrared emitter, children, disabilities.

## 1 Introduction

This study focus on using simple equipment for children with developmental disabilities in their learning needs, the improvement of equipment to meet the needs of teaching. Wii interactive whiteboard in the teaching carried out in the practical results of the promotion, has been very effective and a lot of reports[1].Interactive whiteboard is a new interface on teaching resource, for teaching purposes with a number of software features, offer children a wealth of information and promote the teaching of students in the classroom attention, abstract understanding of memory after school[2].Integration of visual and auditory information has a positive function

G. Lee (Ed.): Advances in Intelligent Systems, AISC 138, pp. 199–206.
springerlink.com          © Springer-Verlag Berlin Heidelberg 2012

for children with special needs[3].Interactive whiteboard to promote the ability of understanding in the teaching materials, links the experience of learning styles [4]. Disable children which prevent them from using base computer control devices, but custom made alternative devices always more expensive [5], one kind of solution is to explore the application of devices used in contemporary gaming technology[6], such as the Nintendo Wii or air mouse[7].Infrared camera is generally used in tracking systems and this leads to costs often not affordable, particular, wii remotes used as infrared cameras[8].Some research combined wii remote and infrared emitter to create low-cost interactive whiteboard[9], so that teachers enable to design teaching materials enhance learning interest for children with developmental disabilities. The interactive technology consists of a wii remote, infrared emitter, a laptop and a projector. Combined low-cost gear to create a cheaper device, the laptop could be controlled by the infrared emitter that functions much like a mouse[10]. Computer-mediated communication facilitates the understanding of communication patterns, forms, functions and subtexts, which can in turn engender an understanding of how to derive meanings within such context [11]. The application of a virtual interface and infrared light emitter is similar like make a mouse in the participant's hand, the design of virtual interfaces adopts an interactive method while the design of teaching materials adopts flash software that could invent interesting display in order to raise children's curious to add the rate of using such interactive teaching materials. In addition, children with developmental disabilities not only could taste new teaching method but also are impressed by them. Virtual interface design has an advantage in that the application is able to make corrections to the teaching materials, while the assistive technology may also be transfer to other training courses. Just like this research as an demonstration, the teaching material displayed via swf, however, through the virtual interface and infrared emitter, the process becomes uncomplicated for the children; only operate the tool using the sense of tuition.

## 2    Method

For children with disabilities, it is required to come to the station to the screen to operate, in fact, the process is difficult, In the experience of many experiments show, the children with developmental took the infrared emitter pen near the projective wall, they always face to the projective screen, often because the operating habits, they covered the infrared transmitter, making the wii remote can not successfully receive the infrared emitter of the message, so that interactive activities can not proceed smoothly. Normal children were asked to hand holding infrared emitter pen as a pc mouse ,and at the same time, pay attention to body position, do not cover the wii remote and the infrared emitter pen transmission direction, for the normal children's understanding, it should be just a simple command, however, is also very easy to implement. But for children with developmental disabilities, even for children with multi developmental disabilities potentially increasing the learning process a lot of trouble with the learning load. Therefore, the first part of this study was improved for the devices, the subjects do not have to come to the wall is projected to operate the IR pen, but rather, against the participation of children in the wii remote, at the time during calibrated moment, the research consider the scope of the children's fingers could stretch as the operation of the virtual range. In this way, the projected image still on the walls, but the subjects can seat in a chair, holding a infrared emitter pen as a pc mouse in the learning process, just as Fig 1.

The purpose of this research is to help children with developmental disabilities have some opportunities enjoy happiness on their learning process. Assistive technology is a helpful method for learning, which has prominent influences on helping teachers explain difficult concepts, giving access to a huge range of examples and resources, and inducing pupils to engage in learning easily. This research showed appreciation with many experts who devote their technology about interactive whiteboards using the wii remote. Wii remote is a handheld device just like a television remote, a high-resolution high speed IR camera and wireless Bluetooth connectivity. Wii remote camera is sensitive only to bright sources of infrared light emitter, tracked objects must emit a significant amount of near infrared light to be detected[12]. With infrared emitter and wii remote could be create an effect just like pc mouse, it's a low-cost and custom-made tool[13]. Therefore, the study is able to integrate the feedback of flash software and power point and to develop teaching materials for children with developmental disabilities.

## 2.1    Hardware Set-Up

A projector projected the images to wall that children with developmental disabilities could control the display by using the relative devices. The devices included infrared emitter and a wii remote, the principle of infrared emitter pen focus on the micro switch, when the children press the micro switch, the pen will emit infrared light, the wii remote will receive the signal. The equipment cost so cheap and portable, so it is useful for teachers to use the devise in different places even they are itinerant teachers who need teach children between different elementary.

The theory applied making use of blue tooth to connect computer, wii remote and infrared emitter. Via infrared emitter, wii remote could track the location of infrared emitter, so, the function of the infrared pen just like a pc mouse, the children with developmental disabilities could use a pen to instead of a mouse as a controller, just as Fig.1. Based on the requirement for children, the weight of the device that children hold

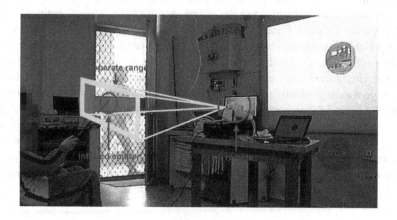

**Fig. 1.** The principle of this research

is only a simple pen with micro switch. The display mode of this method is an intuitive learning tool. Children with developmental disabilities will raise their delight and achievement toward lessen their frustration during their operate processes. Since the wii remote can track sources of infrared light emitter, so the research make a virtual interface, there is a virtual operate range, when the children press the micro switch in this area, the infrared emitter just like a mouse to control to system.

## 2.2    Interaction Design

In this research, the virtual interface displays the function of infrared receiver. It is a simple tool for children to control the micro switch, and the teaching material could be modified for other courses. In the research ,the researcher use flash software to design interactive teaching materials, when the research set up the device in the classroom, the children with developmental disabilities only take an infrared pen behind the virtual interface could use the interactive effect. Participants from the resource classroom could recognize how to operate this virtual interface, because only take one object just like a pen and wave it could see the content.

# 3    Case Study

This 2 cases are custom made for children with developmental disabilities so as to increase their intention. Using virtual interface and infrared light emitter could make an interactive design, which is able to increase the attention of teaching materials, there are 2 cases demonstrate on this study. The research performed how to use the assistive tool, the child was asked for to hold the infrared emitter pen in her hand. When she press the micro switch of the infrared pen and move it in the air, she could see the reflections that the projector projected the images on the wall.

**Case 1.** The participant of case 1 is a girl with intellectual deficit and speech disability who is belongs to moderate multiple retarded. She is a $6^{th}$ grade student at elementary school. In this case, she sat on a chair and hold a infrared emitter pen behind the virtual interface range, when the girl press the micro switch of the infrared emitter pen, just like she operate a mouse. Case 1 is one kind of teaching materials for customer made, pictures of children on the content of the teaching materials, the program set in the beginning is to present the status of invisible, when the girl operate the infrared emitter pen in the virtual interface range, the projector will show the picture that include the girl's image appear gradually projected on the wall, and she know she could control the situation by herself. The case is not only a custom-made teaching materials design but also a real-time feedback. Fig. 2. shows the step by step pictures in the experimental process. This case focuses on using low-cost assistive technology could make interactive effect, children are offered with digital presentation of designing and learning concepts for easier ways to operate, by this way, children to train their body active ability by virtual interface design with interaction.

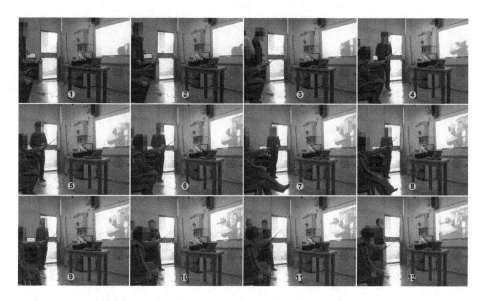

**Fig. 2.** A girl participant a custom-made course (To protect the handicapped child, all pictures have used masaic effect on their face)

**Case 2.** The participant of case 2 is a boy with autism. He is a $3^{rd}$ grade student at elementary school. Case 2 is one kind of teaching materials for erase the white area then could see the clue of the picture. In case 2, the researcher work the interactive interface design by flash software, the research change different pictures from case 1 for the course that the theme of this course on campus scenery. The child with developmental disabilities hold the infrared emitter pen, when he press the micro switch, the function just like we control a computer with left button of a pc mouse, when he moved her hands and kick the object, her images on the screen would change another shape on the projective screen. In case 2, the researcher design different pictures of campus scenery, the child could use a pen to control the content of an interactive interface, in the learning process, the participant didn't use pc mouse, it's means that decrease the load for the child.

It could improve the attention of children's activity; because the real-time feedback could attract children's attention. Furthermore, because the image design used by flash software, the application of flash software is very easy, teachers only take the original design program and change the pictures that could make a new function for different courses, Fig. 3.shows the application on campus scenery unit, when the children with developmental disabilities, they could control the message only by a virtual interface and an infrared pen, the wii remote will receive the singles through blue tooth transfer to computer, then the output will display in the real-time projected on the wall.

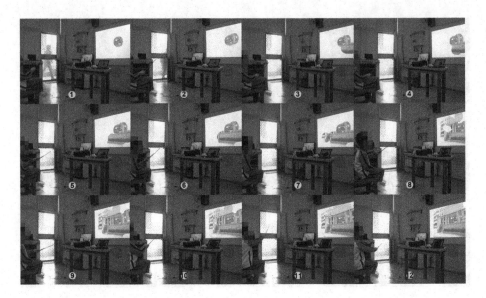

**Fig. 3.** The participant wave the tool to show the campus scenery(To protect the handicapped child, all pictures have used masaic effect on their face)

## 4   Conclusion

Just as the application of low cost interactive technology, the emphasis is placed on aspects such as ease of learning by resource class teachers, low equipment cost and ease of promotion.

The sensor from the infrared emitter could transmit the singles to the computer, and then the projector project the interactive effect on the wall at the same time. To put it simply, we used an infrared emitter pen as a mouse, therefore the continue researchers could design different teaching module courses. The main purpose is to instruct children in the application of physical activities. The research tried to design more different course for different disabled children.

There will be more application on the field of virtual interface, because the virtual interface focus on easy operate and easy use, moreover ,the virtual interface provide not only real-time feedbacks but also lower price could execute the experiment. It is real and quite assistance to design interactive teaching materials for teacher, special for the resource teacher and itinerant teachers. Based on the fundamentals of making the operating interface simpler and burden-free, the assistive technology is applied on elementary school students and kindergarten children to produce fun learning.

The main theme of the project is the assessment of the application of virtual interface for the design of interactive teaching materials; flash is used as the main application software to produce the interactive contents such as videos and animations. The main concern is to induce the interest of the elementary teachers who have participated in the first stage to use the easy-to-learn software. Since they are already equipped with basic skills to use the software while formulating the teaching materials, the teachers will be able to focus their energy on designing the teaching material. In terms of the hardware, the infrared emitter is modified so that the projected images are interactive in real-time, relieving the burden of using keyboards and mice for the children. Furthermore, the introduction of interactive, assistive technology also facilitates the children's learning process.

When teaching children from resource classes, the teaching material is modified according to their needs. The development of the study may cater to more children with diverse needs; by introducing wii's interactive technology into the project, the project is not only academically sound, innovative and flexible, it also has significant influence on the design and development of teaching materials. Consequently, the teachers' teaching materials may be presented in a more diverse manner, in turn facilitating the introduction of 100% interactive teaching materials into elementary schools. If during the children's learning process, instead of employing creed-oriented teaching, knowledge is obtained via interaction, the process will become less burdening for children. The development of digitization possesses high potential; as far as children are concerned, not only is the application of digital content refreshing, since the teaching materials emphasize methods such as interaction and coordination, children will become more interested while learning, which in turn generates a sense of accomplishment. Therefore, the development of interactive interface to assist the children's learning is imperative.

**Acknowledgement.** This work was partially supported by the National Science Council, Taiwan , under the Grant No. 98-2410-H-024-018-and 98-2515-S-024-001.

# References

1. Lin, C.Y., Lin, H.H., Jen, Y.H., Wang, L.C., Chang, L.W.: Interactive technology application program of experience learning for children with developmental disabilities. Key Engineering Materials (accepted, 2011)
2. Coyle, Y., Yanĕz, L., Verdú, M.: The impact of the interactive whiteboard on the teacher and children's language use in an ESL immersion classroom. System 38, 614–625 (2010)
3. Kim, J.K., Zatorre, R.J.: Generalized learning of visual-to-auditory substitution in sighted individuals. Brain Research 1242, 263–275 (2008)
4. López, O.S.: The Digital Learning Classroom: Improving English Language Learners' academic success in mathematics and reading using interactive whiteboard technology. Computers & Education 54, 901–915 (2010)
5. Vidal, J.C., Mucientes, M., Bugarn, A., Lama, M.: Machine scheduling in custom furniture industry through neuro-evolutionary hybridization. Applied Soft Computing 11, 1600–1613 (2011)
6. Wuang, Y.P., Chiang, C.S., Su, C.Y., Wang, C.C.: Effectiveness of virtual reality using Wii gaming technology in children with Down syndrome. Research in Developmental Disabilities 32, 312–321 (2011)
7. Shih, C.H.: Assisting people with attention deficit hyperactivity disorder by actively reducing limb hyperactive behavior with a gyration air mouse through a controlled environmental stimulation. Research in Developmental Disabilities 32, 30–36 (2011)
8. De Amici, S., Sanna, A., Lamberti, F., Pralio, B.: A Wii remote-based infrared-optical tracking system. Entertainment Computing 1, 119–124 (2010)
9. Lee, J.C.: Hacking the Nintendo Wii Remote. Pervasive Computing, 39–45 (2008)

10. Lin, C.Y., Lin, C.C., Chen, T.H., Hung, M.L., Liu, Y.L.: Application infrared emitter as interactive interface on teaching material design for children. Advanced Materials Research (accepted, 2011)
11. Bower, M., Hedberg, J.G.: A quantitative multimodal discourse analysis of teaching and learning in a web-conferencing environment–the efficacy of student-centred learning designs. Computers & Education 54, 462–478 (2010)
12. Standen, P.J., Camm, C., Battersby, S., Brown, D.J., Harrison, M.: An evaluation of the Wii Nunchuk as an alternative assistive device for people with intellectual and physical disabilities using switch controlled software. Computers & Education 56, 2–10 (2011)
13. Lin, C.-Y., Wu, F.-G., Chen, T.-H., Wu, Y.-J., Huang, K., Liu, C.-P., Chou, S.-Y.: Using Interface Design With Low-Cost Interactive Whiteboard Technology to Enhance Learning for Children. In: Stephanidis, C. (ed.) HCII 2011 and UAHCI 2011, Part IV. LNCS, vol. 6768, pp. 558–566. Springer, Heidelberg (2011)

# Leaf Area Measurement Embeded in Smart Mobile Phone

Dianyuan Han and Fengqing Zhang

Dept. of Computer and Communication Engineering,
Wei Fang University, Shandong 261061, China
{wfhdy,zhangfq}@163.com

**Abstract.** This paper concerns the problem of leaf area measurement based on image processing embedded in smart mobile phone. After the leaf image was captured with the camera of the mobile phone and preprocessing to eliminate the noise, the first step is the detection of the referenced rectangle with 2-side scan method. Then the leaf region was separated from the background by comparing the value of 2G-R-B to a threshold which was defined with OTSU method. Next, the leaf area was corrected by morphological image processing. Lastly, the leaf area was calculated according to the pixels proportion between leaf region and the referenced rectangle. Experiment results show that our method has as equivalent accuracy as traditional method with digital camera and computer.

**Keywords:** image processing, image segmentation, leaf area measurement, smart mobile phone.

## 1    Introduction

Leaf is the vital organ of plants that can synthesize organics through photosynthesis. The size of leaf area acts as an important role in forestry resources survey and plants physical ecology research. Leaf area measurement method based on the image processing has attracted increasing attention because of its equipment simple, easy operation and high precision [1]. In this method, a digital camera and a computer are necessary to catch the leaf picture and calculate the leaf area, so it is not convenient in outdoors.

Currently, there is at least one camera in the smart mobile phone, and the preference of the camera is so powerful that it can be comparable with a digital camera. At the same time, the solution of application based on image processing embedded in mobile phone has become basically mature, and there are some image processing chips embedded in mobile device in the market. So the image processing technology can be embedded in the mobile phone easily. After the leaf picture is shot by the phone camera, the leaf area can be calculated with the application we developed. This method has some advantages such as high operation speed, real timing and portability, so it is of high application value. In this paper, the technologies will be reported.

## 2    Builting Development Environment

At present, the mostly used mobile phone operation systems are Symbian, iPhone, Android and Windows Mobile. Among them, Windows Mobile is developed by Microsoft

G. Lee (Ed.): Advances in Intelligent Systems, AISC 138, pp. 207–213.

for mobile devices. Though it has been not long since it was marketed, its development is extraordinarily rapid.

In our work, we selected Windows Mobile  6.5 as the operation system for mobile phone, selected Windows Mobile 6.5 Professional DTK as the development platform and selected VS2008 as the programming language. The software structure was shown in Fig.1.

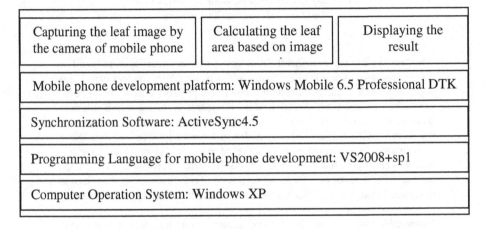

| Capturing the leaf image by the camera of mobile phone | Calculating the leaf area based on image | Displaying the result |
| --- | --- | --- |
| Mobile phone development platform: Windows Mobile 6.5 Professional DTK | | |
| Synchronization Software: ActiveSync4.5 | | |
| Programming Language for mobile phone development: VS2008+sp1 | | |
| Computer Operation System: Windows XP | | |

**Fig. 1.** The sofrware structure of leaf area measurement embeded in smartphone

To start the mobile phone development, the development platform should be built firstly. The main steps are as follows:

step1: Setup Microsoft Visual Studio 2008 + SP1.
Step2: Setup Microsoft ActiveSync 4.5
Step3: Setup Windows Mobile 6 Professional SDK Refresh.
Step4: Setup Windows Mobile 6.5 Professional Developer Tool Kit.

In this paper, the leaf area measurement flowchart was shown in Fig.2.

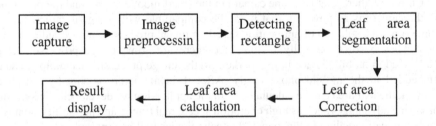

**Fig. 2.** The flowchart of leaf area measurement embeded in smart phone

The first step of the leaf area measurement based on image processing embedded in mobile phone was leaf image acquisition. Then the image was preprocessed to eliminate the noise. Next, the referenced rectangle in the board was detected and the number of pixels in the rectangle was counted. Subsequently, the leaf area was extracted and the number of pixels in the leaf area was counted. Lastly, the leaf area was calculated according to the proportion of the pixel number between rectangle and leaf region. In this work, we select 24-bit true color image as the example for research.

## 3    Acquisiting and Preprocessing the Leaf Image

The image acquisition is basis of the leaf area measurement based on image processing. In this work, we used the camera of the mobile phone to catch the photo of the plant leaf.

Before the leaf photo was shot, a referenced rectangle was drawn on the board with white background, or it could be printed on a white paper. The rectangle should surround the leaf as shown in Fig.3.

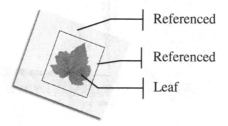

**Fig. 3.** The leaf image acquisition diagram

In order to reduce the geometric distortions and increase the precision, the camera should be perpendicular to the blade when the leaf photo was shot.

The main purpose of image preprocessing is to eliminate the noise. In this work, we used median filter method.

## 4    Detecting the Rectangle

In order to detect the referenced rectangle quickly and accurately, we used oscillation scan method (OSM). As it only need to scan the image for one time, so the speed was improved greatly. The OSM is as fellows.

Firstly, set the element of the binary image scan as 1, and its size was equal to the leaf image.

Then the leaf image was scanned from the first row until to the last row. To a particular row, it was scanned from left to right, and the gray value of every pixel was calculated with Eq. 1.

$$gray = R*0.3 + G*0.59 + B*0.11 \tag{1}$$

If gray>T1, then the element in the corresponding position of binary image scan was set as 0. Else, if gray<=T1, it shows that the current position was the left edge of the rectangle, so the scan direction was changed from the right side of the same row to left side. When gray<=T1 appeared again, the scan was ended. There T1 was an experienced threshold and it was set as 128.

The rectangle was recorded in the binary image scan which was created with OSM. If a pixel value was 1, then it belonged to rectangle and displayed with white color. The number of 1 in scan was the number of the rectangle. The rectangle detected with OSM was shown in Fig.4. There (a) was the original image shown in mobile phone and the area with white color of (b) was the detected rectangle.

(a)The leaf image                (b)The detected rectangle

**Fig. 4.** The detected rectangle by using OSM

## 5   Leaf Image Matting

**The Rough Matting of Leaf Image.** The images captured by the camera of mobile phone are RGB mode. The color of plant leaves is generally green, or mainly green, while the color of referenced board is white, so the color contrast between the object and the background is obvious. To every pixel of the total image, if the value of 2G-B-R [2]is calculated and then compared to a threshold T2, the leaf area can be easily separated from the background.

There the selection of threshold T2 is very important. In our work, we used OTSU method [3]. Otsu's method is an attractive one due to its simplicity in computation. In Otsu's paper, the between-class variance (BCV) is defined, and the gray level at the BCV maximum determines an optimal threshold. In many cases the Otsu's method is used to automatically perform histogram shape-based image thresholding and in this paper it was also used.

The main steps of the rough matting of leaf image are as follows:

Firstly, a binary image rough_leaf was initiated which had the same size with the original leaf image.

After the threshold T2 was calculated, the original leaf image was scanned and 2G-B-R of every pixel was calculated. If |2G-B-R|>T2*2/3, then the corresponding position of rough_leaf was set as 0, else it was set as 1. Experiment showed that if T2 was times by 2/3 it has more good result than T2 only. The segmentation image according to 2G-B-R was shown in Fig.5(a).

(a) 2G-B-R segmentation image   (b) Handled with close morphology operation

**Fig. 5.** The result of leaf area segmentation

**Correcting the Leaf Area by Morphological Image Processing.** Closing is an important morphological operation [4]. It generally smoothes the sections of an object and fuses narrow breaks and long thin gulfs, eliminates small holes and fills gaps in the contour. The closing of set A by structuring element B, denoted A·B, is defined as Eq. 2.

$$A \bullet B = (A + B) - B \qquad (2)$$

Thus the closing A by B is the dilation of A by B, followed by the erosion of the result by B.

After the rough leaf image rough_leaf was got with the foregoing 2G-B-R, it was handled with close morphology operation. Suppose the handled image was leaf_image, The result is shown in Fig.5(b).

## 6    Calculating the Leaf Area

The 1-value pixels in the binary image scan which was got with OSM was rectangle, and when all pixels value were added, the result was the pixel number of the referenced rectangle. Similarly, if all the pixels value in binary image leaf_image were added, the result was the pixel number of the leaf area. On of the experiment result was shown in Fig.6.

We supposed that the plant leaf and the referenced board were at the same plane, and the geometry distortion was consistent. As the area of the referenced rectangle was known, so the leaf area could be calculated with Eq. 3.

$$S_{leaf} = \frac{N_{leaf}}{N_{rec\tan gle}} S_{rec\tan gle} \qquad (3)$$

There $N_{rectabgle}$ represented the pixel number of the referenced rectangle, $N_{leaf}$ represented the pixel number of the plant leaf, and $S_{rectabgle}$ represented the area of the rectangle.

**Fig. 6.** Display the area of the leaf

# 7   Experimental Results

The leaf area can be measured accurately by using of  leaf area measuring apparatus. So the result of area measuring apparatus was set as standard and compared with other methods. In our work, different methods were used to compare the measurement precision.

In our experiment, the mode of leaf area measuring apparatus is X3/M317003, the digital camera is Canon 30D, and the smart mobile phone is HTC HD2 Leo T8585. Some experiment results with different methods were show in Table 1.

Table 1 shows that the measurement result with digital camera and computer method has high precision, and the average relative error is 0.025. The average relative error with mobile phone method is 0.039, so the error is bigger than digital camera and computer method. It is because that the camera of mobile phone has bigger geometric distortions and the image has much more noise.

**Table 1.** Contrast table of leaf area measurement result with different methods

| Test images | Area measuring apparatus (cm$^2$) | Camera and computer (cm$^2$) | Relative error (cm$^2$) | Mobile phone (cm$^2$) | Relative error (cm$^2$) |
|---|---|---|---|---|---|
| image 1 | 26.42 | 25.22 | 0.05 | 24.52 | 0.07 |
| image 2 | 69.45 | 70.55 | -0.02 | 71.55 | -0.03 |
| image 3 | 112.26 | 114.16 | -0.02 | 115.88 | -0.03 |
| image 4 | 80.65 | 81.74 | -0.01 | 82.46 | -0.02 |
| image 5 | 90.42 | 92.44 | -0.02 | 93.96 | -0.04 |
| image 6 | 53.85 | 54.98 | -0.02 | 55.74 | -0.04 |
| image 7 | 77.79 | 78.84 | -0.01 | 78.93 | -0.01 |
| image 8 | 35.98 | 34.38 | 0.04 | 34.14 | 0.05 |
| image 9 | 58.46 | 59.76 | -0.02 | 60.52 | -0.04 |
| image10 | 44.56 | 46.16 | -0.04 | 47.34 | -0.06 |

# 8    Conclusions

The leaf is a vital plant organ which has the functions of photosynthesis and evapotranspiration. The leaf area measurement has significance to study plant characteristics and guide production. The leaf area measuring apparatus which is made based on the scanner principle have higher accuracy, but they are expensive. The digital camera and computer are not convenient for carrying.

In the wild planting resources survey, the measurement equipment portability was particularly important. Smart mobile phones and other small electronic equipment with camera provide prospects for embedded image processing. In our work, the leaf area measurement technology based on the image processing is embedded in smart mobile phones, which provides new methods and procedures for the wild plant resources survey, and has greater practical value.

# References

1. Han, D.: Influence and correction of the shadow and halo of leaf edge to image-based measurements of leaf area. Forest Resources Management 3, 98–103 (2010)
2. Meyer, G.E., Kocher, T.F., Mortensen, D.A., et al.: imaging and discriminant analysis for distinguishing weeds for spot sparying. Transactions of the ASAE 41(4), 1189–1197 (1998)
3. Otsu, N.: A threshold selection method from gray-level histograms. IEEE Trans. Systems, Man, and Cybernetics 9(1), 62–66 (1979)
4. Gonzalez, R.C., Woods, R.E.: Digital image processing 2ED, ch. 9. Prentice Hall Publisher, New Jersey (2002)

# A Study and Design on Web Page Tamper-Proof Technology

Hu Hong-xin

Suzhou Vocational University,
Suzhou, Jiangsu, China
hhx@jssvc.edu.cn

**Abstract.** This thesis presents a web page tamper-proof technology which is proposed to ensure web site security. It can make the web page failback when it is tampered illegally and do real-time monitoring. By integratedly applying the three precautions of Web firewall, web page real-time monitoring and web page flow detection, we can not only well prevent the web pages from being tampered, but also failback the tampered web pages in time.

**Keywords:** Web page tamper-proof, Web firewall, Web page real-time monitoring, Web page flow detection.

## 1 Introduction

With the popularity of internet usage, network threats occur at all times when people are experiencing the inexhaustible shared resources brought by the internet. There also many problems for the websites. The findings from National Computer Network Emergency Response Technical Team and Coordination Center of China shows that the number of the websites being tampered in mainland China had been up to 2748 only in May of 2010, which brings extremely bad effects on the government and enterprises and public institutions. Therefore solving the problem of web page tampering is hot today, and needs to be discussed and further improved.

## 2 Cause and Type Analysis on Web Page Being Tampered

The causes of the web pages being tampered are mainly as follows:

(1) The webmaster is lack of safety consciousness and negligent;
(2) The systems and programs have many vulnerabilities, or cite free and open source code;
(3) No use of website tamper-proof protection technology;
(4) Attacker has brilliant means and advanced tools.

See Table 1 for the current main attack types on website:

G. Lee (Ed.): Advances in Intelligent Systems, AISC 138, pp. 215–220.
springerlink.com          © Springer-Verlag Berlin Heidelberg 2012

**Table 1.** Attack Types on Website

| Threat type | Measurements | Consequences |
|---|---|---|
| Injection attack | Create SQL statements to illegally inquire the data base | Hacker can visit backend database to steal and modify data |
| Drive-by downloads | Embed malicious code into the website's home page | Hacker can control victims' client and steal users' information |
| Upload fake documents | Avoid webmaster's restriction to upload documents in any type | Hacker can tamper web page, pictures and download documents etc. |
| Unsafe local storage | Steal cookie and session information | Hacker can obtain key data of the users and pretend to be them |
| Illegally run script | Run system default or self upload WebShell script | Hacker full controls the server |
| Illegally run system commands | Use Web server vulnerabilities to run Shell command | Hacker obtain server message |
| Source code disclosure | Use Web server or application vulnerabilities to obtain script's source code | Hacker analyses the source code and attack the website more targetedly |
| URL access restrictions failure | Hacker can visit unauthorized resource links | Hacker can forcibly visit some login pages and history web pages |

## 3    Solution and Deployment for Web Page Tamper-Proof

We can use Web firewall, web page real-time monitoring and web page flow detection to protect dynamic and static web pages, and prevent the Web server from operations such as illegal access, malicious scanning, drive-by download, SQL injection attack and web page files tampering etc. When the website is invaded and web page files are successfully tampered by hacker, the system will take self defense for the website, and alert and recover automatically. In addition, web page tamper-proof technology can also monitor situations of resource utilization and system security in site server, and let the webmaster know the operating condition of the current site at any time, and then make website maintenance more convenient and visualized.

Many websites have built Web firewall, which increased the security of the website to some extent. However firewall relies on virus signature or attacking data base to defense, hackers can use the time difference of database update to conduct the attack, so it's not enough to only use firewall to guard against web page tampering, especially the attack to the application layer, which four layers of firewall basically can not do much effect, so we need a set of safer solutions.

Here, web page tamper-proof technology will adopt three precautionary measures to protect the website: first is building Web firewall to intercept regular illegal operations such as attacking, scanning and illegal request to the Web; Second is real-time monitoring the files within the site. When it finds any illegal operations like modifying and deleting the web pages, it can take automatic protection and alert immediately;

Third is to verify web pages outflowed from site. As long as it finds the web page is different from the former backup file, it will intercept, alert and automatically recover the modified file, and preclude the tampered web pages from seeing by the visitors. Web page tamper-proof technology can only achieve comprehensively safety protection for the website by using multilayer and multiaspect measures.

The workflow of web page tamper-proof shows in Figure 1:

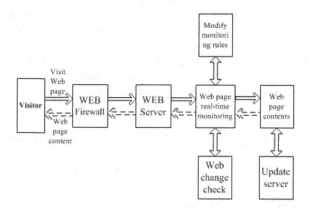

**Fig. 1.** Workflow of Web Page Tamper-proof

The implementation of web page tamper-proof technology doesn't influence the original website mode. According to the whole deployment of the web page tamper-proof technology, the system is divided into three main modules which can be installed into different servers respectively. Usually monitor client is installed on Web server, and management client is installed on management server which has independent deployment. The management client server can correspond and update multiple Web servers at a time. From now on, when you want to do safety protection on every Web server, you only need to install web page tamper-proof monitor client module on every Web server, and then verify its connection with the background management server. By doing this, you can do rule management on background management client server. Network topology shows in Figure 2:

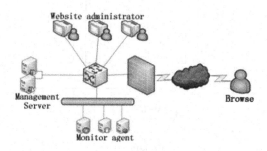

**Fig. 2.** Network Topology

When deploying web page tamper-proof system, you can first build management control end, and then install monitor agent software. In this way, monitor agent server will establish safe connection and make data communication with management control server. Management control server send site safety rules to monitor agent server which will then make safety surveillance and send back alert log to management control server. Through management control server, the webmaster can check the operation condition and log information of the monitor agent server. Update server is responsible for the content updating of all web pages and building watermark (Note:digital watermarking value is the unique tag value generated by doing HMAC-MD5 calculation for 128 bit key for all the web page elements.)backup information for every monitoring end site. When monitoring end software detects that there are web pages being tampered, it can automatically require for watermark contrast with update server, and recover correct web page information in time. Although management control and monitor agent module can be installed in same server, it's best to manage separately for security consideration.

Because management control server has the power of control and site backup, generally you can set an application gateway device at the front of the management control client for site webmaster to connect management control client safely.

## 4    Design of Web Page Tamper-Proof

### 4.1    Web Firewall

Web firewall is mainly used for protecting web page access. It's the first line of defense on analyzing and intercepting illegal users' access request. Web firewall has functions such as preventing malicious scanning, malicious code upload, illegal web page request, drive-by download, SQL injection attack and keywords filtering etc. Its working diagram shows in Figure 3:

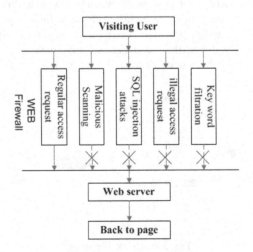

**Fig. 3.** Working Diagram of Web Firewall

(1) Malicious scanning and SQL injection attack: Web firewall makes intrusion detection rules matching on access requests through intercepting every Http request for visiting monitored web page to avoid general attacks such as SQL injection.

(2) Illegal access request: when users want to access the website by using illegal ways, the Web firewall can embed rules according to the user's condition, and open to the access requests and ports which match the rules and shield other access requests.

(3) Keyword filtration: set keyword filtration rules. All the key words, sensitive words involving rules or information including attack code field in the page data submitted by the users will be intercepted by the Web firewall. It will block its connection with the data base, and avoid further attack to the Web server resulted in threat.

### 4.2   Real Time Monitoring

Real time monitoring embeds in the Web server. Applying Web server used for web page content output as an entry point, it embeds tampering detection module into the Web server software, and make real time access monitoring on tampered web page and alert and recover automatically by doing integration verification between every outflowed web page and the real ones, which can timely and efficiently solve the problem of tamper-proof and is an active and direct way to protect website files. By applying this technology, wen can protect website's directory files in advance, and forbid any other process and port accesses to modify the protected directory and files except specified legal process and port service etc., and cut off its connection before illegal process starts to invade the system and forbid its next action.

### 4.3   Outflow Detection

At its first mirroring, the update server has copied all contents in the site and made watermark record for every web page and stored them into watermark library. When web page is modified, the tamper-proof system will get event trigger notification and do watermark detection for current web page. If it finds the watermark is different, it will automatically recover the web page from the mirror. In addition, it will also do watermark detection and keyword filtration when web page outflows. If the system finds unmatched rules, the web page can not be outflowed. Therefore when the web page files in the Web server are modified, the tamper-proof system will immediately substitute the former backup files for the modified files automatically and make the modified files can not flow to the client, which prevents web pages in the site from being modified and ensures its instant recovery and the web site's correctness and authority.

## 5   Conclusion

Web page tamper-proof technology can be implemented not relying on the original Web system framework. It ensures the security and stability of the Web server by multiple protective mechanism and provides web sites with a more secure guarantee.

# References

1. Zhang, J.H., Li, T., Zhang, N.: Mechanism of anti-modification and anti-replacement on WebPages. Journal of Computer Applications 26(2), 327–331 (2006)
2. Gao, Y.L., Zhang, Y.Q., Bai, B.M.: Survey of Webpage Protection System. Computer Engineering 30(10), 113–115 (2004)
3. Liu, J.J., Li, J.H.: The Design and Realization of an Anti-tampering Web Page System. Application Research of Computers (10), 137–139 (2002)

# ETD-MAC: An Event-Tracking Detecting MAC Protocol for Wireless Sensor Network

Ping Liu[1], Yi Chen[1], Mingxi Li[1,2], and Xiaoming Deng[2]

[1] New Star Research Institute of Applied Technology, Hefei, China
[2] School of Computer Science University of Science and Technology of China,
Hefei, China
chenyiaini.student@sina.com

**Abstract.** The TDMA MAC protocol is customarily used in event-driven WSN applications. But it only has good performance in the unmovable event detection, and can't cope with the event whose trend is variable. In this paper, an Event-Tracking Detecting MAC (ETD-MAC) protocol is proposed for event-tracking detection in densely deployed event-driven WSN. In this protocol, an energy-based clustering technique is used to achieve event-tracking detection and a sleeping mechanism is proposed for energy conservation. Moreover, a tight scheduling mechanism is proposed to reduce the latency. The protocol is compared with the cluster-based MAC protocol and S-MAC protocol on NS-2 and Mica2 platform respectively, and shows its superiority in energy, scalability and latency.

**Keywords:** wireless sensor network, MAC protocol, event-tracking, scalability, energy conservation, low-latency.

## 1 Introduction

The applications in wireless sensor network can be classified into four categories in terms of data delivery: continuous, event-driven, query-driven, and hybrid [1]. In event-driven WSN applications, the sensor nodes maintain the network with low data traffic on the regular status; once the event of interest (e.g. magnet) occurs, they will send data to the sink node simultaneously and engender high data traffic.

Efficiency in any wireless sensor based application relies significantly on the medium access control it implements [2]. In densely deployed WSN, especially the magnetic WSN, high data traffic will result in data collision and retransmission wasting time and energy. Thus the energy conservation and low-latency should be achieved in the application. The TDMA MAC protocol is customarily used in event-driven WSN applications for its collision free and no idle listening which achieves energy conservation. To reduce the latency, this protocol is usually integrated with dynamic clustering. But all of these only have good performance in the unmovable event detection, while the event-tracking (e.g. armored target tracking) detection is more complicated.

In this paper, an Event-Tracking Detecting MAC (ETD-MAC) protocol is proposed for event-tracking detection in densely deployed event-driven WSN. The protocol

G. Lee (Ed.): Advances in Intelligent Systems, AISC 138, pp. 221–232.
springerlink.com © Springer-Verlag Berlin Heidelberg 2012

defines two statuses: regular status and event status. On regular status, the S-MAC is executed to maintain time synchronization of the network. On event status, an energy-based clustering technique and a sleeping mechanism are used to achieve event-tracking detection and energy conservation. Moreover, a tight scheduling mechanism is used to reduce the latency.

This paper is organized as follows. Section two introduces related works and points out our motivations. Detail issues in ETD-MAC are described in section three. Section four presents simulation and implementation. Conclusion and future work are given in section five.

## 2    Related Work

S-MAC [3] is a contention-based protocol especially designed for WSN. It forces sensor nodes to operate at low duty cycle by putting them into periodic sleep, and uses RTS/CTS/DATA/ACK to avoid data collision and overhearing. It has good performance when the data traffic is low and can be used on regular status as stated before.

The traditional TDMA MAC protocol is proved to be poorly scalable. But the cluster-based TDMA MAC protocol whose TDMA schedule is integrated with clustering technique is more scalable than the traditional one. In this protocol, the sensor nodes are organized into several clusters and cluster heads are responsible for scheduling their members in a TDMA manner. The ED-SMAC protocol [4] improves cluster-based TDMA MAC protocol by an energy-based clustering algorithm which homogenizes the energy distribution of network, but it can't adjust or form a cluster according to the trend or direction of the event. The EB-MAC protocol [2] is tailored for event-based system and adaptive to the target-tracking detection. But the proposed algorithm is only adaptive to small-scale WSN (e.g. acoustic).

Low-latency is a significant requirement in event-drive WSN. Latency expression for both of traditional TDMA MAC protocol and cluster-based TDMA MAC protocol is the same. The BMA MAC protocol [5] is designed for event-driven WSN applications. As shown in Fig. 1, the operation of BMA is divided into rounds including a set-up and a steady-state phase. It reduces the number of data slots by only assigning them to the source nodes to improve the latency. On the basis of the BMA, a new slot allocation approach in [6] is proposed to decrease the number of source nodes transmitting by only allocating data slots to the ones that have different data. But the occupancy of the idle period in data transmitting period is ignored.

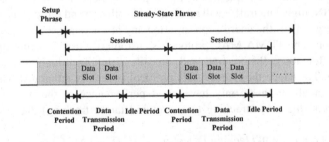

**Fig. 1.** Illustration of single round for BMA [7]

## 3    ETD-MAC Protocol Description

The ETD-MAC protocol which is designed for event-tracking detection in densely deployed event-driven WSN is a cluster-based TDMA MAC protocol. The design objectives can be described as follows. (i) Scalability: it adjusts or forms a cluster according to the trend or direction of the event. (ii) Energy conservation: it not only homogenizes the energy distribution of the network but also saves energy in event-tracking on the premise of satisfying the requirements of event-tracking detection. (iii) Low-latency: it reduces the time from sensor nodes to sink node in event-tracking detection.

Here are some assumptions and notifications which refer to the following discussions.

(i) The sensor nodes are uniformly deployed. The average coverage of the network is $K_{cov}$ and $K_{cov} > k$, where $k$ is the minimum to meet the requirement of event-tracking detection.

(ii) The network time can be synchronized exactly on regular status by S-MAC.

(iii) Each sensor node can receive data from its neighboring nodes correctly, and detect all events inside its detection range.

(iv) There are five energy levels in sensor nodes, where the highest one is 5. The sensor nodes which have a lower energy level than level 2 can't be elected as the cluster head.

(v) The sensor nodes have three states on event status: active state, sleeping state and listening state.

(vi) As shown in Fig. 2, the packet is piggybacked with the packet type, the ID of the source node, the destination of the packet and the data part.

**Fig. 2.** Packet format

**The Operation of ETD-MAC Protocol.** As shown in Fig. 3, on event status, the operation of ETD-MAC is divided into rounds including the setup, the steady-state and the dissolution phrase.

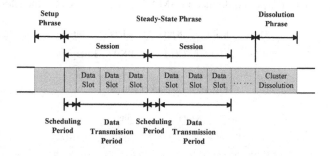

**Fig. 3.** Illustration of single round for ETD-MAC

(i) **Setup Phrase:** When a specific event happens, the sensor nodes which have detected it turn into listening state. If the channel is busy, they will keep silence to wait for an invite from the cluster head. Otherwise, they will check if their energy level is higher than level 2. If so, they will broadcast a CCH (Contest Cluster Head) packet as shown in Fig. 4 for 3 times. The contention nodes also receive CCH packets from others, and check if they can be a cluster head by a contention rule which described as follows. The sensor node that has the highest received signal strength (RSS) readings of the event detected will be given the highest priority, because it can form a cluster with the greatest coverage. If there are some equal ones, the sensor node with maximal remnant energy will be chosen for the cluster head (CH). If there are still some equal ones, the one with the smallest ID will win the election eventually. Then the CH will broadcast a HELLO packet as shown in Fig. 5 to set up cluster. After the sensor nodes have received the packet, they will keep listening and wait for the steady-state phrase.

| 8 bits | 8 bits | 8 bits | 16 bits | 8 bits |
|:---:|:---:|:---:|:---:|:---:|
| CCH | ID | Broadcast | RSS | Energy |

**Fig. 4.** CCH packet format

| 8 bits | 8 bits | 8 bits | 8 bits |
|:---:|:---:|:---:|:---:|
| HELLO | ID | Broadcast | CH |

**Fig. 5.** HELLO packet format

(ii) **Steady-State Phrase:** Steady-state phrase can be divided into scheduling and data transmission period. In the scheduling period, the sensor nodes in listening state will send a RTS packet as shown in Fig. 6 to the CH. The CH will choose $k$ nodes by the contention rule as stated before. $k$ is the number of sensor nodes needed to cover the event with a required quality. Then the schedule will be calculated according to the RSS values of the detected event. The nodes whose RSS value is higher than others will be assigned to earlier slots as they contain a low percentage of error [2]. After that, the CH will broadcast a CTS packet as shown in Fig. 7. The sensor nodes chosen for cluster members will turn off their radio for waking up at their own transmitting time, while others will sleep for $T$ seconds. In data transmission period, the sensor nodes in active state will send a DATA packet as shown in Fig. 8. The idle period defined in BMA is canceled to reduce the latency.

| 8 bits | 8 bits | 8 bits | 16 bits | 8 bits |
|:---:|:---:|:---:|:---:|:---:|
| RTS | ID | CH ID | RSS | Energy |

**Fig. 6.** RTS packet format

| 8 bits | 8 bits | 8 bits | 8k bits | 8 bits |
|:---:|:---:|:---:|:---:|:---:|
| CTS | ID | Broadcast | Schedule | Start-time |

**Fig. 7.** CTS packet format

| 8 bits | 8 bits | 8 bits | 16 bits | 8 bits | 32 bits | 8 bits |
|--------|--------|--------|---------|--------|---------|--------|
| DATA | ID | CH ID | RSS | Energy | TimeStamp | Start-time |

**Fig. 8.** DATA packet format

(iii) **Dissolution Phrase:** At the end of every session, the CH will check if its energy level is lower than level 2 or lose the event. If so, the CH will broadcast a Dissolution packet which has the same format as HELLO packet. If the cluster members have received this packet, they will clear the information about the CH, and come into a new round.

**Energy-based Sleeping Mechanism.** There are two features in the energy-based sleeping mechanism. (i) Minimize idle listening to save energy. The CTS and DATA packet both have Start-time which indicates the start time of next session. The sensor nodes which wake up to send data are ensured to acquire the start-time in time and sleep till then to send a RTS packet without keeping listening. (ii) Minimize contention nodes to save energy. The sensor nodes that lose the election in the scheduling period will be allotted a $T$ sleeping time to reduce the number of nodes which contest to send data in every session. The process of this mechanism is shown in Fig. 9.

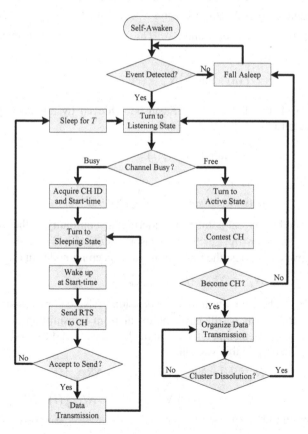

**Fig. 9.** The process of sleeping mechanism

**Deduction of $T$.** $T$ is the key point in this mechanism. If the value is too large, it can't meet the requirement of event-tracking detection. On the contrary, too small one is ineffective on energy conservation. Thus, it is decided by the event velocity, detection requirement $k$ and the average coverage of the network $K_{cov}$.

We convert the event velocity to the number of nodes which detect or lose an event per session, which is denoted as $N_{detect}$ and $N_{lose}$ respectively. And its value is given by $N_{detect} = N_{lose} = \lambda$, where $\lambda$ is assumed to be constant. The number of sensor nodes which detect an event in a specified time is subject to *Poisson* distribution whose parameter is the average coverage of the network [8]. Thus, $K_{cov}$ and $k$ are subject to Eq. 1.

$$poisson\left(Number \geq k \mid K_{cov}\right) \geq 0.95. \tag{1}$$

The time length of one session is $\Delta t$, hence $T$ can be define as Eq. 2.

$$T = n\Delta t, \left(n \geq 1, n \in N\right). \tag{2}$$

As state before, the objective of $T$ is to minimize the contention nodes. The number of nodes which contest to send data in session $i$ is denoted as $N_{contest}(i)$. It is assumed that the average of sessions per round is $s$. Thus the objective function can be defined as Eq. 3.

$$T(n) = \min \sum_{i=1}^{s} N_{contest}(i). \tag{3}$$

As given by Eq. 4, the deduction of $T$ can be achieved by nonlinear programming method according to Eq. 1, Eq. 2 and Eq. 3.

$$T(n) = \min \sum_{i=1}^{s} N_{contest}(i)$$
$$s.t. n \geq 1; \tag{4}$$
$$poisson\left(Number \geq k \mid K_{cov}\right) \geq 0.95.$$

where $n$ is assumed to be real number considering the continuity of the function.

In session $i$, as given by Eq. 5, the sensor nodes which lose the event can be classified into two categories: the transmitting and sleeping nodes in the last session.

$$N_{lose} = NL_{sleep}(i) + NL_{transmit}(i). \tag{5}$$

As given by Eq. 6, the contention nodes in session $i$ also can be classified into three categories: the nodes which still have the event from transmitting nodes in the last session, the new nodes detecting an event in the last session and the nodes that wake up after $T$ seconds sleeping in this session.

$$N_{contest}(i) = NR_{transmit}(i) + N_{detect} + NR_{awake}(i). \tag{6}$$

Since $NR_{transmit}(i) = k - NL_{transmit}(i)$ and $N_{detect} = N_{lose} = NL_{sleep}(i) + NL_{transmit}(i)$, the Eq. 6 can be converted to Eq. 7.

$$N_{contest}(i) = k + NL_{sleep}(i) + NR_{awake}(i). \tag{7}$$

The number of sensor nodes which are allotted a $T$ sleeping time in this session can be defined as Eq. 8.

$$N_{sleep}(i) = NL_{sleep}(i) + NR_{awake}(i). \tag{8}$$

Because of the uniform distribution of the network, the mathematical expectation of $NL_{sleep}(i)$ can be defined as Eq. 9.

$$\overline{NL_{sleep}(i)} = \lambda \frac{K_{cov} - k}{K_{cov}}. \tag{9}$$

The sensor nodes in $NR_{awake}(i)$ are from $N_{sleep}(i-n)$ which decrease in the following $n$ sessions because of event losing. The decrease rule is analyzed as follows.

In session $i-n+1$, the decrease is $N_{sleep}(i-n) \dfrac{NL_{sleep}(i-n+1)}{K_{cov}-k}$.

In session $i-n+2$, the decrease is $\left[1 - \dfrac{NL_{sleep}(i-n+1)}{K_{cov}-k}\right] \dfrac{NL_{sleep}(i-n+2)}{K_{cov}-k} N_{sleep}(i-n)$.

......

In session $i$, the decrease is $\displaystyle\prod_{j=i-n+1}^{i-1}\left[1 - \dfrac{NL_{sleep}(j)}{K_{cov}-k}\right] \dfrac{NL_{sleep}(i)}{K_{cov}-k} N_{sleep}(i-n)$.

Thus, the mathematical expectation of $NR_{awake}(i)$ can be defined as Eq. 10 according to Eq. 9.

$$\overline{NR_{awake}(i)} = \prod_{j=i-n+1}^{i}\left[1 - \frac{\overline{NL_{sleep}(j)}}{K_{cov}-k}\right] N_{sleep}(i-n)$$

$$= \left(1 - \frac{\lambda}{K_{cov}}\right)^{n} N_{sleep}(i-n). \tag{10}$$

Thus, the mathematical expectation of $N_{sleep}(i)$ can be defined as Eq. 11, according to Eq. 9 and Eq. 10.

$$\overline{N_{sleep}(i)} = \lambda \frac{K_{cov}-k}{K_{cov}} + \left(1 - \frac{\lambda}{K_{cov}}\right)^{n} \overline{N_{sleep}(i-n)}. \tag{11}$$

The definition of $\overline{N_{sleep}(i)}$ is provided from session 1 to $i$ as follows.

It's equal to $\overline{NL_{sleep}(i)}$ from session 2 to $n$, except for $K_{cov} - k$ in session 1. Because there is no node wakes up from $T$ seconds sleeping. It's assumed that the $i$ can be denoted as Eq. 12. Here both $a$ and $b$ are integer.

$$i = an + b, a \geq 0, b \in [1, n]. \tag{12}$$

When $b = 1$, it can be converted to Eq. 13.

$$\overline{N_{sleep}}(i) = \lambda \frac{K_{cov}-k}{K_{cov}} + \left(1 - \frac{\lambda}{K_{cov}}\right)^n \times$$

$$\left[\lambda \frac{K_{cov}-k}{K_{cov}} + \left(1 - \frac{\lambda}{K_{cov}}\right)^n\right]^a (K_{cov}-k). \tag{13}$$

When $b \neq 1$, it can be converted to Eq. 14.

$$\overline{N_{sleep}}(i) = \lambda \frac{K_{cov}-k}{K_{cov}} + \left(1 - \frac{\lambda}{K_{cov}}\right)^n \times$$

$$\left[\lambda \frac{K_{cov}-k}{K_{cov}} + \left(1 - \frac{\lambda}{K_{cov}}\right)^n\right]^a \left(\lambda \frac{K_{cov}-k}{K_{cov}}\right). \tag{14}$$

Thus, Eq. 3 can be converted to Eq. 15

$$T(n) = \min \left\{ \begin{array}{l} \sum_{i=an+1}^{s}\left(k+N_{sleep}(i)\right) + \sum_{i=n+2,i\neq an+1}^{s}\left(k+N_{sleep}(i)\right) + \\ \left(K_{cov}-k\right) + \lambda \frac{K_{cov}-k}{K_{cov}}(n-1) \end{array} \right\}. \tag{15}$$

Eq. 4 can be converted to Eq. 16 as $s$ is assumed equal to $cn$, where $c > 0, c \in N$.

$$T(n) = \min \left\{ \begin{array}{l} A_1^n \sum_{i=0}^{c-2}\left(A_1^n + A_2\right)^i \left(A_2 n + A_3 A_1\right) + \\ cn(k+A_2) - nk + A_3 A_1 \end{array} \right\}$$

$$= \min \left\{ \begin{array}{l} A_1^n\left[\left(A_1^n + A_2\right)^{c-2} - 1\right]\left(A_2 n + A_3 A_1\right) + \\ cn(k+A_2) - nk + A_3 A_1 \end{array} \right\} \tag{16}$$

$s.t. n \geq 1;$

$poisson\left(Number \geq k \mid K_{cov}\right) \geq 0.95.$

Here $A_1 = 1 - \frac{\lambda}{K_{cov}}, A_2 = \lambda \frac{K_{cov}-k}{K_{cov}}$ and $A_3 = K_{cov}-k$. It can be solved by using a Math tool like MatLab [9]. An example is given as follows to validate the objective function.

For $K_{cov} = 12, k = 4, \lambda = 4$ and $c = 10$, Eq. 16 can be converted to Eq. 17.

$$T(n) = \min \left\{ \begin{array}{l} 0.67^n\left[\left(0.67^n + 1.33\right)^8 - 1\right] \times \\ (1.33n + 5.33) + 49.3n + 5.33 \end{array} \right\} \tag{17}$$

$s.t. n \geq 1;$

$poisson\left(Number \geq 4 \mid 12\right) \geq 0.95.$

As shown in Fig. 10, the minimum is 265, when $n$ is equal to 4. The number of average contention nodes per session is 6.6, which is smaller than 12 in traditional TDMA MAC.

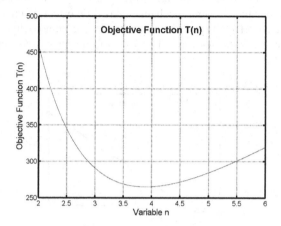

**Fig. 10.** Illustration of $T(n)$

# 4   Simulation and Implementation

The protocol is validated in terms of energy conservation, scalability and low-latency. Its Energy conservation is simulated on NS-2 [10] which has good energy models, and compared with the cluster-based (CB) TDMA MAC protocol. The scalability and latency of the protocol is implemented on Mica2 platform, and compared with S-MAC. All simulations and experiments only refer to the event status.

**Energy conservation Simulations.** 100 sensor nodes are randomly placed in a 100m×100m square region with no mobility. According to the Mica2, the power consumption levels are: 0.06 mW for sleep, 30 mW for receiving, 24mW for idle listening, and 81 mW for transmitting. The simulation parameters are listed in Table 1.

**Table 1.** Simulation parameters

| Bandwidth[Kps] | | 15 |
|---|---|---|
| Event Detection Requirement[nodes] | | 4 |
| ETD-MAC Session | Scheduling Period[s] | 0.4 |
| | Transmitting Period[s] | 0.6 |
| Cluster-based TDMA MAC Session | Scheduling Period[s] | 0.4 |
| | Transmitting Period[s] | 0.6 |
| | Idle Period[s] | 0.5 |

The average coverage of the network which is subject to Eq. 1 increases to simulate the augment of event signal strength, and $\lambda$ increases to simulate the increasing of event velocity. As shown in Fig. 11 and Fig. 12, the energy consumption of ETD-MAC protocol is not sensitive to the event signal strength, but increases sharply when $\lambda$ is more than 4. Because the sleeping mechanism has little effect on energy conservation as the event moves too fast. Then it will operate like a common cluster-based TDMA MAC protocol, but the tight session will result in faster energy consumption than the compared one.

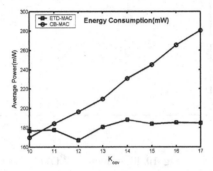

**Fig. 11.** Energy consumption of ETD-MAC vs. CB-MAC as $K_{cov}$ increases

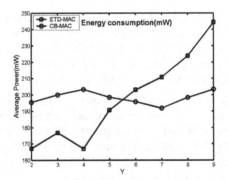

**Fig. 12.** Energy consumption of ETD-MAC vs. CB-MAC as Event velocity increases

**Scalability and Latency Experiments.** To test the proposed protocol, we have simulated an event-detected by storing a buffer of data in different sensor nodes which simulate data being captured from the sensors (e.g. magnet). The cluster head is the sink node and each node can communicate with it directly. The time length of one sleep/wake-up period in S-MAC is 2s.

In the scalability experiment, we use one sink node and 20 sensor nodes which are divided into 4 groups. To simulate the mobility of the event, different node groups will execute the two protocols orderly in a certain time whose length decreases as the event velocity increases. We define the detection that the sink node receives packets form all sensor nodes in the same group as a successful event-tracking detection. Thus the average ratio of successful ones in the experiments is defined as event-tracking probability. Fig. 13 shows the time length increases from 1s to 10s and the probability of event-tracking detection.

In the latency experiment, the number of sensor nodes is varied to simulate the $k$ in different event-tracking detections and the latency is defined as the time from the point of initially detected the event to the point when sink node receives packets form all sensor nodes in the group. The average of latency in the experiments will be calculated for the compare of the two protocols. Fig. 14 shows the number of sensor nodes increases from 2 to 9 and the average latency.

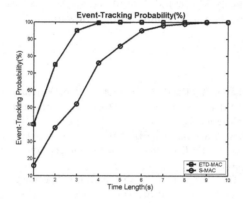

**Fig. 13.** Event-tracking probability of ETD-MAC vs. S-MAC as the time length increases

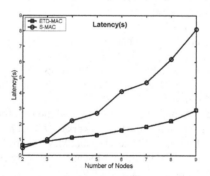

**Fig. 14.** Latency of ETD-MAC vs. S-MAC as number of nodes increases

## 5    Conclusion

In this paper, a new MAC protocol called ETD-MAC is tailored to work with densely deployed event-driven WSN which has the requirement in energy conservation, scalability and low-latency. The sensor nodes at each round can organize and adjust the cluster according to the trend of the event. The latency is reduced by a tight schedule.

Special care is taken to avoid energy waste in transmitting contention by allotting a $T$ sleeping time to the nodes which lose the election. The protocol is not only simulated on NS-2 but also implemented on Mica2 platform, and compared with the cluster-based TDMA MAC protocol and S-MAC respectively. The design shows its superiority in energy conservation, scalability and low-latency and achieves the objectives presented before. Our future work includes studying the algorithm performance for multi-hop communication and high velocity event both in energy consumption and latency, and implementing it in target-tracking system to prove the efficiency of the proposed MAC protocol.

**Acknowledgment.** The authors thank Professor Xunxue Cui and Master Xingyu Chen for their useful comments and help during the course of this work. The authors also gratefully acknowledge the useful comments of reviewers.

# References

1. Stojmenovic, I., Olariu, S.: Data-Centric Protocols for Wireless Sensor Networks. In: Stojmenovic, I. (ed.) Handbook of Sensor Networks: Algorithms and Architectures. John Wiley and Sons, Chichester (2005)
2. Merhi, Z., Elgamel, M., Bayoumi, M.: EB-MAC: An event based medium access control for wireless sensor networks. In: IEEE International Conference on Pervasive Computing and Communications, pp. 1–6 (2009)
3. Heinzelman, W.B., Chandrakasan, A.P.: An Application-Specific Protocol Architecture for Wireless Microsensor Networks. IEEE Transactions on Wireless Communications 1(4) (2002)
4. Hou, Y.X., Wang, H.G., Liang, J.X., Pei, C.X.: A Cross-Layer Protocol for Event-Driven Wireless Sensor Networks. In: 1st International Conference on Information Science and Engineering, pp. 3926–3929 (2009)
5. Li, J., Lazarou, G.Y.: A Bit-Map-Assisted Energy-Efficient MAC Scheme for Wireless Sensor Networks. In: Third International Symposium on Information Processing in Sensor Networks, pp. 55–60 (2004)
6. Sazak, N., Erturk, I., Koklukaya, E., Cakiroglu, M.: An event driven slot allocation approach to TDMA based WSN MAC design and its effect on latency. In: Computer Engineering Conference, pp. 22–25 (2010)
7. Lazarou, G.Y., Li, J., Picone, J.: A cluster-based power-efficient MAC scheme for event-driven sensing applications. Ad Hoc Networks 5(7), 1017–1030 (2007)
8. Vales-Alonso, J., Egea-Lopez, E., Bueno-Delgado, M.V., Sieiro-Lomba, J.L., Garcia-Haro, J.: Optimal p-Persistent MAC Algorithm for Event-Driven Wireless Sensor Networks. Next Generation Internet Networks, 203–208, 28–30 (2008)
9. MathWorks. MATLAB 6.5 (2002)
10. NS-2 on, http://www.isi.edu/nsnam/ns/

# The Research on the Network Switch Intelligent Management<sup>*</sup>

Weichun Lv

Center of Network Management,
Suzhou Vocational University,
Suzhou 215104, China
lwc@jssvc.edu.cn

**Abstract.** By writing VBScript programme  and using SecureCRT terminal tool to perform the function of the batch upgrading of the network switch software version and the switch configuration backup so as to improve the efficiency of the network administrator and achieve the purpose to manage the network equipment quickly and easily.

**Keywords:** Switch, VBScript, Managerment, Configuration.

## 1   The Present Situation Analysis

With the continuous development of the Internet, the size of the LAN is becoming larger and larger, and access layer switches have a wide range of applications in all the industries, it is a very exhausting thing to upgrade the LAN in large qualities and backup the configuration regularly. The campus dormitory network in Suzhou Vocational University has 300 S2026G switches in total. It is estimated to take 2 or 3 days to complete the upgrade by the administrator's continuous operation. It will take more workdays to complete the regular configuration backup. The paper aims to reduce the administrator's tedious labor and duplication of labor by the batch operation. This paper will explain how to automatically complete the LAN upgrade and configuration backup by SecureCRT scripting function.

SecureCRT [1] is a very useful terminal tool, which began to support VBScript [2], two scripting languages of Jscript from Version 5.0 and it can easily do some repetitive operation by writing the scripts.

## 2   Programme Design

Through the design process to achieve a reasonable batch upgrade and configuration backup function, see Figure 1 .

---

<sup>*</sup> This work is partially supported by Suzhou Vocational University Research Foundation (SZD09L21).

G. Lee (Ed.): Advances in Intelligent Systems, AISC 138, pp. 233–238.

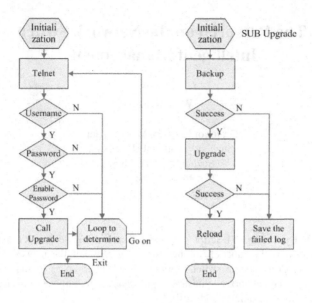

**Fig. 1.** Program Chart

## 3    Program Design and Implementation

### 3.1    Main Function Programme

The major functions of scripting are related to two subs, one of which is Jude Function which judges the functions of various conditions, the other of which Upgrade Function which fulfills the upgrade and backup function.

```
    Sub Judge
  crt.Screen.Synchronous = True
    Set objFSO = CreateObject("Scripting.FileSystemObject")
    Set objTextFileRead = objFSO.OpenTextFile ("ipfileurl", ForReading)
      Do While objTextFileRead.AtEndOfStream <> true
    ipaddress = objTextFileRead.ReadLine crt.Screen.Send "telnet " & ipaddress &
  vbCR   result = crt.Screen.WaitForString ("Username:",30)
      If result = -1 Then
            crt.Screen.Send "***" & VbCr
      result = crt.Screen.WaitForString ("Password:",30)
      If result = -1 Then
      crt.Screen.Send "***" & VbCr
      result = crt.Screen.WaitForString (">",30)
      If result = -1 Then
            crt.Screen.Send "enable" & VbCr
                crt.Screen.Send "***" & VbCr
                  result = crt.Screen.WaitForString ("#",30)
                  If result = -1 Then
                  Call upgrade
```

```
            Else
                crt.Screen.Send "exit" & VbCr
                crt.Screen.Send VbCr
                Set     objTextFileWrite     =     objFSO.OpenTextFile
            ("upgrade_log.txt", ForAppending, True)
                objTextFileWrite.WriteLine Now() & " @" & ipaddress & "
            Device Enable password wrong or not set. The upgrade failed."
                objTextFileWrite.Close
            End If
        Else
            crt.Screen.Send "exit" & VbCr
            crt.Screen.Send VbCr
            Set objTextFileWrite = objFSO.OpenTextFile ("upgrade_log.txt",
        ForAppending, True)
            objTextFileWrite.WriteLine Now() & " @" & ipaddress & " Device
        Telnet password error. The upgrade failed ."
            objTextFileWrite.Close
        End If
    Else
        Set objTextFileWrite = objFSO.OpenTextFile ("upgrade_log.txt",
    ForAppending, True)
        objTextFileWrite.WriteLine Now() & " @" & ipaddress & " Device
    telnet username input error, and log.The upgrade failed."
        objTextFileWrite.Close
    End If
Else
    Set objTextFileWrite = objFSO.OpenTextFile ("upgrade_log.txt",
ForAppending, True)
    objTextFileWrite.WriteLine Now() & " @" & ipaddress & " Device can
not be a Telnet connection, please make sure device is working or is configured
correctly. The upgrade failed. "
    objTextFileWrite.Close
End If
Set    objTextFileWrite    =    objFSO.OpenTextFile    ("upgrade_log.txt",
ForAppending, True)
objTextFileWrite.WriteLine VbCr
objTextFileWrite.Close
Loop
objTextFileRead.Close
MsgBox " Upgrade script is finished, for more information, please see the
scripts directory upgrade_log.txt file "'script is finished, pop-up balloon.
    crt.Screen.Synchronous = False
End Sub


Sub Upgrade
    crt.Screen.Send "copy flash:config.text tftp://IPserver /" & ipaddress &
"config.bak" & VbCr
```

```
result = crt.Screen.WaitForString ("Success : Transmission success",30)
If result = -1 Then
    Set  objTextFileWrite  =  objFSO.OpenTextFile  ("upgrade_log.txt",
ForAppending, True)
        objTextFileWrite.WriteLine Now() & " @" & ipaddress & " Device
configuration backup successed."
        objTextFileWrite.Close
            crt.Screen.Send "copy tftp:// IPserver /rgnos.bin flash:rgnos.bin" & VbCr
        result = crt.Screen.WaitForString ("SUCCESS: UPGRADING OK",300)
    If result = -1 Then
        Set  objTextFileWrite  =  objFSO.OpenTextFile  ("upgrade_log.txt",
ForAppending, True)
            objTextFileWrite.WriteLine Now() & " @" & ipaddress & " Device
upgrade is successful, it will reload after 5 minutes."
            objTextFileWrite.Close
            crt.Screen.Send "reload in 00:05" & VbCr
            crt.Screen.Send "exit" & VbCr
            crt.Screen.Send VbCr
    Else
        Set objTextFileWrite = objFSO.OpenTextFile ("upgrade_log.txt",
ForAppending, True)
            objTextFileWrite.WriteLine Now() & " @" & ipaddress & " Device
upgrade file failed, please make sure the TFTP server is on, the address is correct
and the upgrade file rgnos.bin have been placed to the root directory.The upgrade
failed."
            objTextFileWrite.Close
            crt.Screen.Send "exit" & VbCr
            crt.Screen.Send VbCr
    End If
Else
    Set  objTextFileWrite  =  objFSO.OpenTextFile  ("upgrade_log.txt",
ForAppending, True)
        objTextFileWrite.WriteLine Now() & " @" & ipaddress & " Device
configuration backup failed,please make sure the TFTP server is on and the address
is correct.The upgrade failed "
        objTextFileWrite.Close
        crt.Screen.Send "exit" & VbCr
        crt.Screen.Send VbCr
End If
End Sub
```

## 3.2   Specific Implementation

Before running the program at first you must assure the current windows screen resolution should be 1024*768 or above to ensure that the software version from the terminal to obtain accuracy. If the windows screen resolution is less than 1024*768, it is prohibited to use this script to upgrade. Specific implementation steps are as follows:

1) Open the SecureCRT, and login to a terminal, where the terminal can be a network device, windows system, linux system, but you need to ensure that the terminal supports Telnet command [3].
2) In the SecureCRT menu, click "Script" - "Run ...", and select the script to "upgrade and backup switch batch script. Vbs".
3) Click "Run" button to start the script, the script after the prompt implementation of the following: Figure 2.

**Fig. 2.** The tip window after the prompt script

4) After the upgrade is complete, open the script directory upgrade_log.txt file, view the detailed upgrade information; Figure 3.

```
                         Upgrade_log.txt
2010-6-4 7:43:49 @10.0.14.51 Device configuration backup successed.
2010-6-4 7:45:35 @10.0.14.51 Device upgrade is successful, it will
reload after 5 minutes.

2010-6-4 7:46:35 @10.0.14.52  Device can not be a Telnet connection,
please make sure device is working or is configured correctly. The
upgrade Failed.

2010-6-4 7:47:09 @10.0.14.53 Device configuration backup successed.
2010-6-4 7:48:48 @10.0.14.53 Device upgrade is successful, it will
reload after 5 minutes.
```

**Fig. 3.** Upgrade Log

## 4    Conclusion

The S2026G switch upgrade and configuration backup script in this paper is written or edited  by using VBScript, mainly to achieve the following functions:

1) The bulk upgrade S2026G switch;
2) The text of papers custom S2026G switch IP information;
3) Automatic identification S2026G switch connectivity, abnormal situation will be recorded in the log;
4) Automatic identification S2026G switch username, telnet, enable the password of the correctness of abnormalities recorded in the log;
5) automatically determine the current switch model S2026G, abnormalities will be recorded in the log;
6) Automatic stack environment after the upgrade waiting for synchronization;

7) Automatic identification TFTP server and file transfer of the correctness of abnormalities will be recorded in the log;
8) automatic backup before upgrading the current configuration;
9) Automatic identification device is working properly after reboot, and before and after the upgrade version or upgrade failure information recorded in the log;
10) log support for timestamp.

If there are multiple access layer LAN switching equipment, can be the basis of this script slightly modified version of the judge to increase the switch also enables batch configuration upgrades and backup functions. Similarly, after some modifications this script alone script to achieve the switch configuration, with windows scheduled tasks to configure the switch to achieve a regular backup.

# References

1. Zhang, X.-P.: Using SecureCRT to Implement the Automatic Management of Network. Journal of Neijiang Teachers College 20(02), 48–50 (2005)
2. Kingsley-Hughes, Kingsley-Hughes, Read: VBScript Programmer's Reference. Wrox, pp.11:15–11:696 (1999)
3. Lv, W.-C., Dong, J.: The Reserch of Dormitory Network Switch Security Configuration. Computer Knowledge and Technology 5(36), 10203–10204 (2009)

# Discussion on Computer Network Base Course Reform

Yin Tang

Office of Academic Affairs, Suzhou Vocational University, Suzhou, China
ty@jssvc.edu.cn

**Abstract.** Conduct a series of reforms on teaching contents, teaching methods, and teaching means in Computer Network Base, and run through the project learning, inquiry learning, and collaborative learning into the whole process of reform. Cultivate the students' subject consciousness, participation consciousness, independence consciousness and comprehensive executive ability for future professional change by adopting opening teaching, practical teaching, "team" teaching, "project" teaching and so on. And therefore, realize the goal of nurturing the advanced technical talents that can work in front line position in computer network enterprises and fields with their combined network base theory knowledge and network practical ability.

**Keywords:** Computer network base, teaching reform, local colleges.

## 1 Introduction

Computer network is an ongoing subject that combines tightly the computer development and communication technology. Its theoretical development and application level, an important sign of degree of modernization and comprehensive national strength, directly reflects a country's high-tech development level. In the process of propelling industrialization by informatization and promoting informatization by industrialization, computer network plays a more and more important role. In order to accommodate the talents cultivation demand of information society, computer network is not only an important course for the computer major, but also for non-computer majors. The goal of opening this course is to cultivate high-quality applied talents; the key point of it is to improve the students' practical application ability, emphasizing on meeting the cultivation demand of all kinds of students.

Right now, Computer Network Base course has become the core major course for many high vocational colleges' computer major and the corresponding majors. This course, in accordance with the feature and demand of vocational education "enough theories, more practices", introduces the basic knowledge of computer network. Starting from the basic conceptions of computer network, it explains the basic communication theory, computer network architecture, Internet and TCP/IP, local area network conception and its composition, network design and networking technology, operation and maintenance of network operation system, knowledge on computer network security and application. It attaches more attention on course experiment and curriculum design to develop students' application ability. Furthermore, it introduces the modern education technique to teaching, demanding the teachers to change the conception and transform the role to adapt vocational education reform.

G. Lee (Ed.): Advances in Intelligent Systems, AISC 138, pp. 239–244.
springerlink.com         © Springer-Verlag Berlin Heidelberg 2012

# 2    Teaching Content Organization and Arrangement

The current society not only requires the graduates to have solid theoretical knowledge, but also demands them to have certain practical ability to solve the problems. The teaching arrangement of Computer Network Base should not be confined in knowledge system completeness, and it has to have pertinence and practicability. During the teaching process, the students' professional post skills and quality should be fostered to accommodate the employment demand after graduation. This could help the students to fulfill the goal of practicability after learning.

## 2.1    Theoretical Teaching Content Arrangement

By combining computer network teaching practice, Computer Network Base class teaching content can be divided into four learning units with the principal line of Internet technology and high speed network technology, as well as the latest development fruit of network development. The specifics are as follows: first unit mainly discusses the basic conception, development and application of computer network, including: Ethernet structure, switch, datagram, virtual circuit, protocol and other network basic knowledge. The second unit has the discussion on basic conception and protocol of wide area network physical layer and data link layer. On the basis of introducing media access control method, it helps the students to learn the local area network technology development and application. It introduces the basic networking technology of LAN, VLAN, and WLAN. The third unit undergoes the systematic discussion on network layer, transport layer and application layer of TCP/IP protocol. Moreover, by taking typical example of analysis on application layer protocol, it helps the students to digest the knowledge and reinforce the understanding on network working principle and implementation technique. The fourth unit introduces the technology development hotpot issues, such as network management technology, encryption, digital signature, security e-mail, intrusion detection and firewall, VPN and other knowledge on network security. This will enhance the learning initiative of students.

## 2.2    Experiment Teaching Content Arrangement

In practice, the experiment teaching falls into basic content and advanced content. Basic content includes network cable making, network addresses distribution, network routing configuration, switchboard configuration, network application erection, network traffic monitoring and so on. The purpose of it is to cover all contents from first level to seventh level in OSI reference model. Advanced content selectively conducts network design planning, maintenance management, network security, network protocol analysis, network malfunction diagnosis, network programming, etc. The purpose of network experiment teaching is to promote further understanding on theories of network courses of the students.

# 3   Teaching Methods and Teaching Means Change

## 3.1   Teaching Methods Change

*1) Adopting opening teaching to promote the interactive cooperation between learning and teaching*

Opening teaching emphasizes on interaction between teaching and learning. Teachers and students should interact, communicate, enlighten, complement with each other. During the teaching process, teachers and students should share the thoughts, experience and knowledge with each other to communicate each others' feelings, experiences and notions, enriching teaching content and seeking new discoveries, and therefore, reaching consensus, sharing, and mutual progress. This will realize the common development of teaching and learning. In the teaching process of Computer Network Base, we adopt heuristic method, discussion method, interaction method and other teaching methods to invigorate the classroom atmosphere, arouse the learning interest of students, and promote their active thinking. In the class, the teachers should pay more attention on handling relationship between key points and difficult points, conception and application, theory and operation. The main content should be spoken pithier and practiced more. Some content have to be spoken and practiced together. The past completion project cases should be made into courseware, and the teachers could explain them to the students. Besides, the students should be encouraged to take part in the analysis. This will arouse the students' learning initiative to a large extent.

*2) Adopting practical teaching to advance the professional ability of students in an around way*

Practical teaching set adheres to the design principle of ability oriented, placing improving students' technology application ability in an outstanding seat. The department has set up advanced network labs, equipping route, switchboard, server and network software. It can conduct the practical training of several projects, effectively supporting the autonomous and inquiry study of students. We have also designed several experiments on the basis of current computer network application situation, giving the basic operation skill training to students. All of these experiments are separated. The technical practice activities completed in the labs include common network settings and network checking, network cable making, WWW server establishment, FTP server establishment, etc. Through these practical teaching, the students initially master the basic operation skills of computer network, get familiar with the common network products and basic techniques. Meanwhile, the intertwining of theoretical teaching and practice training improves the integration between them, reinforcing the students' understanding on course content.

*3) Adopting "project" and "team" teaching to enhance the students' comprehensive execution*

The theories in books and practical teaching are far from enough to cultivate the students' operation ability and application skills for this course. The reasons for this are that: on the one hand, this is a course with strong practicality; on the other hand, the course requires a lot in comprehensive quality. The students face the problem of how to synthesize the software, hardware and communication knowledge. This synthesis is not the simple overlying, but integration. It is the mutual penetration of computer software, hardware and communication basic knowledge, forming an integration conception. This requires students to have the corresponding comprehensive ability. "Project" and "team" teaching mean to inspire the students' learning initiative by planning, design, realization, test and evaluation of practical project in the unit of project team. While practicing the students' organization ability, communication ability, cooperative ability and team spirit, their innovative awareness, spirit and ability have been fostered, completely enhancing their task-completion ability under restricted situation and their cross-disciplinary and cross-departmental comprehensive execution. This has laid a solid foundation for them to face the changing working environment in the future career. Besides, part of the students are arranged to participate the enterprise network construction in the internship. This task-driven method helps the students to transform the knowledge into practical skill and makes the students experience the ongoing medium and small sized enterprises and companies' network integrated project scenes. In the project realization teaching process, the students also get to know the comprehensive quality required in engineer implementation, project management and practical work. Therefore, this reinforces the synthesis effect of teaching, better realizing "short-distance interface" of classroom teaching and future working posts.

### 3.2    Teaching Means Change

*1) Multi-media teaching means*

Course content basically makes multi-media demonstration courseware, showing the former abstract and complex theoretical content by vivid pictures and animations. By doing this, the students can intuitively understand the teaching contents, inspiring their learning interest. Teachers separate the courseware to students, so that it is needless to take notes and the students can focus on classes. Meanwhile, this facilitates the reviewing. In a word, the teachers for this course take full advantage of modernization teaching means and gain excellent teaching effect.

*2) Network teaching means*

Computer Network Base is a practical technology which is used widely. This represents the application in reality of network theories. So, real-time display is of great importance. We put in the Internet with the corresponding knowledge teaching and explain the knowledge with real network simultaneously. Through this, the students will not feel boring and they can understand the importance and practicability of network.

*3) Theory and practice combination teaching means*

Change thoroughly the traditional teaching method of "theoretical classes first, then experiment classes". The training content, using the object teaching and field operation as the teaching aid, adopts "teaching while practicing" teaching method. The teachers demonstrate on the teaching computer, and the students will follow the teachers' schedules. This kind of process is normally two times, and the next is self-practicing by students. After doing it, the students will remember the operation. When they are operating by themselves, they could learn while exploring. This improves the learning efficiency greatly.

*4) Keeping up with the time and adjust the teaching content*

Most of the teaching content of this course is the relatively stable knowledge points. In recent years, on the one hand, we strengthen the contact and cooperation with this industry to establish practice training bases, getting to know and mastering the application situation of computer network and demand for graduates. Part of the teaching module will be readjusted in line with enterprises' "demand" to meet the knowledge structure requirements on students by enterprises. On the other hand, introduce the teaching reform and research fruits or discipline latest published fruits into teaching on time, adjusting teaching content appropriately and keeping up with the global computer network technology development. The teaching content should pay attention on enough basic knowledge, but also reflect the industry advancement and sense of times, helping the students to master practical and useful knowledge.

## 4    Final Assessment Methods

In order to strengthen the students' self-learning, innovation and practical ability, we have also conducted the related reform on examination system. The constitutions of attainment test are as follows: experiment performance takes up 25%, mainly including test report and final curriculum design; class performance takes up 15%; final exam score 60%. The last two weeks will be allowed to do curriculum design. Give the students an assignment related to enterprise LAN internet, and ask them to propose a network construction plan with learned knowledge, considering network system security and reliability. For the students who have finished open innovative experiment or hardware networking experiment, their experiment performance proportion can be increased to 50%, and final exam score 50%. This assessment will encourage the students' learning passion, and provide an opportunity for excellent students to show their innovative ability.

## 5    Conclusions

Rearrange the course teaching content organization and arrangement in accordance with the teaching goal of Computer Network Base. Use the teaching method that combines the multi-media assistant teaching and experiment teaching. And at last, formulate the flexible final assessment standard on the basis of teaching program. Experiment teaching can effectively reinforce the students understanding on network technology and improve their practical ability and problem-finding and problem-solving ability in practical setting. For several years' experience of teaching Computer Network Base in computer major, while passing on the theoretical knowledge, I also give the students proper guidance in project practice and the good results are gained.

# References

1. Zhu, T., Gao, J.: Technology and Programming of Digital-controlled Processing course teaching reform on the basis of interactive teaching in network environment. China After School Education (August 2010)
2. Zhao, J., Chen, M.: Exploration on computer network practical teaching reform. Science & Technology Information (March 2010)
3. Huang, Y.: Network technology lab construction and opening experimental teaching reform. Journal of Liaoning Institute of Science and Technology (April 2008)
4. He, J., Wu, L.: Teaching reform and practice on high vocational colleges "Computer Network Base" course. Fujian Computer (August 2009)

# Economic Analysis and Realization of the Information Sharing in Supply Chain Production System

Wei Zhang[1] and HuiFang Li[2]

[1] Department of Management Engineering,
Suzhou Vocational University, Suzhou, China
[2] Department of Computer Engineering,
Suzhou Vocational University, Suzhou, China
{zw,lhf}@jssvc.edu.cn

**Abstract.** From the angle of supply chain production system, this article has analyzed the economic benefit and game of information sharing among members in it as well as the realization of information sharing inside. We need to establish enterprise strategic alliances from organizational means and establish centralized information system from technological means.

**Keywords:** Supply chain production system, Information sharing, Strategic alliances, Information system.

## 1 Introduction

Supply chain refers to the network of supply and demand which is consisted of raw materials suppliers, manufacturers, wholesalers and final consumers involved in the course of production and circulation of products, namely from the processing and manufacturing of raw materials to the finished products in customers' hands. It is also the network composed of enterprises and their departments involved in the entire process. According to the system theory, supply chain logistics management makes the upstream and downstream firms in the supply chain as a whole, through mutual cooperation and information sharing, which could realize reasonable disposition of resources throughout the supply chain as well as enhance the fast-reacting ability of logistics management.

From the angle of the supply chain operation, the supply chain is consisted of four subsystems: production system, consumption system, social system and environmental system. And the production system is consisted of suppliers, manufacturers, distributors and retailers; consumption system is consisted of consumers; Social system is consisted of social factors of supply chain operation; Environmental system is consisted of resources and living beings and so on. As shown in figure1.

Due to three obstacles in the supply chain operation, namely externalities, asymmetric information and interestedness among the members in supply chain, which leads to its low efficiency accompanied with the difficult realization of the supply chain operation objectives. In order to improve each enterprise's competitive ability and economic benefit in the supply chain, it is necessary to realize information sharing in the entire supply chain.

G. Lee (Ed.): Advances in Intelligent Systems, AISC 138, pp. 245–250.
springerlink.com          © Springer-Verlag Berlin Heidelberg 2012

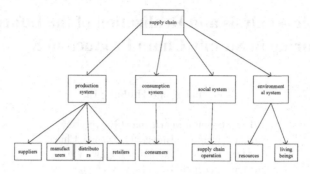

**Fig. 1.** Structure of supply chain

From the angle of supply chain production system, this article has analyzed the economic benefit of information sharing among members in it as well as expounded the realization method of information sharing.

## 2    Type Style and Fonts the Game Analysis of Information Sharing among Members Inside the Production System

Assuming there are two behavioral agents in supply chain production system, namely a manufacturer and a supplier. The supplier decides price of the intermediate products, and the manufacturer decides the price of the final products. In order to facilitate the deduction and assume that a unit of the final products needs a unit of intermediate products, we suppose the inverse demand curve that the production system would be confronted as follow:

$$p2 = a - bq \qquad (a > 0, b > 0)$$

Here    $p_2$ is the price of the final products provided the manufacturer

q is the quantity of products consumed by consumers

Then suppose the supplier's unit production cost as c 1, and the manufacturer's production cost as c2. x   i is the technical information for reducing the cost.

So the supplier's unit production cost is: c1 – c 0 1 = x1

And the manufacturer's production cost is: c 2 =c 0 2 +p1 – x 2

here : x i = x 0 i +Δx j

p 1 is price of the intermediate products provided by the suppliers ;

x 0 i is the technical information owned by the enterprise itself ;

Δx j is the technical information that enterprise i gets from enterprise j ;

According to the above assumptions, now discuss the profits of the two in the situations of information sharing and not sharing.

### 2.1   Not Share the Information

The game process is as follows:

**First stage:** the manufacturer decides the market price and the corresponding production of the final products. Manufacturer's goal is to seek profit maximization, max($\pi2$)

q1, p2

here: $\pi2 = p2\, q - c2\, q$

$$\frac{d\pi_2}{dq} = 0$$

Get the optimal solution:

$$q^*_{1N} = \frac{a + X_{02} - C_{02} - p_1}{2b}$$

$$p^*_{2N} = \frac{a + C_{02} + p_1 - X_{02}}{2}$$

Second stage: the supplier's decision-making. As far as the supplier is concerned, the goal is to seek profit maximization.

namely max $(\pi1)_o$

Because of the previous hypothesis that the entire supply chain has the same quantity in market demands and the final products. A final product requires an intermediate product, so as far as the supplier is concerned, its variable of decision-making is the price of the intermediate product, and satisfies

$$\frac{d\pi_1}{dp_1} = 0$$

The optimal solution can be obtained:

$$p^*_{1N} = \frac{a + X_{02} + C_{01} - C_{02} - X_{01}}{2}$$

According to the above results, the profit level of both the manufacturer and supplier can be obtained.

$$\pi^*_{2N} = \frac{(a - C_{01} - C_{02} + X_{01} - X_{02})^2}{16b}$$

$$\pi^*_{1N} = \frac{(a - C_{01} - C_{02} + X_{01} - X_{02})^2}{8b}$$

## 2.2   Information Sharing

When the two behaviors all share information, suppose each behavior agent gave the same information to the other through agreement

namely:        $\Delta x_i = \Delta x_j = \Delta x$

$\Delta x \in [\, 0, \quad \min(x_i, x_j)\,]$

According to the above solving process we can get:

$$q^{\ast}_y = \frac{a - (c_{02} - x_{02} + p_1 - \Delta x)}{2b}$$

$$p^{\ast}_{1y} = \frac{a + x_{02} + c_{01} - c_{02} - x_{01}}{2}$$

$$p^{\ast}_{2y} = \frac{a + (c_{02} - x_{02} + p_1 - \Delta x)}{2}$$

Further we can work out the supplier and manufacturer's profit level when sharing information

$$\pi^{\ast}_{1y} = \frac{(a - c_{01} - c_{02} + x_{01} + x_{02} + 2\Delta x)^2}{8b}$$

$$\pi^{\ast}_{2y} = \frac{(a - c_{01} - c_{02} + x_{01} + x_{02} + 2\Delta x)^2}{16b}$$

The manufacturer and supplier's difference in profit level before and after information sharing

$$\Delta \pi_1 = \pi^{\ast}_{1y} - \pi^{\ast}_{1N} = \frac{\Delta x^2}{2b}$$

$$\Delta \pi_2 = \pi^{\ast}_{2y} - \pi^{\ast}_{2N} = \frac{\Delta x^2}{4b}$$

Therefore we can get the following conclusion: the manufacturer and supplier can improve their profit level together when share information with each other.

# 3   Realization of Information Sharing within the Supply Chain Production System

Information sharing can improve the profit level as well as promote the members to improve production activities and technology level in the whole supply chain.

However, because of the members' self-interest in the whole supply chain, every enterprise attempts to increase its own profits through giving less information to the other enterprise. The result is that profits of each enterprise become lower and the efficiency decreases in the whole supply chain. To achieve information sharing within the production system, we should start from two aspects. On the one hand, we could establish the enterprise strategic alliance, promoting the information sharing among enterprises by organizational means, and managing the information, storage and logistics from the angle of large system; On the other hand, we could establish a centralized information system in the supply chain.

As for many problems involved in setting up the business enterprise strategy alliance, there have been already many studies and discussions, so here I would only talk about the jobs should be done in terms of three aspects:

### 3.1 Trust in the Enterprise Strategic Alliance

Taking the reliability prediction as standard, emphasizing the concept of honest, reliability, and credit among partners, believing in each other, caring for the overall interests including their own interests in the entire supply chain, we say trust is the foundation of cooperation. Cooperation makes the partners all have their benefit. Distrust naturally will increase the cost of each other, even bring in harm. The prisoner's dilemma model of game theory has explained this point very well.

### 3.2 Reasonable Share of Information Sharing Cost and Reasonable Distribution of the Extra Profits

It is well known that the "bullwhip effect" exits in supply chain. Just because we failed to forecast the terminal clients' effective demand accurately enough, the safety stock in the supply chain is amplified. Another reason is that we are not clear enough about the existent cargo information. The advanced prediction technology and information technology are important solving methods which however will cost a lot. Therefore it is necessary to share the cost with all partners in the supply chain scientifically and reasonably; meanwhile, information sharing is generally launched by the core enterprise or enterprises that will get many profits after this action, so it certainly will cause the unequal distribution of profits. Supply chain emphasizes the rapid response to demand. Therefore, information mainly comes from the downstream enterprises, while the increase of profits mainly embodies in upstream enterprises. Enterprises all take self-interest maximization as their ultimate goal. If the overall increase of extra profits failed to be allocated reasonably among members in the supply chain, which inevitably will lead to the resistance of some enterprises, even destroy the cooperative relations of supply chain enterprises.

### 3.3 Information Sharing and the Enterprise's Own Commercial Secrets

Some information and original data of the enterprise are considered to be highly confidential business secrets, even inside the enterprise people are controlled when access to, let alone disclose to anybody outside the enterprise. Every enterprise worries its partners about abusing information and possessing the extra profits, therefore, it is necessary for the enterprise to keep its own costs, output and purchase price and other information as secrets to keep the information superiority. In addition, excessive information sharing might leak the enterprise's commercial secrets which may cause great loss of the enterprise. Supply chain is a dynamic organization structure, and environmental change could undermine the partnership between enterprises. Once the supply chain is disintegrated, enterprises exposed more information may lose their competition superiority in the fierce competition in the market later. Therefore, it is necessary to correctly handle the relationship between information sharing and protecting the enterprise's own commercial secrets in the supply chain.

To realize the information sharing among each behavior agent in the supply chain, we have special requirements to the information management technology which requires information management system of the supply chain to meet the following four points;

(1) Interoperability: Because members of the supply chain include manufacturers, suppliers, vendors, transporters and consumers, the equipments used by each behavior agent differs. And the differences in information platform make the behavior agents fail to share information in the supply chain. So when choosing computer software and hardware platform, the behavior agents must choose the platforms with good mutual operation.

(2) Safety: In the supply chain operation process, because of changes in external environment, consumer's preferences, the reliable technology and manufacture, today's partners may become competitors tomorrow. This requires members to share private information which may turn into the dynamic change. Then it is necessary to adopt appropriate safety measures to avoid affecting the enterprise information exchange or leaking each other's commercial secrets.

(3) Dynamic allocation: As for the supply chain operation process, it will change with the changes of external environment, workmanship, manufacturing process and the consumer preferences. Then as for the management of information sharing, it should also satisfy the requirement of being changed or adjusted at any time to support the dynamic allocation capability.

(4) Standardization technology: The construction of information system involves lots of standards, such as STEP(standard for the exchange of product model data), CORBA(common object request broker architecture) ,VRML(virtual reality modeling language) ,TCP/IP and so on. These standards closely related to the express and exchange of product information in different fields. So we must solve problems in terms of mapping and coordination among the standards to make sure information of different areas in the supply chain could be satisfied.

# References

1. Huang, X.: Supply chain operation - coordination, optimization and control. Science press (2007)
2. Sun, Y.: Principle of supply chain management. Shanghai university of finance and economics press (2004)
3. Tan, L.: Commodity Science. Science press (2008)
4. Peng, Z.: Modern logistics and supply chain management theory. Publishing House of Shandong university (2004)
5. Wang, N.: Green supply chain management. Tsinghua university press (2005)

# Experiment of Soil Thermal Properties Based on Thermal Probe Method

Yuanyuan Gao, Jun Zhao, Jiuchen Ma, and Xinyang Song

Department of Thermal Engineering,
Tianjin University, TJU
Tianjin, China
feifeilong520@163.com,
innermongolianma@yahoo.com

**Abstract.** The design of the ground source heat pump system (GSHPS) is the key point of the whole system application, whereas the correct measurement of the thermo- physical properties of the soils affects the design process in a significant way. Therefore, an experiment on the thermo- physical properties of soils using the modified thermo-probe method was carried out in order to improve the precision of result of the in-situ thermal response test. With the equations of Kelvin line source model, the experiment data was processed and the soil thermal conductivity and diffusivity were obtained at the same time. The change law of the thermo-physical properties of the soils was derived based on the analyses of the experiment result. This convenient method plays an active role in the reasonable design of GSHP, mainly manifesting in the in-situ thermal response test.

**Keywords:** ground source heat pump system (GSHPS), soil thermal properties, thermal probe, in-situ thermal response test, soil physical properties.

## 1   Introduction

In recent years, ground-coupled heat pump technology has been a new energy technology with much worldwide attention, which has been identified as one of the best energy-saving technologies for space heating and cooling in residential and commercial buildings. Its essence is a recycling operation mechanism of cross-seasonal energy storage and releasing.

As the critical step in design of GHPS, to make accurate measurement of soil thermal properties directly influences the calculation of the heat exchanger area, energy balance in ground surface, the energy storage in soil, temperature distribution characteristics and other basic parameters, etc. Therefore, the accurate measurement of soil thermal properties is an important guarantee to develop its superiorities in energy-saving, environmental protection and economic. However, it should be noticed that conventional thermal probe method [1] based on five assumptions has principle errors. This paper is concerned with the high accuracy experiment on the basis of above method.

G. Lee (Ed.): Advances in Intelligent Systems, AISC 138, pp. 251–261.
springerlink.com          © Springer-Verlag Berlin Heidelberg 2012

## 2    Experimental Apparatus and Method

Based on the normal thermo-probe experiments, this paper proposes an improvement of the experimental apparatus, so that the thermo-physical properties of typical soil are measured more accurately. The structure of normal thermal probe was simplified. The thermocouples were moved from the thermo-probe tube to the soil in order to directly measure oil own temperature. In addition, the conventional single temperature measuring point was changed into a number of temperature measurement points burying in different positions, which can measure soil temperature in different positions in the process of heating soil by thermal probe.

The experiment built a small cylindrical tube-like laboratory table. 15 group experiment soil samples of shallow soil were collected from Tianjin. Each soil sample was closely filled in the cylindrical sleeve in sequence. In a standard circular section at the upper height of the sleeve, the thermocouples were symmetrically distributed centering on the circle center, as shown in Figure 1. The thermocouples were fixed to the rigid plastic panel to ensure that the positions of measuring points were not undermined when filling soil. The thermal probe was inserted into axis position of the cylindrical soil whose heated part was completely covered by experimental soil samples.

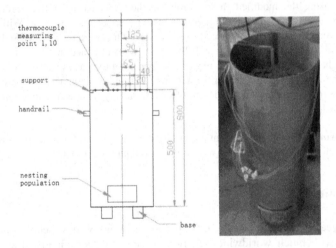

**Fig. 1.** The plan of main structural and physical map of experimental device

This study on improving the normal thermo-probe method is positive. The way of burying the thermocouple is optimized to obtain the parameter of soil conductivity and diffusivity at the same time. The temperature changes of different position in the soil were monitored all the time to control the start and stop of the experiment. Make sure that the experiment simulates the infinite boundary condition to satisfy the hypothesis, in order to get the more accurate results. Therefore, the errors caused by the nonuniform initial temperature and nonuniform thermal properties can be avoided. Compared with the conventional method, there is a certain distance between thermocouples and the probe so that the contact resistance is reduced and the impact of ignoring internal resistance on the experimental is lowered.

# 3   The Process and Phenomenon of the Experiment

Take sand 1 for example. Open data acquisition system, and read out initial temperature recorded at soil points. After the initial soil temperature to be stable, turn on the power and regulate the voltage regulator to set thermal probe's heating power 100W. The temperature collector began to record the temperature of measuring points. When the temperature of the edge measuring point 1 and point 10 marked warming, stop the probe heating, adjust voltage regulator to zero and cut off the power. As there was still some distance between the measuring point of the outer edge and the outer edge of the soil, it was the critical moment that heat transfer can be considered not to be affected by the environmental. In that moment stop heating to satisfy the hypothesis of soil infinite boundary condition and a set of soil's thermo-physical properties experiment had been completed. As the difference of the symmetrical thermocouple temperatures on both sides of the thermal probe was small, the average of symmetrical thermocouple readings on both sides was equal to the temperature of the column soil at equal radius. Figure 2 is the radial equivalent temperature of sand one over time. The first half is soil temperature rise curve when to heat thermal probe and the second half is the curve of temperature recovery when to stop heating.

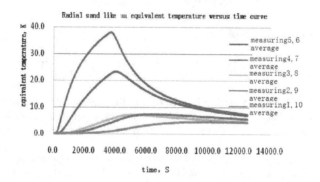

**Fig. 2.** The curve of the radial equivalent temperature of sand one over time

From Figure 2, the nearest location 20mm from thermal probe, namely measuring point 5 and point 6, is the most sensitive in the thermal response test, where the temperature begin to rise after having been heated for 60s, the temperature rising rate over time is the fastest, and its temperature rise reaches the maximum 37.6K when to stop heating. In the location 40mm from the thermal probe, namely measuring point 4 and point 7, the thermal response time of heating is turned about 300s after having been heated, where the temperature rising rate over time becomes slower, and the maximum temperature rise decreases. The locations 65mm and 90mm from the thermal probe, respectively start heating up after a longer time where the rates of temperature rising are relatively lower. In the location 125mm from the thermal probe, namely measuring point 1 and point 10, the temperature does not change significantly. The main purpose of monitoring this temperature is to control heating time of thermal probe.

The other experiments of 14 types of soil are similar to sand one and the temperature trend is also consistent with the sand one. The same law of temperature variation can be obtained.

## 4    Data Analysis

Suppose the initial temperature and structure of the experimental soil even are uniform, and ignore contact thermal resistance between experiment soil and thermal probe. Under these hypotheses, the mathematical description of the experimental principle is:

$$\Delta T(r,t) = T(r,t) - T_0 = \frac{Q}{4\pi\lambda} \int_{\frac{-r^2}{4at}}^{\infty} \frac{e^{-y}}{y} dy \tag{1}$$

As previously mentioned, when the time "t" is large and the radius "r" is small, the exponential integral can be expressed by approximate formula, that is:

$$\Delta T(r,t) \approx \frac{Q}{4\pi\lambda}[\ln(\frac{4at}{r^2}) - c] \tag{2}$$

Restructure the formula 2 to obtain:

$$\Delta T(r,t) = \frac{Q}{4\pi\lambda}[\ln(\frac{4a}{r^2}) - c] + \frac{Q}{4\pi\lambda}\ln t \tag{3}$$

Compared with Mogensen derived formula[2], the above formula reduces various thermal resistances between the fluid within heat exchanger and the wall in the actual GSHP, and therefore simplifies the calculation so that the final derived formula is simple.

In the experiments, the heating power of thermal probe is controlled constantly, and thus the experimental theoretical model, equation 3 has only one variable ln (t). Therefore, according to Mogensen's thinking[2], the equation is simplified to the linear equation using ln (t) as the independent variable x and $\Delta$ T (r, t) as the dependent variable y:

$$y = mx + b \tag{4}$$

Where:

$$b = \frac{Q}{4\pi\lambda}[\ln(\frac{4a}{r^2}) - C] \tag{5}$$

Thermal conductivity and diffusivity are as follows:

$$\lambda = \frac{Q}{4\pi m} \tag{6}$$

$$a = \frac{r^2}{4}\exp(\frac{b}{m} + C) \tag{7}$$

Through the analysis of the time-varying temperature data during the experiment, linear curves of temperature rise on different locations to the logarithm of time can be fitted and the fitting formula has been gotten with Formula Fitting method including the slope "m" and intercept "b". Then according to formula (5) and (6) thermal conductivity "λ" and thermal diffusivity "a" can be solved.

This article, sand one as an example, explains the specific steps of fitting method. Ignore the first 10 minutes experimental data to fit the requirements of using approximate formula in the theoretical model. This paper mainly analyzes the temperature change in the location of radius 20mm, and then makes a linear curve of the average temperature rise to the logarithm of time on measuring point 5 and point 6 shown in Figure 3. According to the fitting formula with the linear coefficient of 0.997, the slope m equals 16.89 and the intercept b equals −100, so that we can calculate the thermal conductivity and diffusivity of sand one. Similarly, the thermo-physical parameters of the remaining 14 experimental soil samples can be calculated.

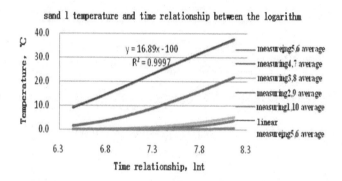

**Fig. 3.** The formula curve fitting with experimental data of sand one

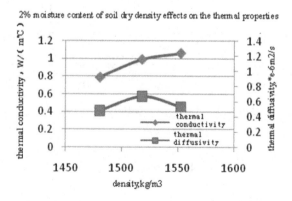

**Fig. 4.** Dry density's effect on soil thermal properties

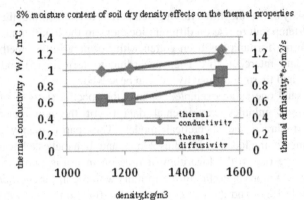

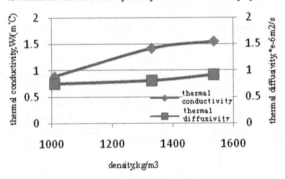

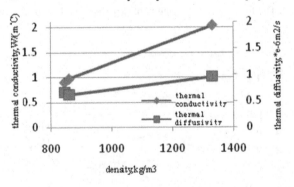

**Fig. 4.** (*continued*)

# 5    Soil Physical Parameters' Effect on the Thermo-Physical Properties

A few samples were presented from each group test soil, cutting Ring method to measure the density and drying method to measure the moisture content. According to the relationship between moisture content and dry and wet density, the dry density can be calculated, so as to obtain the physical parameter values of various soil samples.

## 5.1    Soil Dry Density's Effect on the Thermo-Physical Properties

According to the moisture content, the test soil samples were divided into four groups of 2%, 8%, 13% and 20% to analysis the effect of soil dry density on soil thermo-physical, shown in figure 4.

From the four curves, it can be intuitively seen that the soil thermal conductivity increases with dry density values. When the moisture content is 20%, with the increase of the dry density the soil thermal conductivity adds more largely, whose growth trend is very steep, mainly because of larger span of its dry density value. However, when the moisture content is 2%, 8% or 13%, the increase in thermal conductivity of soil gradually slows down with increasing of the dry density, which is in agreement with the theoretical foundation. When the density value increases, the soil porosity decreases with denser soil particles and thermal performance enhancement, and when the density continues to increase, the ability of soil thermal conductivity reaches the upper limit so that the gradual growth trend becomes flatter.

Among the four curves, except for the curves of the soil with 2% moisture content whose thermal conductivity tends to decrease with the augmentation of dry density, the thermal conductivity of the soil samples with 8%, 13% and 20% moisture content adds with the increase of dry density. As a result, when the moisture content is small, thermal conductivity fluctuates unstably with the dry density, but the thermal conductivity takes on an increasing trend with the dry density.

## 5.2    The Moisture Content's Effect on Soil Thermo-Physical Properties

According to the experimental results of test soil samples with similar dry density, the effect of the moisture content on soil thermal conductivity and diffusivity can be analyzed. The following figure 5 includes the experimental data image of the sand-like series with dry density about 1530 kg/m$^3$ and mixed sand-earth series with dry density about 1329 kg/m$^3$, where the curves show the effect of different moisture content on soil thermo-physical properties.

From the figure 5, in the two comparative experiments, the thermal conductivity increases steadily with the moisture content increasing. The thermal diffusivity of soil also keeps an increasing trend with the moisture content, but its growth rate is not very regular. The figure shows the thermal diffusivity becomes significantly large with the increase of the moisture content when the moisture content is in the range of 0-8%, declines slightly when the moisture content 8-13%, and then increases lightly afterward.

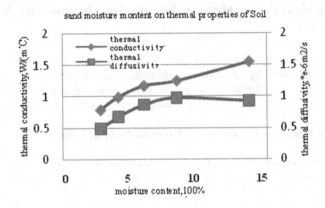

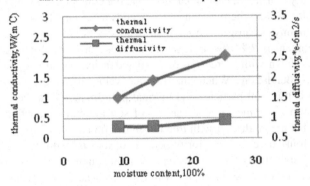

**Fig. 5.** The effect of the Moisture effects on thermo-physical properties of soil

## 5.3    The Combined Effects of the Physical Parameters of Soil on Thermo-Physical Properties

According to the physical and thermo-physical parameters obtained by experiments, the effect of the moisture content and dry density on soil thermo-physical properties is comprehensive analyzed. Cloud images of the moisture content and dry density to the thermal conductivity and diffusivity are drew respectively in figure 6. The increase of thermal conductivity with the moisture content can be intuitively seen. Similarly, the thermal conductivity also shows a rising trend. The results are consistent with the single physical parameter's effect on the thermal conductivity. At the same time, the results shows that it has little effect on the conductivity when the value of the moisture content and the density is small, and as the value of the two parameters grows, the effect becomes significant.

According to the result, the basic form of fitting curves can be set:

$$\lambda = (A_1 - B_1 \times C_1^{\omega})(A_2 - B_2 \times C_2^{\rho}) \tag{8}$$

Based on the high fit precision, by adjusting the various coefficients, the formulas of the thermal conductivity to the moisture content and dry density can be obtained, as follows:

$$\lambda = -1.04192 + 2.05079 \cdot 1.025^{\omega} + 3.9647 \cdot 0.99815^{\rho}$$
$$-6.73781 \cdot 1.025^{\omega} \cdot 0.99815^{\rho}$$ (9)
$$R^2 = 0.94556$$

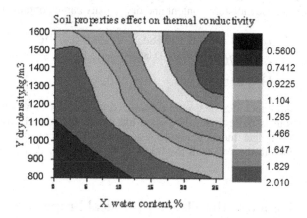

**Fig. 6.** Cloud image of the combined effects of soil physical parameters on the thermal conductivity

From the figure 7, the increase of the temperature diffusivity with the moisture content increasing can be intuitively seen. Similarly, as the dry density increases, it also shows a rising trend. Unlike the thermal conductivity, its change is not stable, fluctuating within a certain range. The results are also consistent with the single physical parameter's effect on it. The effect of the moisture content is the leading

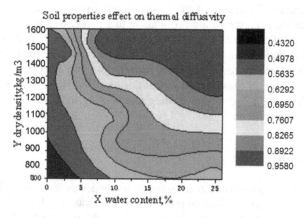

**Fig. 7.** Cloud image of the combined effect of soil physical parameters on the temperature diffusivity

factor when the value of the moisture content is under 8%, and the dry density has very limited impact on it. As the value of the moisture content grows, the dry density has a more important effect, while moisture content is gradually weakening.

According to the result, the basic form of fitting curves can be set:

$$\alpha = (A_1 - B_1 e^{C_1 \omega})(A_2 - B_2 e^{C_2 \rho}) \tag{10}$$

Based on the high fit precision, by adjusting the various coefficients, the formula of the thermal diffusivity to the moisture content and dry density can be obtained, as follows:

$$\alpha = 0.4268 - 0.15718 e^{-0.15767\omega} + 0.06701 e^{0.001487\rho}$$
$$-0.05768 e^{0.001487\rho - 0.1567\omega} \tag{11}$$
$$R^2 = 0.92202$$

The soil heat transfer includes heat convection except for heat conduction when the soil moisture content is more than 50%. Then the theoretical model, just considering heat conduction, no longer meets computing accuracy. Therefore, the conclusions of this study only apply in the soil with the 800 ~ 2000 kg/m$^3$ dry density and 0 ~ 40% moisture content.

# 6    The Application of the Thermo-Physical Properties' Variation Rules

According to the foregoing analysis, the thermo-physical properties of the soil with known moisture content and dry density can be initially determined. This convenient method of quickly getting soil thermo-physical properties plays an active role in the reasonable design of GSHP, mainly manifesting in the in-situ thermal response test.

First of all, based on preliminary soil thermo-physical parameters, heat exchange amount in per unit depth of well can be estimated, thus determining the heating power value of a borehole so as to guarantee a reasonable in-situ thermal response test to be carried out. Second, the results of the in-situ thermal response test can be compared with these obtained from the fitting formulas in this paper, which can be as the design basis in the range of permitted error. If the deviation is greater, there may be a big error in the in-situ thermal response test, and then a second test is needed. In this way, the unreasonable operations, test equipment failure, the big miscalculation and other serious error, which can cause the irrational design of GSHPS, can be found so as to avoid the inadequate or overloaded heating and cooling load and the waste problems during run of actual systems.

# 7    Conclusion

1) An innovation to the normal thermo-probe method was made in order to improve the experimental precision. Thermal probe's structure was simplified. The temperature measurements were arranged directly in the soil. By optimizing the layout of the thermocouples, soil thermal conductivity and temperature diffusivity can be measured at the same time.

2) Using the fitting formula derived by Kelvin line heat source model, the experimental data can be analyzed in order to get more reliable thermal conductivity and diffusivity.

3) In this paper, the relations of the thermo-physical properties of the soils with the easy-to-measure parameters of the density and the moisture content were concluded and the equations to calculate the thermo-physical properties were obtained which only apply in the soil with the 800 ~ 2000 kg/m$^3$ dry density and  0 ~ 40% moisture content. This work can provide a guide for proposing theoretical relationship between the thermal properties and the physical properties.

4) The inner mechanism of the thermal properties with moisture content and density should be further researched so as to obtain the theoretical relationship.

**Acknowledgment.** The work is supported by National Major Project of Scientific and Technical Supporting Programs of China during the 11th Five-Year Plan Period (Grant NO.2006BAJ03A06) and The Comprehensive Research and Application of Thermal Energy Storage Technology Using Superficial Underground Saline Water in Binhai New Area in Tianjin Project (Grant NO.08ZCGYSF02400).

# References

1. Choudhary, A.: A approach to determine the thermal conductivity and diffusivity of a rock in situ. Ph.D. dissertation. Oklahoma State University. (1976)
2. Mogensen, P.: Fluid to duct wall heat transfer in duct system heat storages. In: Proceedings of the International Conference on Subsurface Heat Storage in Theory and Practice, Stockholm, Sweden (June 6-8, 1983)
3. Gehlin, S.: Thermal response test method development and evaluation, doctoral thesis, Department of Environment Engineering Division of Water Resources Engineering, LULEA university of technology (2002)
4. Mattsson, N., Steinmann, G., Laloui, L.: Advanced compact device for the in situ determination of geothermal characteristics of soils. Energy and Building 40, 1344–1352 (2008)
5. Georgiev, A., Busso, A., Roth, P.: Shallow borehole heat exchanger: Response test and charging–discharging test with solar collectors. Renewable Energy 31, 971–985 (2006)
6. Zhao, J., Duan, Z., Song, Z.: A method for in situ determining underground thermal properties based on the cylindrical heat source model. Acta Energiae Solaris Sinica 09 (2006)
7. Yu, M., Peng, X., Fang, Z.: A Simplified Method for In-situ Measurement of the Thermal Conductivity of Deep-layer Rock Soil. Journal of Engineering for Thermal Energy and Power 05 (2003)

# The Precise Design of Bandpass Filter Miniaturization Using Microstrip SIR

Luo Hui[1], Li Weiping[1], and Hu Rong[2]

[1] EastChina JiaoTong University,
Nanchang, China
[2] Jiangxi Industry Polytechnic College,
Nanchang, China
{lh_jxnc,xuefeixu}@163.com,
liweiping@ecjtu.jx.cn

**Abstract.** Coupled resonator circuits are of importance for design of RF/microwave filters, in particular the narrow-band bandpass filters that play a significant role in many applications.

Microstrip stepped impedance resonator with inner coupling are used to realize miniaturization. The coupling coefficient k and external quality factor Q of the filter are obtained by full-wave analysis. According to the extracted curves, the geometrical dimension of the filter is figured. At last, a 1.52 GHZ microstrip band-pass filter is designed and the transmission characteristic is measured by experiment to validate the design.

**Keywords:** SIR, band-pass filter, coupling coefficient, external quality factor.

## 1   Introduction

Microstrip band-pass filter is one of the main segments of communication system. When the frequency is low, the size of the circuit that using harf-wavelengh resonators is large, thus not suitable for integration. But the is size of the circuit can be half decreased by using microstrip stepped-impedance-resonator. From the previous work, we can see a lot of the formal design didn't consider the input and output coupling and the influence of microstrip T-type connector and circuit corner. Practically the factors above have great influence to the capability of the filter. In this letter, we propose the precise design method for miniaturization, then make the design to execution just validating the correctness of our design method.

Microstrip line resonator is adopt above VHF frequency band, half wavelength stepped-impedance-resonator (SIR) shown in Fig. 1 is frequently used. By using SIR, sufficient freedoms on the structure and design will be obtained without decrease the unload quality facor.

For the half wavelength SIR, the parameters Yi and Zi are set to represent the input admittance and impedance respectively.

Without considering the edge discontinuity and fringe influence, the equation of $Z_i$ is as below:

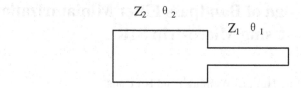

**Fig. 1.** Half wavelength SIR

$$z_i = jz_2 \frac{z_1 \tan \theta_1 + z_2 \tan \theta_2}{z_2 - z_1 \tan \theta_1 \tan \theta_2} \tag{1}$$

The resonance condition is obtained when Yi  is zero, which is as:

$$z_2 - z_1 \tan \theta_1 \tan \theta_2 = 0 \tag{2}$$

$$\tan \theta_1 \tan \theta_2 = \frac{Z_2}{Z_1} = R_z \tag{3}$$

From(1)~(3), it's easy to find that the resonance condition is determined by electrical length $\theta_1$  $\theta_2$ and the impedance ratio $R_z$ .  The resonance condition of uniform impedance resonator (UIR) is only determined by the length of transmission line, while to SIR we must consider both length and impedance ratio. Hence, SIR has one more design freedom to UIR by comparison.

$\theta_T$  is set as the whole electrical length of SIR[2]:

$$\theta_T = \theta_1 + \theta_2 = \theta_1 + \arctan \frac{R_z}{\tan \theta_1} \tag{4}$$

The curve of (4) is shown in Fig. 2, the figure express that the electrical length is max. when $RZ \geq 1$ while min. when RZ<1.

$0 < \theta_T < \dfrac{\pi}{2}$  when   $0 < R_Z < 1$, it represents that the dimension of SIR is smaller than UIR, this is why we use SIR to realize miniaturization.

## 2    Design Theory

The equations above are obtained without considering the higher mode of the microstrip line and discontinuity of step plane, only based on the TEM mode. So there is difference between the theory and experiment.

Consequently, the design in this article first extracts the curve that represent the relationship between the coupling coefficient, external quality factor and coupling gap, using full-wave analysis. Then get the number "n "of the resonators and the element value of the low-pass prototype. Finally the physical structure based on the curve and external quality factor curve is confirmed.

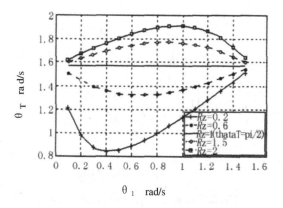

**Fig. 2.** Electrical length

The design of band-pass filter can be confirmed by coupling coefficient $k_{j,j+1}$ and external quality factor $Q_e$, $Q_{e1}$ ($Q_{e2}$) is used to represent the relationship between the filter and input circuit (output circuit), follow the equations below:

$$Q_{e1} = \frac{g_0 g_1}{FBW} \qquad (5)$$

$$Q_{e2} = \frac{g_n g_{n+1}}{FBW} \qquad (6)$$

The relationship between the resonators is determined by the coupling coefficient, $k_{j,j+1}$

$$k_{j,j+1} = \frac{FBW}{\sqrt{g_j g_{j+1}}} \qquad (7)$$

In the equations above, $g_j$ is used to represent the value of low-pass prototype; FBW is Fractional Band Width; $Q_{e1}$ and $Q_{e2}$ are external quality factors of input circuit and output circuit respectively, $k_{j,j+1}$ is the coupling coefficient between level j and level j+1.

### A.  The coupling between resonators

When two resonators with the same resonance frequency are coupled together, their resonance frequency will separate, as shown in Fig. 3.

f1 and f2 are introduced to represent the two separate resonate frequencies , the coupling coefficient can be defined as:

$$k = \frac{f_2^2 - f_1^2}{f_2^2 + f_1^2} \qquad (8)$$

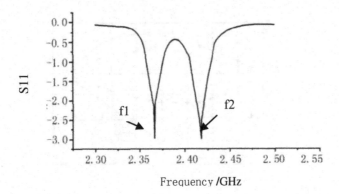

**Fig. 3.** Resonate frequencies separate apart

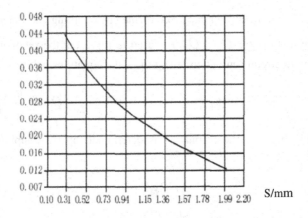

**Fig. 4.** Relationship curve of "k" and "s

$f_1$ and $f_2$ can be extracted from full-wave analysis. Fig.4 express the relationship curve which describe the relationship between coupling coefficient "$k_{j,j+1}$" and coupling gap "s". From the figure it can be found that coupling coefficient "k" is decreased by increasing the value of coupling gap "s".

## B.   *The Coupling between Resonators and Out circuit*

The external quality factor $Q_e$ is defined by equation:

$$Q_e = \frac{f_0}{BW} \tag{9}$$

in which BW represents the 3dB bandwidth.

Two typical input/output (I/O) coupling structures for coupled microstrip resonator filters, namely the tapped line and the coupled line structures, in this letter, the tapped line structure is adopt, just shown in Fig.5. For the tapped line coupling, usually a 50 ohm feed line is directly tapped onto the I/O resonator, and the coupling or the external quality factor is controlled by the tapping position t, the smaller the t, the closer is the tapped line to a virtual grounding of the resonator, results in a weaker coupling or a larger external quality factor.

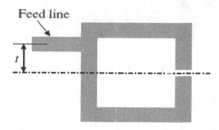

**Fig. 5.** Tapped-line coupling

The relationship between $Q_e$ and the tapping position "t"of input(output) feedline can be obtained by using Full-wave analysis, as shown in Fig. 6.

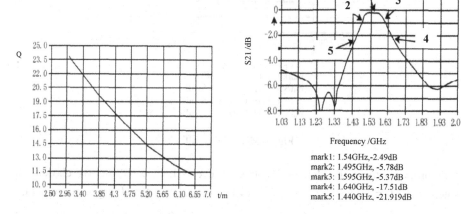

mark1: 1.54GHz,-2.49dB
mark2: 1.495GHz, -5.78dB
mark3: 1.595GHz, -5.37dB
mark4: 1.640GHz, -17.51dB
mark5: 1.440GHz, -21.919dB

**Fig. 6.** Relationship curve of "Q" and "t"          **Fig. 8.** Simulated result

## 3    Practicla Design and Result

The target is to design a band-pass filter the middle frequency of which is 1.52GHz; BW=100MHz; insertion loss $\leq$ 3dB; on the cut-off frequency point, the attenuation $\geq$ 20dB.

**Fig. 7.** Practical Design

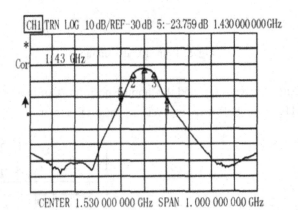

CH1 TRN LOG 10 dB/REF −30 dB 5: −23.759 dB 1.430 000 000 GHz

CENTER 1.530 000 000 GHz SPAN 1. 000 000 000 GHz

mark1: 1.530GHz,-2.529dB
mark2: 1.484GHz, -5.490dB
mark3: 1.574GHz, -5.651dB
mark4: 1.630GHz, -21.846dB
mark5: 1.430GHz, -20.816dB

**Fig. 9.** Measured result

In order to meet the demands above, 3-level Chebyshev filter is chosen, in which $Q_1 = Q_2 = 19.68$, $k_{1,2} = k_{2,3} = 0.0433$. From Fig.4 and Fig.5 the structure parameters can be obtained : $S_{1,2} = S_{2,3} = 0.3mm$, $t = 3.8mm$. The BPF is fabricated on the substrate of Rogers with dielectric constant $\varepsilon_r$ =3.38 and thickness h=0.813mm. The circuit structure is shown in Fig.7, the size of practical design is smaller than 30mm×25mm.

Finally, Fig.8 and Fig.9 are used to describe simulated result and measured result, the simulated result is done by EM simulation software Sonnet and the measured result is obtained using network analyzer Agilen8720ET. It's clear that they inosculate very well.

# 4   Conclusion

In this letter, the microstrip stepped impedance resonator with inner coupling is adopt to realize miniaturization, extract the relationship curve that describe the relationship between coupling coefficient k, external quality factor Q and physical structure with the help of full-wave analysis software. Finally, the theory is well validated by practical design.

The theory which is mentioned above is very helpful to synthesize waveguide filters, dielectric resonator filters, ceramic combline filters, microstrip filters, superconducting filters, and micromachined filters. In the future, more works will be done to discover novel and compact filters for realizing miniaturization.

# References

1. Hong, J.S., Lancaster, M.J.: Microstrip Filters for RF/MicrowaveApplications. Wiley, New York (2001)
2. Matthaei, G.L.: Interdigital band-pass filters. IRE Trans. Microwave Theory Tech., 479–491 (1962)
3. Makimoto, M., Yamashita, S.: A design method of band-pass filters using dielectric-filled coaxial resonators. IEEE Trans. Microwave Theory Tech. MTT-33, 152–157 (1985)
4. Advanced Design System (ADS_), Version 1.3. Agilent Technologies Inc., Palo Alto (1999)
5. Matthaei, G.L., Young, L., Jones, E.M.T.: Microwave filters, Impedance matching networks, and coupling structures. McGraw-Hill, New York (1964)
6. Wong, J.: Microstrip tapped-line filter design. IEEE Trans. Microwave Theory Tech. MTT-27, 44–50 (1979)
7. Dishal, M.: A simple design procedure for small percentage bandwidth round-rod interdigital filters. IEEE Trans. Microwave Theory Tech. MTT-13, 696–698 (1965)
8. Cameron, R.J.: General coupling matrix synthesis methods for Chebyshev filtering functions. IEEE Trans. MTT-47, 433–442 (1999)
9. Hong, J.-S.: Couplings of asynchronously tuned coupled microwave resonators. IEE Proc. Microwaves, Antennas and Propagation 147, 354–358 (2000)
10. Montgomery, C.G., Dicke, R.H., Purcell, E.M.: Principle of Microwave Circuits, ch. 4. McGraw-Hill, New York (1948)
11. EM User's Manual. Sonnet Software Inc., New York (1993)

# The Simulation, Optimization and Modeling of Mirostrip Filter Using ADS2008

Li Weiping, Luo Hui, and Guan Xuehui

EastChina JiaoTong University, Nanchang, China
lwp8277@126.com, lh_jxnc@163.com, xh_guan@hotmail.com

**Abstract.** Microstrip filter is used widely in wireless communication area, especially in WLAN. Compared to the conventional manufacturing techniques using lumped element, we can get extremely better characters in high frequency wave band using microstrip lines. ADS2008 (Agilent Advanced Design System 2008) is an excellent software platform for high frequency and high speed EDA manufacturing. In this letter, the modeling of microstrip filter which works at appointed frequency is got by transforming the lumped elements to microstrip elements. Finally, the microstrip filter simulated, optimized by ADS2008.

**Keywords:** ADS, microtrip filter, lumped elements, Kuroda rule.

## 1   Introduction

Filters play important roles in many RF/microwave applications. They are used to separate or combine different frequencies. The electromagnetic spectrum is limited and has to be shared; filters are used to select or confine the RF/microwave signals within assigned spectral limits. Emerging applications such as wireless communications continue to challenge RF/microwave filters with ever more stringent requirements—higher performance, smaller size, lighter weight, and lower cost. Depending on the requirements and specifications, RF/microwave filters may be designed as lumped element or distributed element circuits; they may be realized in various structures, such as lumped elements and microstrip.

Filters which are realized by lumped elements are fit for the frequency band that is lower than 500MHz. Their advantages are small volume, easy installment, no autoecious passband and flexible design. But they are not fit for higher frequency band especially when the bandwidth is narrow because the working wavelength is very close to their physical dimension and brings large insertion loss.

Microstrip filters are used widely in the millimeter range such as for wireless communication because of their distributed parameter structure. With the advantages of smaller volume, light weight, broad band and easy matching, microstrip filters are applied frequently in the design of microwave circuit system. In this letter, the design method of microstrip bandpass filter is demonstrated for example.

G. Lee (Ed.): Advances in Intelligent Systems, AISC 138, pp. 271–277.
springerlink.com          © Springer-Verlag Berlin Heidelberg 2012

## 2    Basic Theories of the Filter Design

### 2.1    The Low Pass Prototype Filter

Filter syntheses for realizing the transfer functions, usually result in the so-called lowpass prototype filters.

A lowpass prototype filter is in general defined as the lowpass filter whose element values are normalized to make the source resistance or conductance equal to one,denoted by g0 = 1, and the cutoff angular frequency to be unity, denoted by $\Omega_C = 1$ (rad/s). Fig.1 demonstrates two possible forms of an n-pole lowpass prototype for realizing an all-pole filter response, including Butterworth, Chebyshev, and Gaussian responses. It should be noted that in Fig.1, gi  for i = 1 to n represent either the inductance of a series inductor or the capacitance of a shunt capacitor; therefore, n is also the number of reactive elements. This type of lowpass filter can serve as a prototype for designing many practical filters with frequency and element transformations.

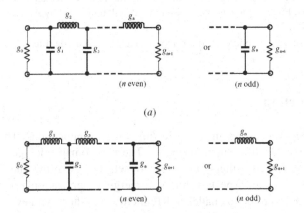

**Fig. 1.** Low pass prototype filters with a ladder network structure

### 2.2    Bandpass Transformation

Assume that a lowpass prototype response is to be transformed to a bandpass response having a passband  $\omega_2 - \omega_1$ , where $\omega_1$ and $\omega_2$ indicate the passband-edge angular frequency. The required frequency transformation is:

$$\Omega = \frac{\Omega_c}{FBW}\left(\frac{\omega}{\omega_0} - \frac{\omega_0}{\omega}\right) \tag{1}$$

with

$$FBW = \frac{\omega_2 - \omega_1}{\omega_0}$$

$$\omega_0 = \sqrt{\omega_1\omega_2}$$

(2)

where $\omega_0$ denotes the center angular frequency and FBW is defined as the fractional bandwidth. If we apply this frequency transformation to a reactive element g of the lowpass prototype, we have

$$j\Omega_g \rightarrow j\omega\frac{\Omega_c g}{FBW\omega_0} + \frac{1}{j\omega}\frac{\Omega_c \omega_0 g}{FBW}$$

which implies that an inductive/capacitive element g in the lowpass prototype will transform to a series/parallel LC resonant circuit in the bandpass filter. The elements for the series LC resonator in the bandpass filter are:

$$L_s = (\frac{\Omega_c g}{FBW\omega_0})\gamma_0 g$$

$$C_s = (\frac{FBW}{\omega_0\Omega_c})\frac{1}{\gamma_0 g}$$

(3)

for g representing the inductance.

The impedance scaling has been taken into account as well. Similarly, the elements for the parallel LC resonator in the bandpass filter are:

$$C_p = (\frac{\Omega_c g}{FBW\omega_0})\frac{g}{\gamma_0}$$

$$L_p = (\frac{FBW}{\omega_0\Omega_c})\frac{\gamma_0}{g}$$

(4)

for g representing the capacitance.

It should be noted that $\omega_0 L_s = \frac{1}{(\omega_0 C_s)}$ and $\omega_0 L_p = \frac{1}{(\omega_0 C_p)}$ hold in (3) and (4). The element transformation in this case is shown in Fig. 2.

## 3   Realization of Bandpass Filter

The design of a practical bandpass filter used in wireless communication area (802.11b) is considered in this letter, the description of the parameters is shown in Fig. 3.

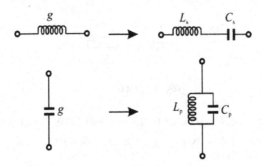

**Fig. 2.** Low pass prototype to bandpass transformation

| | Fs1 | Fp1 | Fp2 | Fs2 | As | Ap |
|---|---|---|---|---|---|---|
| Input Parameters | 2.0E9 | 2.2E9 | 2.6E9 | 2.8E9 | 20.0 | 3.0 |

| | CF-Des | CF-Actual | Gain Dev (dB) | MA-LSB | MA-USB | |
|---|---|---|---|---|---|---|
| Performance | 2.400E9 | 2.400 GHz | 3.000 | | | |

| | F | S11 (dB) | | S21 (dB) | | Delay (ns) |
|---|---|---|---|---|---|---|

| CF: Center Frequency (Desired or Actual) | Fs1: Lower Stopband Edge |
|---|---|
| Dev: Deviation in Passband | Fp1: Lower Passband Edge |
| MA: Minimum Atten. Lower/Upper Stopband | Fp2: Upper Passband Edge |
| F: Frequency | Fs2: Upper Stopband Edge |
| 1/2: Input/Output Ports | Ap: Atten at PB Edge or Ripple |
| Spec: Frequency Specification | As: Atten at SB Edge |

**Fig. 3.** Parameters of practical bandpass filter for 802.11b wireless communication

## 3.1    Realization of Lumped Element Filter

First the filter design guide is used to fulfill the design of lumped element filter. The bandpass filter model and the its corresponding subcircuit is shown in Fig. 4, the entire process is implemented by Advanced Design System 2008(ADS2008).

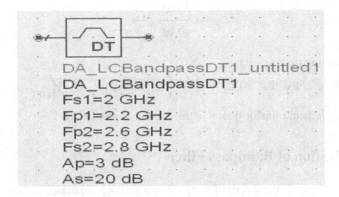

**Fig. 4a.** The practical bandpass fiter model

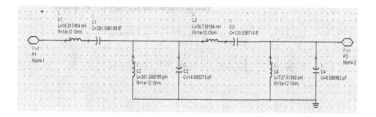

**Fig. 4b.** The corresponding subcircuit

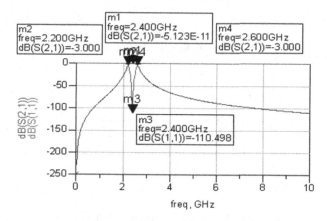

**Fig. 5.** Full wave simulation response of lumped element bandpass filter

After the structure of the lumped element bandpass filter is confirmed, simulation is taken to validate the correctness of our design., the full wave simulation response is shown in Fig. 5. It is noticed that the results satisfy our demands.

## 3.2    Transformation from Lumped Element Filter to Microstrip Filter

The measurement response of lumped element filter is not good in wireless communication frequency area but better results can be obtained by using microstrip filter, so the transformation from lumped element filter to microstrip filter is needed. Kuroda identities which are shown in Fig.6 are used to fulfill the transforming process. Zc and Yc are used to represent the characteristic impedance and admittance of transmission line, Zu is used to represent the characteristic impedance of unit element.

Finally the schematic of the practical microstrip filter is obtained by filter assistant and also the correlative parameters are optimized in ADS 2008, just as Fig. 7 shows.

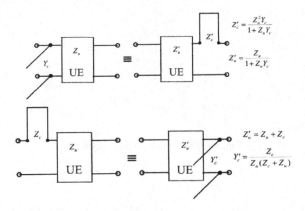

$$Z'_c = \frac{Z_u^2 Y_c}{1 + Z_u Y_c}$$

$$Z'_u = \frac{Z_u}{1 + Z_u Y_c}$$

$$Z'_u = Z_u + Z_c$$

$$Y'_c = \frac{Z_c}{Z_u(Z_c + Z_u)}$$

**Fig. 6.** Kuroda identities

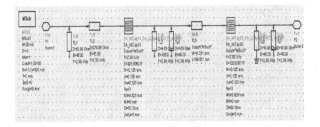

**Fig. 7.** Schematic of microstrip filter

## 4    Conclusion

Comparing to the conventional methods, it is much more efficient to complete the filter design process with the help of ADS2008, in this letter, the basic work is the design of lumped element filter, then the microstrip filter is synthesized by filter assistant using Kuroda rules, it is noticed that the designing time is obviously decreased and the accuracy is increased by using software. Although the whole designing process is simplified by ADS2008, there are some work to do such as matching work during the fabrication, use value will be obtained by combing the software design and practical realization together.

## References

1. Ludwing, R.: RF Circuit Design:Theory and Applications. Publishing House of Electronics Industry (2001)
2. Gupta, K.C.: Microstrip Lines and Slot lines. Artech Hourse, Ded-ham (1979)
3. Atwood, W., Stinehelfer, H.E.: Multi stub Filter for Microstripline. IEEE MTT MTT-16(7), 177–180 (1968)
4. Rhodes, J.D.: Theory of Electrical Filters. Wiley, New York (1976)

5. Helszajn, J.: Synthesis of Lumped Element, Distributed and Planar Filters. McGraw-Hill, London (1990)
6. Weinberg, L.: Network Analysis and Synthesis. McGraw-Hill, New York (1962)
7. Saal, R., Ulbrich, E.: On the design of filters by synthesis. IRE Trans. CT-5, 284–327 (1958)
8. Ozaki, H., Ishii, J.: Synthesis of a class of strip-line filters. IRE Trans. Circuit Theory CT-5, 104–109 (1958)
9. Mattaei, G., Young, L., Jones, E.M.T.: Microwave Filters, Impedance-Matching Networks, and Coupling Structures. Artech House, Norwood (1980)
10. Matthaei, G.L., Hey-Shipton, G.L.: Novel staggered resonator array superconducting 2.3-GHz bandpass filter. IEEE Trans. MTT-41, 2345–2352 (1993)

# The Design of People Management System in the Campus Base on RFID

Hu Rong[1], Luo Hui[2], and Li Weiping[2]

[1] Jiangxi Industry Polytechnic College, Nanchang, China
[2] EastChina JiaoTong University, Nanchang, China
{xuefeixu,lh_jxnc}@163.com, liweiping@ecjtu.jx.cn

**Abstract.** With recent advances in wireless technologies, Radio frequency identification (RFID) becomes an important enabling technology for the management system. RFID system is a high efficient information collection system. It consists of two major components, one is reader and the other is tag. If it can be used in the management of the people in the campus, the effect is great. In this paper, the hardware and software design of the RFID system used in people management in the campus are given, the reader, the tag and wireless network connection scheme are designed in details.

**Keywords:** RFID, People Management, Reader, Tag, Wireless network.

## 1 Introduction

RFID technology is starting from the 20th century. It is a Automatic Identification Technology. With contact identification technologies in the same period or the early stage, it succeeded to combine RFID and IC card technology. Using a two-way radio frequency communication realize the recognition and data exchange of the kinds of objects, equipment or personnel in the different state (moving or stationary), by the non-contact manner. With the expansion of the enrollment of students in colleges, the difficulty of the management is increasing, and the level of management needs to be enhanced. Radio frequency identification system can finish a long-range identification and collection of information without human intervention. The collected data can be analyzed in a management system.

## 2 Feature of RFID System

In addition to inheriting their storage capacity, easy to read and write, good privacy, intelligent, etc, RFID also has the following characteristics:

(1) Easy to carry, simple, reading the identification card in the range of about 7.5cm (do not plug card);
(2) No hardware access, avoiding mechanical failure, long life;
(3) Radio frequency identification cards has not exposed metal contacts. the card fully enclosed, has a good waterproof, dustproof, anti-fouling, anti-magnetic, anti-static properties;

G. Lee (Ed.): Advances in Intelligent Systems, AISC 138, pp. 279–286.

(4) For the wireless transmission, the data must be the random encrypted, following a sound, secure communication protocols. Card serial number is unique, the manufacturers solidify this serial number on the card factory curing, high security;

(5) The cards have anti-collision mechanism, prevent the interference of data between the card and its read and write time is less than 10ms. Authentication, data exchange are within 100ms;

(6) High signal penetration ability (penetrate walls, roads, clothing, etc.), a small amount of data transmission, anti-jamming, sensor sensitivity, ease of maintenance and operation.

# 3     Work Principle of RFID Systems

A typical RFID system consists of two parts: radio frequency identification card and reader.

Reader sends a certain frequency RF signals by the transmitting antenna. When the electronic tag gets into the transmitting antenna work area, it generates induced current, accesses energy and is activated. Electronics tag sends out encode information by built-in antenna. When the system receives the carrier signal from the tag antenna, and sends it to the reader by the antenna controller. The reader demodulates and decodes the received signal, and then sends it to the main system for processing. The main system judges the legitimacy of the card under the logical, makes the appropriate settings for different treatment and control, and sends command signal to control the executing agency.

On the other hand, when the radio frequency identification card is within the scope of induction of the reader, the card's coil produces a weak current under induction of the "excitation signal". The current is the card's power of the integrated chip. After resetting the card, The hibernation card is activated and will carry its own ID mark, the manufacturer logo and other information code out in modulation way by the card antennas. Reader module will send wireless signals to on-site controller, the site controller implements signal processing and issues commands.

There are two types of RF signals coupling between the electronic tag and reader.

## 3.1     Inductance Coupling

Inductance coupling generally applies identification system working at close range in the IF and LF radio frequency. The typical working frequencies: 125KHz, 225KHz and 13.56MHz. Recognition distance is less than 1m, the typical distance is 10 ~ 20cm.

## 3.2     Electromagnetic Backscattering Coupling

Electromagnetic backscattering coupling uses radar theory model. Based on the propagation of electromagnetic waves in space, electromagnetic launched out hit target and reflects, and bring back the target information. Electromagnetic backscattering coupling generally applies to radio frequency identification system working at remote range in the high-frequency, microwave radio frequency. The typical working frequencies:     433MHz, 915MHz, 2.45GHz and 5.8GHZ. Recognition distance is greater than 1m, the typical role of distance is 3 ~ 10m.

# 4   System Classification

## 4.1   According to the Label of the Power Supply System, the System Is Divided into Active System and Passive System

RFID tags can be divided into two kinds, active and passive. Active RFID tags use the energy within the cell. Their identifying distance can be up to tens of meters or even 100 meters. But their lives are limited, and they are in higher prices. With the battery, therefore, active tags have relatively large size, and they cannot be made as thin cards. In other way, without batteries, passive RFID tags use reader coupling electromagnetic energy emitted by reader as its energy. They are light, small size, very cheap, and their lives can be very long. They can be made into a variety of thin card, but the launch distance is restricted, usually several centimeters to tens of meters, and the greater readers transmit power are needed.

## 4.2   According to the Label Data Modulation, the System Is Divided into Active, Passive and Semi-active System

Active tags have internal battery power to supply power, the reliability is high, signal transmission distance is far. In addition, active tags can be limited in the use of time or number of times by the design of battery life. It can be used in the place which needs to restrict the use of data or restrict data transfer. But active RFID tags' lives are limited, and with the tag battery power consumption, data transmission distance will be shorter and shorter, thus affecting the system work properly.

Passive tags have not internal battery, and work properly by provided the energy outsiders. Passive tags' typically generate power device is the antenna and the coil. When the label works in the system's work area, the antenna receives a particular wave, the coil will be induced current in the circuit. After rectification, it activates micro-switch circuit, and gives a label supply to the tag. Passive tags have the permanent use lives, and are used to be read and written every day, and passive tags support data transmission at long time and permanent data storage. The main disadvantage of passive tags is the data transmission distance is shorter than the active tag, data transmission distance and signal strength was limited.

In addition, semi-active RFID system is also known as battery support type backscattering modulation. Semi-active tag with the battery plays the role in the internal digital circuitry, but the tag does not send data initiatively through its own energy.

## 4.3   According to the Label's Operating Frequency, the System Can Be Divided into Low Frequency, High Frequency, Very High Frequency and Microwave Systems

Reader sends wireless signals in the frequency known as RFID system frequency. The frequency is basically divided into: low frequency (30-300kHz), high frequency

(3-30MHz), VHF (300MHz-3GHz) and microwave ( 2.45GHz and above). Low-frequency systems can be used in short distance, low-cost applications, high-frequency system can be used in access control and the need to transfer large amounts of data, very high frequency systems can be applied to a longer distance literacy and high literacy rate.

# 5    Key Technologies

## 5.1    RF Beam Power Technology

The energy Require of passive RFID tags are directly from the radio frequency electromagnetic beam. Comparing with active radio frequency identification systems, the passive system requires a larger transmission power, radio frequency electromagnetic waves in the electronic tag turn to voltage required by the electronic labeling.

## 5.2    Backscattering Modulation

A radio frequency identification system is used backscattering modulation in data communication. Backscattering modulation is a communication way the passive RFID tag sends data back to reader. According to different data to be sent, by controlling the electronic tag antenna impedance, making the carrier rate reflected has small changes, so the reflected signal carries the data.

# 6    Design of Management System

## 6.1    Antenna Area

Electronic tag and reader build up the transmission channel between them by their antennas, and the spatial transmission channel is entirely decided by the antenna characteristics.

At some distance, the Distribution Error between the angular distribution of radiation field and the angular distribution of infinite Far is within the allowable range, the region from such point to infinite Far is known as the antenna far-field zone. R is recognized near-field radiation zone and far-field zone boundaries:

$$R = \frac{2D^2}{\lambda}$$

The formula, D is the antenna diameter, $\lambda$ is the wavelength of electromagnetic wave, $D \leq \lambda$.

For radio frequency identification systems, in general, because of the limited label size and the limited size of reader antenna, the antenna structure model is $L / \lambda \ll 1$ or $L/\lambda > 1$. So, the antenna's reactive near field and far field zone can be estimated according to wavelength.

## 6.2    Design of Hardware System

①Design of reader

In the RFID system, the circuit design for readers and tags must include resonant circuits for high-frequency signal transmission, and it asks the two LC resonant circuit loops tuned in the same resonance frequency. After the system design, the antenna inductance is fixed; the resonant frequency of LC circuits can be changed by adjusting the capacity of the circuit.

Series resonant circuit is used in Reader antenna. These three parameters should be taken into account in design such as its resonance frequency $f_0$, bandwidth and quality factor Q. resonant frequency $f_0$ is decided by the $L_R$ circuit inductance and capacitance circuit $C_R$. The specific relationship is as follows:

$$f_0 = \frac{1}{2\pi\sqrt{L_R C_R}}$$

The relationship between Q value, the antenna bandwidth B and the resonant frequency $f_0$ is B = $f_0$ / Q.

In order to ensure the requirements of transmission bandwidth and to improve the reliability of the signal transmission, a suitable Q value should be chosen. In the debugging process, the scope of Q value can be fixed by adjusting loop resistance $R_R$ value. After the experimental test, Q value should be confined within 5 to 15, generally Q = 12.

The relationship between Q value and RR is:

$$R_R = \frac{2\pi f_0 L_R}{Q_R}$$

According to the signal transmission, the receivable power of the tags in the electromagnetic fields can be:

$$P_R = \left(\frac{\lambda}{4\pi D}\right)^2 P_T G_T G_R$$

Of which:

$P_R$ is receptive power of an antenna (w);
$P_T$ is the transmitting power of the transmitter (W);
$G_R$ is the enhancement of a transmitter antenna in the isotropic media;
$G_T$ is the enhancement of a tag antenna in the isotropic media;
D is the distance between the transmitter and tag (m)

When RFID system uses passive tags, the identification distance depends on tag activation energy. The results show that as the activation energy for the RFID is 715uW, at different frequencies, the typical relationship for the passive system functional distance is shown as in table 1.

**Table 1.** The relationship between the typical range, the radiated power and the frequency

| EIRP (W) | 434MHz $G_R$ =1.4, $\lambda$=0.7 | 869MHz $G_R$ =1.64, $\lambda$=0.34 | 2.45GHz $G_R$ =1.64, $\lambda$=0.12 |
|---|---|---|---|
| 0.5 | 1.4m | 0.73m | 0.26m |
| 1.0 | 2.0m | 1.03m | 0.37m |
| 2.0 | 2.8m | 1.44m | 0.52m |
| 5.0 | 4.5m | 2.3m | 0.84m |
| 10 | 6.3m | 3.2m | 1.20m |
| 20 | 9.0m | 4.6m | 1.68m |

From the analysis in Table 1, we can see for the school management, the distance for their tags to identify should be less than 1 meter. Taking into account factors, such as allocation of frequency resources, the working frequency of the RFID system can be 2.45GHz.

Reader can be placed in the entrance of classrooms, dormitories, halls and so on for data collection.

②The choice of tags: because the system is used for personnel management, tags are passive and read-only.

Figure 1 illustrates how these sidebands look, in relation to the reader-generated carrier frequency. The comparatively tiny sidebands have approximately 90 decibels less power than the reader-generated carrier signal, and this is the reason why RFID tag responses often have such a limited transmission range.

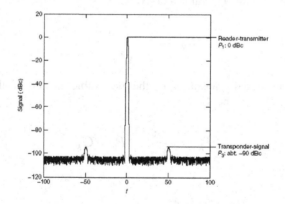

**Fig. 1.** Normal RFID Tag Signal

## 6.3 Software System Design

Software system operates on the WINDOWS XP platform and SQL Server 7.0 databases. It uses the object-oriented VB integrated development tools for the development and design. There are several functional modules as follows:

Control module. Include management of all personnel, grouping according to the functions of different personnel classification authority, setting the control periods and holiday control.

Video monitoring module. Include video switching matrix of automatic / manual switch control, on-site the attitude control of head, video recording and playback.

The alarm module. Include the alarm function setting, the scene plane structure mapping, setting the system input and output field position and interrelated components, alarm monitoring and recording of the state.

Information management module. Include dynamic displaying access data recording system input and output components of the event information, database management, retrieval and query, automatically generate reports, etc..

System maintenance module. Including system initialization, setting the system operation and system management authority, data storage and playback, card and cardholder personal information management, setting control subsystem of each scene, recording of all equipment in the state of real-time, automatic generation system running log, etc..

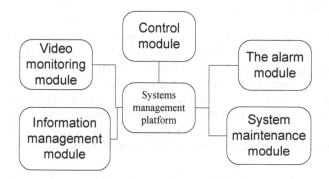

**Fig. 2.** System Software Platforms

# 7 RFID System Security Solutions

The technology of existing security and radio frequency identification systems privacy can be divided into two categories: one is to prevent tag and reader to communicate by physical methods, the other is to enhance the security label in the logical way.

Here are four kinds of physical methods:

## 7.1 Killing Tag

The principle of killing is to let tag RF tag lose function and prevent the reader to track the tag.

## 7.2 Faraday Nets

Radio frequency tags are placed in containers made of conductive material, so that readers cannot communicate with the tag. Therefore, using Faraday nets can prevent criminal access to label information.

## 7.3  Active Interference

The user of RFID tags can take a device to broadcast a radio signal used to prevent or destroy the near RFID reader to identify the tag. But this method is not desirable, because it may interfere with the legitimate reader identification for RF tags. Seriously, it might block the other wireless systems.

## 7.4  Preventing Tag

The principle of this method is to use a special anti-collision algorithm to prevent reading tag. Reader reads the tag which must get the same response data every time, otherwise the label will not be identified.

# 8   Conclusion

Currently, RFID technology has developed rapidly in China and abroad. RFID products have been used in the high-security areas, bank, a general office building, ect.. Combinating alarm systems and video surveillance system is a intelligent management. It strengthens regional security. The system of radio frequency identification systems has traditional access control, alarm, video surveillance and other organic, and achieves a multi-stage operation, real-time monitoring, alarm, joint control and other functions. School personnel management system is the application of RFID technology in the realization of the school management automation and intellectualization.

# References

1. Hind, D.J.: Radio Frequency Identification and Tracking Systems in Hazardous Areas. Electrical Safety in Hazardous Environments 4, 215–227 (1994)
2. International Organization for Standardization. 1S0/IEC 18000-3. Information Technology AIDC Techniques-RFID for Item Management (March 2003)
3. Zhang, Y., Zheng, M., Zhang, Q.: Design of Long Range Passive RFID System. Remote Control and Remote Measurement 25(4), 45–49 (2004)
4. Cheng, X., Fang, S.: Design of Meeting Register System Based on RFID. Electronics Engineer 31(7), 68–71 (2005)
5. Basat, S., Lim, K., Kim, I., Tentzeris, M.M., Laskar, J.: Design and Development of a Miniaturized Embedded UHF RFID Tag for Automotive Tire Application. Electronic Components and Technology (1), 867–870 (2005)
6. Diugwu, C.A., Batchelor, J.C., Langley, R.J., Fogg, M.: Planar Antenna for Passive Radio Frequency Identification (RFID) tags. In: 7th AFRICON Conference in Africa, vol. (1), pp. 21–24 (2004)
7. Strassner, B., Chang, K.: Integrated Antenna System for Wireless RFID Tag in Monitoring Oil Drill Pipe. In: Antennas and Propagation Society International Symposium, vol. (1), pp. 208–211 (2003)

# Kernel PCA and Nonlinear ASM

Liu Fan[1], Xu Tao[2], and Sun Tong[2]

[1] Information and Network Center of CAUC
Tianjin, China
liufan@cauc.edu.cn
[2] College of Computer Science and Technology of CAUC
Information Technology Research Base of CAAC
Tianjin, China
{txu,t-sun}@cauc.edu.cn

**Abstract.** As a nonlinear Principal Component Analysis (PCA) method, Kernel PCA (KPCA) can effectively extract nonlinear feature. For the object image which includes more nonlinear features, traditional Active Shape Model (ASM) couldn't obtain a good result of localization. Concerning this, an extending research on nonlinear-ASM is brought here, and an algorithm of object localization based on nonlinear-ASM is proposed. In the research of nonlinear-ASM, the problem of high dimensionality caused by nonlinear mapping has been solved effectively by the kernel theory. Besides, KPCA can not reconstruct the pre-image of the input space, thus prior model is hardly constructed by the method of the nonlinear-ASM. For solving this problem, the theory of multi-dimensional scaling is researched in the paper. The validity of the proposed method is demonstrated by the results of experiments.

**Keywords:** Kernel Principal Component Analysis, Multi-dimensional Scaling, Active Shape Model, Nonlinear, Object Localization.

## 1 Introduction

Kernel Principal Component Analysis (KPCA) has been powerfully proven as a preprocessing step for classification algorithms. Moreover, KPCA can also be considered as a natural generalization of linear principal component analysis (PCA). It can consider more nonlinear variations, and is better for extracting the nonlinear feature. In recent years, there has been a lot of interest in the study of KPCA, such as face recognition [1] and a pre-processing step in regression problem [2].

However, KPCA is mainly based on kernel method, which induces that KPCA can hardly reconstruct the pre-image, so one can only settle for an approximate solution. In this paper, we research and analyze the idea of the multi-dimensional scaling (MDS) [3] to address this problem, and then use it to construct a non-linear Active Shape Model (ASM). This paper is arranged as follows. First we outline KPCA in Section 2. In Section 3, we first construct prior model for object localization through the MDS method[4], and then propose a localization algorithm of non-linear ASM. For demonstrating the effectiveness of the algorithm, localization experiment on images of hand are presented in Section 4 before conclusions are drawn in Section 5.

G. Lee (Ed.): Advances in Intelligent Systems, AISC 138, pp. 287–293.
springerlink.com      © Springer-Verlag Berlin Heidelberg 2012

## 2    Kernel PCA

KPCA is a nonlinear PCA method which is recently introduced by Sholkopf et al [5], and based on Support Vector Machines (SVM) [6]. The essential idea is both intuitive and generic. In general, PCA can only be effectively performed on a set of observations that vary linearly. When the variations are nonlinear, KPCA utilizes SVM to find a computational tractable solution through a simple kernel function which intrinsically constructs a nonlinear mapping from the input space to feature space. As a result, KPCA performs a nonlinear PCA in the input space.

Given a set of patterns $\{X_1, X_2, ..., X_L\} \in R^N$ , and select the Gaussian kernel function $k(x, y) = \exp(-\|x - y\|^2 / c)$, the kernel matrix $\mathbf{K}$ is computed by $\tilde{\mathbf{K}} = \mathbf{HKH}$ ,

where $\mathbf{H} = \mathbf{I} - \dfrac{1}{L}\mathbf{11}^T$ , $\mathbf{I}$ is the $L \times L$ identity matrix, $\mathbf{1} = [1, 1, \cdots, 1]^T$ is a $L \times 1$ vector.

The kth orthonormal eigenvector of the covariance matrix in the $F$ can be shown to be

$$\mathbf{V}_k = \sum_{i=1}^{L} \frac{\alpha_i^k}{\sqrt{\lambda_k}} \tilde{\Phi}(X_i) , \tag{1}$$

And $\tilde{\Phi}(X_i) = \Phi(X_i) - \dfrac{1}{L}\sum_{i=1}^{L} \Phi(X_i) = \Phi(X_i) - \overline{\Phi}(X)$ .

Then, denote the projection of the $\Phi -$ image of a pattern $X_i$ onto the kth component by $\beta_k$ , and,

$$\beta_k = \tilde{\Phi}(X_i)^T \mathbf{V}_k = \frac{1}{\sqrt{\lambda_k}} \sum_{i=1}^{L} \alpha_i^k \tilde{K}(X, X_i) , \tag{2}$$

where:

$$\tilde{K}(X, Y) = \tilde{\Phi}(X)^T \tilde{\Phi}(Y) \qquad = K(X, Y) - \frac{1}{L}\mathbf{1}^T \mathbf{K}_X - \frac{1}{L}\mathbf{1}^T \mathbf{K}_Y + \frac{1}{L^2}\mathbf{1}^T \mathbf{K}\mathbf{1},$$

and $\mathbf{K}_X = [K(X, X_1), \cdots, K(X, X_L)]^T$ .

Denote

$$\tilde{\mathbf{K}}_X = [\tilde{K}(X, X_1), \cdots, \tilde{K}(X, X_L)]^T$$

$$= \mathbf{K}_X - \frac{1}{L}\mathbf{11}^T \mathbf{K}_X - \frac{1}{L}\mathbf{K}\mathbf{1} + \frac{1}{L^2}\mathbf{11}^T \mathbf{K}\mathbf{1}$$

$$= \mathbf{H}(\mathbf{K}_X - \frac{1}{L}\mathbf{K}\mathbf{1}) . \tag{3}$$

Finally, the projection $P\Phi(X)$ of $\Phi(X)$ onto the subspace spanned by the first t eigenvectors is

$$P\Phi(X) = \Sigma_{i=1}^{t}\beta_k V_k + \bar{\Phi} = \Sigma_{i=1}^{t}\frac{1}{\lambda_k}(\alpha^k) + \bar{\Phi} = \tilde{\Phi}M\tilde{K}x + \bar{\Phi} , \qquad (4)$$

where $M = \Sigma_{i=1}^{t}\frac{1}{\lambda_k}\alpha^k\alpha^{kT}$ is symmetric.

## 3    Nonlinear ASM

ASM consists of the Point Distribution Model (PDM) aiming to learn the variations of valid shapes. While this approach can be used to model and recover some changes in the shape of an object, it can only cope with largely linear variations. If the data of the input space have much complex nonlinear relativity, the ASM could ignore their nonlinear feature. Therefore, this paper will transform ASM from linear to nonlinear through KPCA. However, nonlinear ASM still has one problem that KPCA can hardly acquire the pre-image in the input space (Figure 1), so we research the MDS and use it to approximate the reconstruction image. In this section, we will introduce the MDS method, and then describe an algorithm of nonlinear ASM based on KPCA, where the prior model is constructed by the MDS method.

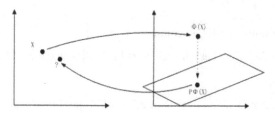

**Fig. 1.** The pre-image problem

*A)  MDS method*

For any two points $X_i$ and $X_j$ in the input space, we can obtain their Euclidean distance $d(X_i, X_j)$. Analogously, we can also obtain the feature space distance $\tilde{d}(\Phi(X_i), \Phi(X_j))$ between their $\Phi$-mapped images. Moreover, for many commonly used kernels, there is a simple relationship between $d(X_i, X_j)$ and $\tilde{d}(\Phi(X_i), \Phi(X_j))$ [7]. For example, for the Gaussian kernel $k(x, y) = \exp(-\|x - y\|^2 / c)$, there is:

$$d_{ij}^2 = -c\log(\frac{1}{2}(K_{ii} + K_{jj} - \tilde{d}_{ij}^2)) . \qquad (5)$$

The idea of the MDS method is as follow (Figure 2). We could require the approximate pre-image to satisfy these constraints.

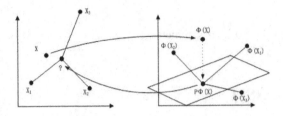

**Fig. 2.** Basic idea of the MDS method

*B)  Algorithm of nonlinear ASM based on KPCA*

In this section, we attempt to develop the nonlinear transformation of ASM based on KPCA and apply it to object localization. First, the training data will be obtained by labeling a set of landmark points, and constructing PDM. Then, extract the feature of training patterns by KPCA and construct prior model through the MDS method. Finally, after locating the initial model near the test image object, localization algorithm will be carried out through an iterative process.

The algorithm is performed as follow:

Step1: choose the initial model, in common $X = \overline{X}$ ;

Step2: in this model, find best fitting point $\left(x_i', y_i'\right)$ along the normal to the shape (Figure 3);

Step3: the new pattern which consists of all best fitting points $\left(x_i', y_i'\right)$ is mapped into the KPCA space to construct prior model. After aligned the    position and shape, the prior model will be new current model X;

Step4: repeat Step2、Step3 till convergence.

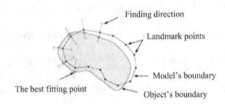

**Fig. 3.** Finding the best fitting points

# 4    Experiments

The data of experiments are selected from a group of image of hand. There are 125 images in all, of which 120 selected as training image, and 5 selected as test image.

These experiments choose hand image as the basic data is due to its unique shape. It is the non-rigid target, not only the profile is changing, but also every finger has a variety of changes. Therefore, the training set formed by such image must contain more nonlinear feature, and it would be better to prove our argument.

In every training image, the landmark points express as a vector of $85 \times 2$: $X_i = \left( x_1, y_1, x_2, y_2, ...., x_n, y_n \right)^T$. For evaluating the performance of the algorithm, the experiments adopt the mean Euclidean distance error as the evaluation criteria (unit: pixel). Define $E'$ as:

$$E' = \frac{1}{n} \sum_{i=1}^{n} \sqrt{(x_i - x_i')^2 + (y_i - y_i')^2},$$  (6)

where $(x_i, y_i)$ is the ith landmark point of the test image; $(x_i', y_i')$ is the ith landmark point of the prior model.

## A) Object localization in the HSV space

This section showed two experiments of object localization in the HSV space to compare the linear ASM and non-linear ASM. First, test image was randomly selected in the set of test images, and then place the initial model near the object (figure 4). Based on the same test image and initial model, two experiments were respectively carried out (figure 5).

As can be seen from the figure 5, nonlinear ASM just need 15 iterations to accomplish localization, and acquired better effect of localization, but linear ASM achieved localization after 20 iterations. Although the initial model was placed near the object, every figure of initial model has a certain angular deviation with object, this is the reason why linear ASM needs more iterations. In the figure 5(a), prior model reconstructed by linear ASM also has much difference with object after iterating one time, while prior model reconstructed by nonlinear ASM has been close to the shape of the object. Figure 6 shows the convergence process of the two experiments. The nonlinear ASM obviously has more fast convergence rate. These are due to the fact that nonlinear ASM algorithm has more advantage to extracting nonlinear feature.

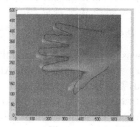

**Fig. 4.** Initial model

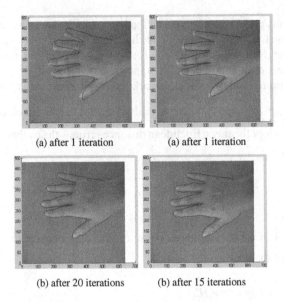

(a) after 1 iteration                    (a) after 1 iteration

(b) after 20 iterations              (b) after 15 iterations

**Fig. 5.** Localization process of linear and nonlinear ASM (The left column images show the localization of linear ASM, and the right column images show the iteration process of nonlinear ASM)

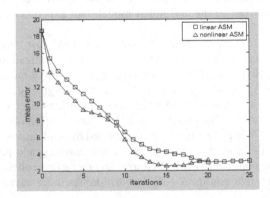

**Fig. 6.** Convergence process of linear and nonlinear ASM (unit: pixel)

## B)   Object localization when object was partly covered

In this experiment, test image was respectively covered the part of palm and finger, and then researched localization based on nonlinear ASM. After 13 iterations, model could fit well the object, and spend 70.25s achieving localization (Figure 7(a) (b)). For comparison, object uncovered was also do experiment (figure 7(c)), and time was 70.21s in the same condition. As can be seen, when the object was covered, the effect of localization based on nonlinear ASM was still good, and time is similar with when object was uncovered. These are due to the outstanding advantage of nonlinear ASM. It could fit the shape of object which is partly covered, and recover the covered part.

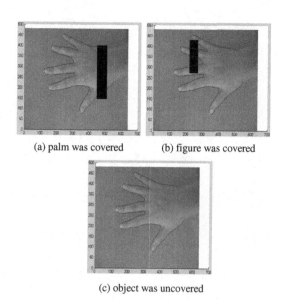

(a) palm was covered        (b) figure was covered

(c) object was uncovered

**Fig. 7.** The effect of localization on object covered and uncovered

## 5    Conclusion

In this paper, KPCA is the focus of the study. As the typical extraction method of non-linear feature, KPCA can extract the nonlinear feature more effectively, so we use this method to improve the traditional ASM, and propose a nonlinear ASM algorithm. However, KPCA raise a difficult problem of pre-image, so the idea of MDS is used to address it. In the end, the validity of this algorithm is demonstrated by two sets of experiment.

## References

1. Kim, K.I., Jung, K., Kim, H.J.: Face Recognition Using Kernel Principal Component Analysis. IEEE Signal Processing Letters 9(2), 40–42 (2002)
2. Rosipal, R., Girolami, M., Trejo, L.J.: Kernel PCA for Feature Extraction and De-noising in Non-linear Regression.Technical Report No.4, Department of Computing and Information Systems, University of Paisley (2000)
3. Cox, T.F., Cox, M.A.A.: Multidimensional Scaling, 2nd edn. Monograghs on Statistics and Applied Probability, vol. 88. Chapman & Hall/CRC (2001)
4. Kwok, J.T., Tsang, I.W.: The Pre-Image Problem in Kernel Methods. IEEE Transactions on Neural Networks 15(6), 1517–1525 (2004)
5. Scholkopf, B., Smola, A., Muller, K.: Nonlinear Component Analysis as a Kernel Eigenvalue Problem. Neural Computation 10(5), 1299–1319 (1998)
6. Vapnik, V.: The Nature of Statistical Learning Theory. Springer, Heidelberg (1995)
7. Williams, C.K.I.: On a Connection between Kernel PCA and Metric Multidimensional Scaling. In: Leen, T.K., Dietterich, T.G., Tresp, V. (eds.) Advances in Neural Information Processing Systems, vol. 13, pp. 675–681. MIT Press, Cambridge (2001)

# Impact on the Pace of the Game of Basketball Analysis of the Factors

Zhang Nan

Shandong Institute of Business and Technology Department of PE,
264005, Yantai, Shandong

**Abstract.** The rhythm game is a game of the basketball soul of a direct impact on the market changes and developments in the war, in the race to control the game tempo of the game will be able to get the initiative. The author of the game affect the rhythm of the major factors, how to control the game tempo discussed.

**Keywords:** rhythm, basketball game, factor.

## 1 Research Purposes

The rhythm of nature in a variety of things to show contains a strong and weak, fast and slow, long and short-term, regular and harmonious movement phenomenon. Basketball is more of the above distinctive characteristics of the basketball game between the pace of technology compact and reasonable use of tactical coordination with tacit, against fierce competition, the level of rational layout and overall performance of the various forms of movement. Modern Basketball, extremely offensive and defensive combat, control and anti-control are pushing very aggressive, reflecting the modern features of a basketball game. Basketball competitions have been revising the rules on the offensive and defensive balance played a role in the promotion. In recent years, the aggressive defense of the extensive use of attack has changed the dominant pattern.

## 2 Research Methods

Literature, observation, logical analysis, comprehensive research methods.

## 3 With the Results of the Analysis

### 3.1 Paced Game of Basketball the Impact a Major Factor

#### 3.1.1 Psychological Adjustment Is to Grasp the Rhythm of the Game Leading

Basketball has always been accompanied by changes in the rhythm of the psychological adjustment, psychological adjustment is to grasp the rhythm of the race leader. Psychological adjustment is only appropriate, in a timely manner, can have an accurate grasp the rhythm of race, reasonable. Match the pace of change is the psychological adjustment to reflect the specific. Basketball game, alternately offensive

G. Lee (Ed.): Advances in Intelligent Systems, AISC 138, pp. 295–299.

and defensive frequent, fast-changing market circumstances, how the game needs, assess the situation and grasp the right tempo game, to give full play to their strengths, and effectively restricted the other side? This is the first psychological factors continue to adjust and Dominated only continue to improve the psychological adjustment, with improved user-friendly in order to grasp the rhythm of the game, controlling the race initiative. In fact the right to match the rhythm of each is a member of the monitoring and regulation of psychological tests and psychological factors also put forward a more specific and more accurate.

### 3.1.2    Offensive and Defensive Technology Is a Reasonable Use of Race to Grasp the Rhythm of the Heart and Foundation

Offensive action and the use of technology in the process of performance rhythm of the major changes in the start-up, running, jumping, stop, from the mass, catch, shoot, dribble, and so on breakthrough technology proficiency, the speed of change, as well as the use of techniques Offensive technology convergence and the use of sham of the time, and so on. If these technologies have a good rhythm, using a timely and reasonable, we can out of defense, its offensive power of technology. For example: out of time, the defender can change the focus of the rhythm of movement, effective from the grasp the favorable opportunity, take advantage of individual tactical purpose. Shooting rhythm with the use of technology to grasp the opportunity well, will be able to raise the norms of shooting, the sham of the use of life-like, will be able to properly control the tempo to avoid aggressive defense, gaining the initiative, to create offensive opportunities, especially in the offensive Technology The pace of convergence in the use of hands and more deterrent. For example: the ball-breaking technology and the rapid convergence of technology, accurate and can be a perfect match in the fast-paced ball in the wrong formation, the defense will be able to mobilize, to facilitate convergence feint to form a rapid breakthrough. Korean women's basketball this study is quite technical, skilled use of on-the-spot, accurately grasp the rhythm, they form a unique fast-paced attack with technical and tactical style. Another example: the ball-shooting techniques and technology convergence in the pace of change has vigorously promoted the development of a variety of shooting technique, it shows a perfect combination between technology and the pace of harmonization. Another example: a breakthrough technology and technology convergence shot in the rhythm of change, and the promotion of the shooting techniques of variability. After the break, such as jump shots in the air, according to the perspective of defensive cover and the rhythm of re-adjustment of shooting rhythm, to avoid capping, changing hands shot. Tupofenqiu convergence technology is also clever use of rhythm changes, diversionary, with the understanding reached. All these have fully demonstrated that each of the offensive actions of the completion of the technical and technological convergence of action there are a rhythm, and offensive moves and techniques used in a reasonable manner, to grasp the rhythm of the game to attack the core and foundation.

Defensive action and the use of technology in the process of change in the rhythm of the main sliding in the mobile, ball, play, steals, blocks, and looting after the market rebounds, as well as a variety of defense technology between the interface between stages. Attack, anti-technology is another constraint, but also promote each other, in order to meet the technical constraints of each other's attack, first of all must be familiar with and master the technology to attack the rhythm of action and the use of the opportunity and take appropriate defensive measures and countermeasures in order flexibility in the use of defense The pace of technology, technical effective in inhibiting the offensive tempo, aggressive defense to meet specifications. For example, defensive shooting, the shooting was based on the general who used different methods of shooting, the selective use of capping of technical, technological constraints of the shot to play. If the clever use of methods of destruction of the shooting rhythm of action, such as the destruction of the ball technology and technology convergence of shooting rhythm, jump stop and the pace of technological convergence, technology and dribble shot between the rhythm of shooting techniques, reducing the effectiveness of the shooting. Another example: Anti-break, not only to stay in step-defense and take the initiative to force the sliding passive and active defense, but should be determined in accordance with rules to limit foot center, and then based on those used by the breakthrough in the cross-step, or step-shun The method of action to effectively grasp the breakthrough in the direction of its movement and rhythm, so that a breakthrough will be reduced by anti-blindness, and greatly enhanced the anti-break and targeted offensive

### 3.1.3    Is the Tactical Ability to Grasp the Rhythm of the Game Reflected in a Comprehensive

Technology is the basis of tactics, technology convergence and cooperation between the organization and tactics and methods and tactics between the conversion is a tactical change in the form of the performance. Technology and skilled with a variety of ways, contributed to the tactical flexibility and variability, however, the tactical ability to cope with the standards, not only depend on the diversity of tactics, and to a greater extent depends on timely and reasonable and accurate selection of different tactics Form, to meet the actual game.

### 3.2    How to Control the Game Tempo

Competition from both offensive and defensive conversion and the composition of the essence in the attack for more scoring shot, when the defense is trying to stop shooting each other's scores in contradiction of the transformation in how to control the rhythm game for game of this initiative?

### 3.2.1    Fighters Determine the Speed of a Rhythm

Contemporary play basketball based on the fast-paced, without a doubt, but the timing wrong, it will be counter-productive, and that haste makes waste. In the race to effectively seize each other's vulnerabilities, to choose a good use of time. When the other party, such as lack of physical strength, fatigue at a time; or the other core members of the main players appear when the mood swings; the other change, suspend, and the tactics do not change at that time; the other main players foul 4 or 8 team fouls when; the other side Lead, paralysis occur when slack; before and after the market rebounds dominant; rapidly accelerate the speed of attack or change the offensive tempo, to play a "climax" to fight the expansion of the score or results, in order to lay the foundation for the victory.

### 3.2.2    The Proper Use of Slow-Paced

Size up the situation, the proper use of slow tempo is also an effective way. When this team feeling irritable or errors in a row, through the slow-paced in order to stabilize the morale of the troops, the attitude adjustment, when the strength of this team have been exhausted, and the lack of reserve forces should be based on the slow pace of relief which will benefit future battles in the fast-paced counter When the other party itself clear upper hand, should switch to slow the pace.

When the leading score, time is running away from the end, eager to draw each other's score, can be used to control the ball to attack, attack to reduce the speed to control the game, playing to a high success rate, so eager to win the other side of the game have shaken confidence, distracted, thinking of the fluctuations Has led to quick results, and its rhythm from the blind chaos to achieve this and to win leader purposes.

### 3.2.3    Match the Height of the Emergence of Rhythm Control

Have a climax in the game, how to maintain the advantage and disadvantage into the game for this initiative?

(1) for the wind force, as far as possible to maintain control of the rhythm game, each team must first clearly aware of this, to minimize the errors so that no other opportunity, one of a relaxation, while fast, slow for a while, so as far as possible with each other Its own rhythm, a rhythm when the other side was immediately after the transformation to adapt to the new rhythm, such as if the whole region was tight after the break the other side, rapidly changing for the half-court defense, the line-up with the changes will lead to changes in rhythm.

(2) on the down side of the team, as far as possible to undermine each other's rhythm has been to seize the aircraft for the climax of their own making. At this time players and coaches will change the rhythm of attack and defense, upset the rhythm of each other's established, can not let the other led by the nose. For example, when playing each other again and again scoring handy, so that the coach may request a suspension of play interrupted, so that an opponent's momentum cold Cold in order to break the rhythm of their play smoothly. In the suspension, the coach should also be targeted set of new tactics and new line-up, destruction of the opponent's rhythm, such as playing on the other side of the main fast-break team, they have to destroy the other side of a mass interfere with each other, and other team members soon to change The other side of the fast-paced, it is difficult to adapt.

### 3.2.4    At the End of the Match Rhythm Control

Offensive team wins control of the speed advantage to play for time to keep the score, teams can defeat the other side so that the press pass, dribble or shoot mistakes, and then seize the aircraft, using reasonable tactics, team formed to fight a losing battle in the end Liwan momentum To speed up the pace of the game, the game for the victory.

## 4    Conclusions and Recommendations

Basketball distinct rhythm, and the team is well-trained team members and the overall level is all about. His team skills and tactics of normal play and psychological adjustment to play an active role. The mainstream of the modern game of basketball, and fully reflects the fast-paced, but it can not ignore the role of the slow-paced, this is a positive adjustment and transition. Rhythm and is closely related to many factors, it is necessary to raise the overall level of basketball training, basketball must be trained into the rhythm of basketball training as a whole, and to constantly improve the level of theory, a better understanding of rhythm in the game of basketball.

## References

1. Chen, Y.: School of Physical Education. People's Sports Publishing House, Beijing (1991) 9621111
2. Liu, Y.: Modern basketball teaching and training. People's Sports Publishing House, Beijing (1992) 582771
3. Wang, S.: Basketball. People's Sports Publishing House, Beijing (1991) 25023011

# Make a Preliminary Research and Analysis on the Psychological Conditions of the College Students in PE Department

KongYang

Binzhou Medical University Mathematic Department, Shandong Yantai 264005

**Abstract.** By methods of questionnaire investigations .this paper intends to make a preliminary research and analysis on the psychological conditions of the college students in PE Department. Which is about the two main question of mental health and mental disposition. It give a basis for providing specific education to suit individual cases. Strengthening the management and improving the teaching quality.

**Keywords:** mental health, mental disposition.

## 1 Introduction

With the establishment of socialist market economy and rapid economic development, their mental state is also constantly changing and becoming more diversified situation. At present, students of sports psychology research is still rare. The topic to answer the questionnaire, physical education students on mental health and psychological conditions of psychological tendencies that made a more detailed investigation to understand. Through the analysis of survey materials, a preliminary understanding of their psychological condition. In order to strengthen ideological and political education of students, psychological education, improving teaching methods and teaching management to provide a theoretical basis.

## 2 Research Subjects and Methods

### 2.1 The Object of study

Physical Education Grade Physical Education Majors 07,08 150 students, of whom 90 boys, 60 girls, aged 20-24 years.

### 2.2 Research Methods

#### 2.2.1 The Questionnaire Design "Mental Status Questionnaire" 150 Questionnaires. 147 Valid Questionnaires. 93.3% in

#### 2.2.2 Number of Questionnaire and Statistics on the Number of Management Information, According to Statistics Required to Do the Processing

G. Lee (Ed.): Advances in Intelligent Systems, AISC 138, pp. 301–307.

### 2.2.3   Literature: Read about Psychological and Theoretical Aspects of Sports Literature, Magazines and Newspapers

## 3   Results and Discussion

### 3.1   The Mental Health of Students

World Health Organization defines health as "the human body, mental and social best, rather than simply the absence of disease", that is that health is not just physical health, including mental health and social harmony. With the accelerated pace of reform and opening up, development of market economy, the increase in the amount of information dissemination, social competition, changes in the pace of life, one can stimulate the students of knowledge, enthusiasm and desire to become a useful person, on the other hand to bring students considerable psychological pressure. Some poor students often can not be self-regulating mental stimulation, can easily cause mental fatigue, poor attitude, that rhythm of cortical excitation and inhibition disorder. Many studies show that about 10% -30% of the students there are different degrees of mental health problems. Physical education students in both the common features of modern college students. At the same time as the special nature of the content of their learning in the physical and mental side is also showing some characteristics of its own.

### 3.1.1   The Findings of the Questionnaire, Nearly 31% of Students Consider Themselves Associated with Stress, Sources of Stress

50% of graduation assignments, and economic difficulties, 10% each empty life, emotional stress, 14%, with more than stress, inevitably have a negative impact on students. That they have the mentality which accounted for 64% of inferiority, anxiety accounted for 18% of the state of mind. Cole had 17 million U.S. college students sample, 23.9% of the students showed an obvious psychological pressure. Pressure mainly from the "unemployment" and "spiritual crisis." My stress comes from the students a larger proportion of total emptiness of their lives, according to material reflects 52% of the students in the void when shopping, listening to music, and study only 29%, 15% go out drinking again, that I students a culture of learning needs to be strengthened, especially in today's market economy, human resources and as a commodity in the market competition, survival of the fittest inevitably produce. So should study hard, including the theory and technology in order to graduate as well as the allocation of future life in an invincible position

### 3.1.2   According to Another Survey 65% of the Students to Keep Their Minds at Ease, If Not Happy, They Have 62%

Students are able to regulate their own, out of the woods. This is more related to the professional students. Sports medicine and exercise physiology study found that people in the movement of the brain can lead to a happy and release some endorphins the body's production of larger, faster speed, so the ability to meet the daily mental distress is relatively strong. In terms of interpersonal social relations, 65% of the students thought that a more coordinated relationship, 34% of students believe that in general,

**Table 1.** Psychological Health Statistics

| Survey questions | Number | Percentage% | Survey questions | Number | Percentage% |
|---|---|---|---|---|---|
| **Spirit stress** | 46 | 31 | **Do not feel comfortable:** | | |
| Sources of stress | 73 | 50 | | 1 | 1 |
| | 26 | 18 | 1Talk to find his brother | 55 | 37 |
| | 26 | 18 | 2Looking for friends to talk | 91 | 62 |
| | 22 | 14 | 3Self-regulation | | |
| 1:Graduation assignments | 94 | 64 | **Interpersonal :** | 96 | 95 |
| 2:Economic difficulties | 26 | 18 | 1 : More coordination | 50 | 34 |
| | | | 2 : General | 1 | |
| 3:Emptiness of their lives | | 29 | 3 Uncoordinated | 28 | 19 |
| 4:emotion | 42 | 52 | 1 : More than mild mental weakness | | |
| | 77 | 15 | 2 : More than mild mental disorder | 35 | 24 |
| Inferiority Complex | 22 | 4 | 3 : Mental and physical health-related | 117 | 80 |
| | 6 | | 4 : Ideological education with psychological counseling | 61 | 42 |
| Anxious state of mind | 95 | 65 | 5 : the Importance of seting up psychological counseling | 128 | 87 |
| **Emptiness of lives:** | | | | | |
| 1;Learning 2:Shopping, listening to music 3: Drinking 4:make Friends | | | | | |
| Keep their minds at ease | | | | | |

significantly higher than that in this regard non-sport students. We know sports and competition, especially in some team sports on the relationship between athletes and act in harmony and coordination of high demands, and in competition or training, the team members can achieve tactical and technical understanding co-ordination, and their mutual relationships is directly related to coordination. A lot of practice has proven that athletes or a sport between the groups in the coordination of relationships whether exercise directly affects their level of technology to improve athletic performance and the acquisition, we can see their sport specific requirements, but also exercise and the psychological development of their ability.

### 3.1.3    Students Faced with Life Choices, Prone to Psychological Disorders. College Students Suffering from Mental Weakness Than

Cases as well. It is due to excessive tension in the brain dysfunction caused by excitation and inhibition of disease. The spirit of psychological factors in the conflict over the ability of the system to withstand the limit, it will lead to a breakdown. Jiangsu Province, the decade 1978-1987 University of the twelve were systematically investigated. Drop out rate due to illness rose from 1.9% to 6.5%, of which mental illness, mental disorder among the top place. In 1998, acceptance of the Tianjin Physical health statistics, the city's fifty thousand students in more than 16% in different degrees of mental disorders, about eight thousand students of psychology unhealthy. According to the survey 19% of material reflect more than the students with mild mental decline and 24% more than the students with mild mental disorder, which is a figure can not be ignored, that the mental health of college students sports low level of students in colleges and universities mental health. 80% of the students thought that mental health and physical health-related, 58% of the students that different ideological education and psychological counseling, or 42% of students believe that the same, indicating that students and thoughts on education, psychological counseling is not clear enough understanding. The results of this study to the management of schools and students such a problem that in strengthening the ideological education of students and moral training at the same time, the mental health of students should not be ignored. The root of the problem caused by different solutions adopted by different means and methods to better guide and help achieve targeted. The premise of doing this work is the officer must have some psychological knowledge, especially basic knowledge of mental hygiene aspects, should also acquire a certain amount of psychological control and treatment techniques. Only in this way can the training, education, management of students in the process, the ability to detect differences between the correct and scientific "psychological problems" and "moral issues" to truly take preventive measures. Counseling can not be attributed to a way of ideological education, psychological counseling can not attributed to a way of ideological education, not only can psychological counseling as unhealthy psychological approach. Consulting Psychologists Li Siman said the U.S. "advisory is to assist the process through interpersonal relationships, educational process, and the growth process." 87% of the students that the Faculty has established sport psychological consultation is necessary, that students recognize the significance and importance of psychological counseling. Through psychological counseling to help students develop mental potential, enhance the ability to adapt, improve relationships, improve quality of life. The above description of college students in the Sports Institute of Psychological consultation to help them scientific and effective regulation of psychological problems, to adapt to students learning and life is very necessary and urgent.

**Table 2.** Statistics survey of psychological tendencies

| questions | Number | Percentage% | questions | Number | Percentage% |
|---|---|---|---|---|---|
| **Students interest** | | | **Students interest** | | |
| 1: Learning | 14 | 10 | 1: Complete, more consistent | 87 | 60 |
| 2: Sing and dance | 20 | 13 | 2: Not completely consistent | 60 | 40 |
| 3: Video movies | 25 | 17 | | | |
| 4: Other | 88 | 60 | **Graduate destination** | | |
| **The needs of students** | | | 1: Find a good job | | |
| 1: Abundant spare time | 26 | 18 | 2: Further studies | 77 | 52 |
| 2: Make close friends | | | | | |
| 3: Improve the professional standards | 35 | 24 | 3: Let matters drift | 58 | 40 |
| 4: Other | 71 | 48 | | 12 | 8 |
| | 15 | 10 | | | |

## 3.2   The Psychological Tendency to Regard the Students

**3.2.1** student interest   Interest is the positive perception of a thing or psychological tendency to engage in certain activities. This tends to pay attention and emotion in close contact with, which always give priority attention to certain things and have feelings of longing. Students interested in physical education is very broad, according to materials reflect, sing, dance 13%, movies, video 17%. In contrast, the proportion of students prefer to learn only 10%. To give people the feeling of playing heavy heart. Professional knowledge of the breadth and depth of the better basic conditions for professional work. Therefore, we should strengthen the ideological education of students so that they become a high theoretical knowledge.

**3.2.2** students need Need is the basis for generating motivation, but motivation is to promote all activities of the internal motivation of students, is the personality and enthusiasm of the internal sources, it is the cause of human behavior and the behavior of people toward a certain goal, in order to their own satisfaction, the results from the survey analysis, full of students learning to improve professional standards is the first requirement, accounting for 48%, followed by 24% to make close friends, once again, 18% of the rich leisure life. Other 10%, indicating that students realize that I learn. It is not difficult to see, along with social development, in particular the establishment of a market economy, the value orientation of students and social needs of the great changes taking place, how to guide students to correctly understand the social division of labor and social integration, that is, raising the level of sports professionals, but also to improve social needs of a variety of ability is very important.

**3.2.3** student aptitudes Students learn professional and personal goals of relevance. Students have learned from the material to see exactly the same profession and inclination, and relatively consistent 60%, indicating that most students choose physical education is more in line with the aspirations of individual interest. While 40% of the students selected professional and personal inclination is not exactly the same, indicating that this part of the PE students are forced to choose so to do, most of them turned out to be learning science and engineering, cultural performance is better, but up to less than the minimum score of science and engineering, had to temporarily find a way out, although personally do not like sports, but after a "crash course" just to achieve a minimum score of sports, they were admitted to the higher cultural achievements of professional sports. Most of these students physical education and overall physical fitness poor skills base, plus do not like or even hate sports, to professional teaching in more difficult, facing this part of the students and the future of the extensive needs of the market economy, schools and teachers should through ideological education and strict organization of teaching, scientific selection tools and methods of teaching. Students interested in this part of the professional learning guide to learn the sport.

**3.2.4** students graduate destination Department of Education is primarily a sports training physical education teachers and special education all kinds of sports talent, training objectives very clear. 52% from the survey reflect the students graduate, hope to find a good job. However, due to the different values of individuals, each person in the eyes of the "good" work vary widely, especially with the development of market economy and the distribution mechanism of the adjustment, social distribution become more inclined to "hard work and" the principle of fairness, therefore, to education students to recognize the return of the community mainly depends on individual personal contributions to the community, and be rewarded. 40% of the students to continue their studies, students should be given for this part of the understanding and support, as far as possible to create certain conditions for them to meet their needs for knowledge, for the country develop a higher level of talent. With economic development, the high degree of social increasing demand, high-level talents in the field of sports is no exception. That society is progressing, personnel standards are improved.

## 4    Conclusions and Recommendations

**4.1** The study's findings show that the Mental Health Physical Education Department's overall mental health of university students was lower than the average level of mental health problems show different professional institutions of higher learning in our universities have a certain universality, so Institute of Current Psychological Counseling in sports it is very necessary and urgent.

**4.2** Physical Education Department in the mental health of students tend to stress and psychological problems and needs of students interested in the results of analysis showed that poor students active learning culture must be strengthened, otherwise it will cause difficulties in school and usually assigned a "spiritual crisis".

**4.3** Physical Education Department in the mental health of students interpersonal and self-regulation is better, there a high level of health, their learning is directly related to the special nature of the content.

**4.4** Most of the students selected professional and personal hobbies and interests, 40% of students with different aptitudes and selected professional schools and teachers must be thinking, psychological education, science teaching methods and methods of selection, guiding them to the sport of professional learning in the past. 52% of students eager to find a satisfying job after graduation, 40% of the students to continue their studies, the school should try to guide them correctly and create favorable conditions for the pursuit of new knowledge to meet their needs. I believe that this will greatly change the students attitude towards learning, improve student learning initiative.

# References

1. Jia, G.: Physical and Health Education. Sports Culture Guide (December 2005)
2. Qi, Y.: Mental Health Education Research Approaches. Xi'an Institute of Physical Education (January 2005)

# Transient Impact Signal's Detection Based on Wavelet Transformation

Xu Cao[1] and Hua-xun Zhang[2]

[1] Center of Modern Education Technology, Changchun Chinese Medicine University,
Changchun, 130051, China
zyxycx@163.com
[2] Electronic Engineering College, Changchun University, Changchun, 130022, China
ccdxzhx@163.com

**Abstract.** The use of wavelet transformation in detecting transient impact signal is mainly discussed in this paper. It has been validated through experiment of axletree's rolling body broken. For standard cycle broken signal, we can't find out its obviously frequency on frequency spectrum, but wavelet transformation can large particular signal which contained malfunction. The result will be known quickly from it. So wavelet transformation fit to detect transient abnormality signal in natural signal.

**Keywords:** signal processing, wavelet transform, MATLAB, signal's detection.

## 1 Introduction

For the moment, technology of diagnose malfunction mainly based on Fourier transformation, so the essential contradiction of Fourier analysis on time domain and frequency domain will be faced, and that stationary signal is a precondition for Fourier analysis, but most of broken signal in control system is contained in transient signal and time-varying signal. Time and frequency analysis can offer both signal and variational degree of signal in some time. This is a new method named scale analysis to analyse signal. This method inherit idea of Fourier analysis to use harmonic function as primary function in order to approach random signal. At the same time, primary function of wavelet analysis is a series function of its scale can be transformed. This make wavelet analysis have good characteristic of time-frequency positioning and ability of signal self-adaption. So it can be used to decompose all kinds of time-varying signal, no distracter. So wavelet transform can be used to detect and fault diagnosis technology for control system in mechanical equipment.

## 2 Statement of Problem

In actual fault diagnosis system for control system, especially in abundant impact signal, signal characteristic of frequency is very important in random time, for example reciprocating machinery, axletree in running period, and so on. A result is concluded

G. Lee (Ed.): Advances in Intelligent Systems, AISC 138, pp. 309–313.
springerlink.com      © Springer-Verlag Berlin Heidelberg 2012

wavelet transform is a set of filter, be made of a low-pass filter and a series of band-pass filter. Through these filters, input signal is split different frequency, include high frequency and low frequency. High frequency can show detail, low frequency can show general, but signal character is not changed in time. Because axletree's fault is usually transient impact belong to periodic signals in rotary machine, obviously frequency is not found on frequency spectrum, but wavelet transform which can depict signal's local quality through multi-resolution can show its transient abnormal phenomenon imbed in normal signal.

In many literatures, signal local character is detailed by Lyapunov exponent, form of Lyapunov exponent follow as:

$$|x(t_0 + \delta) - f_n(t_0 + \delta)| \leq A|\delta|^\beta, n < \beta \leq n+1 \tag{1}$$

where $\beta$ is Lyapunov exponent of signal $x(t)$ at time of $t_0$, $f_n(t)$ is polynomial of degree n across $x(t_0)$, $\delta$ is a submin-measure.

At scale $a = 2^j$, if inequation (2) exist, then $\tau_0$ is called local maximum of model under scale of wavelet transform $a = 2^j$.

$$|WT_x(a,\tau)| \leq |WT_x(a,\tau_0)|, \tau \in (\tau_0 - \sigma, \tau_0 + \sigma) \tag{2}$$

$|WT_x(a,\tau_0)|$ is maximum of module of wavelet transform. Lyapunov exponent $\beta$ of signal $x(t)$ and maximum of module of wavelet transform is satisfied with expression of (3),

$$Log_2|WT_x(a,\tau)| \leq Log_2 K + \beta j \tag{3}$$

$K$ is a constant about wavelet basis. Therefore we obtain a relation between scale of wavelet transform $a = 2^j$ and Lyapunov exponent, as well as evolution between $a = 2^j$ and transform $d_k^{(j)}$

Where $\beta > 0$, maximum of $d_k^{(j)}$ will increase along with scale $a$, where $\beta = 0$, maximum of $d_k^{(j)}$ will immovability, where $\beta < 0$, maximum of $d_k^{(j)}$ will decrease when scale $a$ increase.

From that, we know vibratory signal effects wavelet coefficient $d_k^{(j)}$ more at bigger scale $a$. And that noise effects $d_k^{(j)}$ more at scale $a$ less. So we can transform vibratory signal with noise, in order to separate vibratory signal and noise, and enlarge noise, then use proper measure to filter every subband signal.

## 3    The Actual Application of Wavelet Transform

Axletree is the most important components in rotary engine. Hardware of axletree will strike scathing dot seasonal when axletree appears local scathe in loading operation, a series weakening shake will come into being, frequency of happening weakening shake is malfunction character frequency. We can detect axletree malfunction and position through malfunction character frequency .

Experimental design: An acquisition system for experimental data is established on a monitoring system about reducer experiment. Reducer's fact running state is simulated, control system is master of its ride, change its all kinds of status, monitoring system complete gather sample data. Local malfunction of axletree is put up, malfunction frequency will be calculate from (4):

$$f = \frac{2 \cos \alpha \sin \alpha \sin \beta}{\sin^2 \alpha - \sin^2 \beta} |f_o - f_i| \qquad (4)$$

$\alpha$ is contact angle of coniform ball bearing and outer-race ball track, $\beta$ is contact angle of coniform ball bearing and in-race ball track, $f_o$ is gyro frequency of outer race, $f_i$ is gyro frequency of inner ring.

In expression $\alpha$ =15.5°, $\beta$ =12.5°, based on experimental data, character frequency of malfunction is calculated:

$$f = 238.19Hz$$

## 4    Simulation

Utilize MATLAB programme simulate, result of wavelet transform is proved. Through 10000 Hz sampling, axletree rolling body malfunction signal is analyzed and refine.

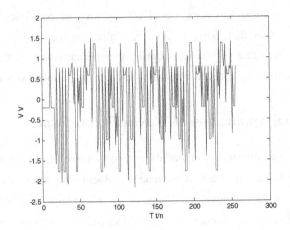

**Fig. 1.** Malfunction signal of axletree rolling body

Fig 1 is the time domain oscillogram, we can't know if malfunction is exist from it.

Use 10 db orthogonal wavelet basis to analyze signal of Fig 1, four layer wavelet detail signal is obtain, look Fig 2. Through Hilbert envelope and spectrum analysis to the first detail signal, we will see Fig 3.

From Fig 3 we find special frequency exist at 200Hz- 250Hz, contrapose to malfunction character frequency, we know axletree malfunction happened.

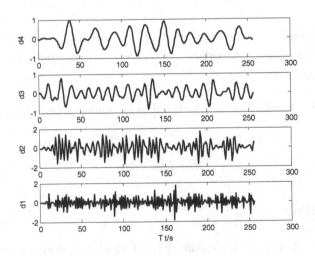

**Fig. 2.** Result of wavelet decomposition

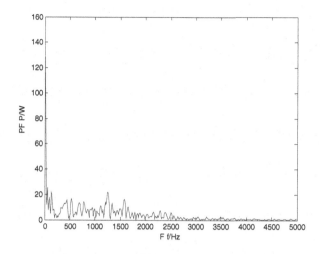

**Fig. 3.** Power spectrum of rolling body malfunction

## 5    Conclusion

Taking advantage of wavelet analysis to transform detecting signal, then reconstruct signal having fault character, in order to demodule and refine frequency analysis. It has been found a good method. To the quasiperiodic signals in rotary mechine. malfunction frequency can't be found obviously on frequency spectrogram, then wavelet transformation can enlarge detail of impulse signals, malfunction can be found quickly, contrast this frequency and malfunction frequency under all kinds of circs, we even find malfunction position.

## References

1. Donoho, D.L.: De-noising by soft-thresholding. IEEE Transactions on Information Theory 41(3), 613–627 (1995)
2. Wavelet Toolbox User's Guide. Mathworks Inc. (2004)
3. Altmann, J., Mathew, J.: Multiple band-pass autoregressive demodulation for rolling-element bearing fault diagnosis. Mechanical Systems and Signal Processing 15(5), 963–977 (2001)
4. Fan, H., Peng, Y.: Interpreting Scattering Mechanism of Radar Target by Wavelet Transform (1995)
5. Daubechies, I.: Orthonormal Bases of Compactly Supported Wavelets (1988)
6. H H Szu.Brian Telfer.shubha Kadambe Neural Network Adaptive Wavelets for Signal Representation and classification 1992(09)
7. Mallat, S.: Singularity Detection and Processing With Wavelets (02) (1992)
8. Kumar, P., Foufoula-Georgiou, E.: A multicomponent decomposition of spatial rainfall fields 1. Segregution of Large-and Small-Scale features using Wavelet transforms. Water Resources Research 29(8), 2515–2532 (1993)
9. Venckp, V., Foufoula-Georgiou, E.: Energy decomposition of rainfall in the time-frequency-scale domain using wavelet packets. Journal of Hydrology 27(3), 3–271 (1996)

# The Research of Log-Based Network Monitoring System

Li Zhang

College of Computer Science and Technology Changchun University
Changchun, China
ccuzl@126.com

**Abstract.** Log-based Network Monitoring System was researched and designed for network security which is aimed at gathering and counting device logs, and at the same time, persisting the final results of the analysis to provide network traffic and intrusion detection and the generation of other relevant reports. LNMS consists of log receiver, log processor, report engine, report scheduler, Web Services interface and web application.

**Keywords:** Network Monitoring, LNMS, Log.

## 1    Introduction

Computer network technology with the Internet era has been a rapid development and application, people can get more and more network resources and services. Along with the increasing number of packets of network services, a variety of network vulnerabilities which are targeted by hackers and viruses are transmitted into the network. Therefore, the monitoring of the network has become one of the key issues in network management. Network monitoring is aimed to provide network managers the way to analyses statistical information of network traffic and intrusion detection, in order to improve security protection and contribute to better use of network.

Network monitoring, in terms of basic functions, typically include three aspects: (1) total network bandwidth usage (easy to analyze the network using the peak traffic bottlenecks, usage of various services and other issues); (2) intrusion detection; ( 3) user-defined situation monitoring.

Network monitoring system is mainly implemented in two ways: (1) based on packet capture technology; This technique is also known as network sniffer (Sniffer), it captures all the network packets and analyzes the data and provide statistical conclusion; (2) Gateway-based log analysis technology. This method is combined with the gateway device, this technology analyzes the logs from the built-in firewall.

User's internal network is usually connected through the gateway device and connected Internet, the gateway device is usually built into the network packet processing system and firewall systems, by analyzing and filtering through the gateway and provides corresponding log that includes network traffic, intrusion detection and other information, the log reflects the basic situation of the current network, all network data is concentrated. Therefore, we can build a log-based analysis of network monitoring system (Log-based Network Monitor System, LNMS), by acquiring and analyzing these logs to achieve the purpose of network monitoring [1-2].

G. Lee (Ed.): Advances in Intelligent Systems, AISC 138, pp. 315–320.
springerlink.com          © Springer-Verlag Berlin Heidelberg 2012

## 2    LNMS General Design

### 2.1    LNMS Design Goals

LNMS is based on the device logs the network monitoring system by analyzing the log generated by the firewall in the gateway device, it generates the final status of the network with various statistics and reports according to user's needs which determines the system's basic design goals: (1) system functions to support multiple network devices, with Disassembling and configurable, each module can be individually configured and loaded, meanwhile its core functions  are called by other applications; (2) system compatibility, stable running, easy to install,  friendly use; (3) supports a variety of modes, not only as a standalone application, but also to provide external interfaces ; (4) The current system needs to support several common log format, and the need to fully consider the possible future of the other log formats for easy system upgrades [3].

### 2.2    LNMS Functional Design

LNMS as an independent system, needs to provide the following functions: (1) user management function, the practical application of LNMS access is limited, different levels of users can access different functions; (2) system backup and recovery function, the system will backup stored data to prevent system crash while cause data loss, mean while  a recovery feature that enables a system restore to the point of working condition; (3) data import and export functions, it can use the existing log into the LNMS, it can export data to a general format file as well which is compatible to other applications in the system; (4) report customization capabilities, according to the user's requirements the system can generate customized reports  and send to the user at a specified time; (5) the system configuration, LNMS have many operating parameters to be configured, and therefore a corresponding module is required; (6) External interface functions, to provide Web Services, EJB, etc. to facilitate the interface with other systems of communication and integration. LNMS general structure is shown in Figure 1 [4].

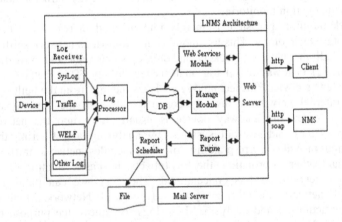

**Fig. 1.** LNMS general structure diagram

According to the system frame diagram, it is shown that the LNMS system consists of the following function modules:

(1) log receiver module (Log Receiver): responsible for receiving all kinds of logs from the device and redirects to the log file;

(2) log processing module (Log Processor): responsible for receiving the log from the computer where log receiving program bases, statistically analyzing log files , and finally achieving persistence;

(3) report generation engine (Report Engine): is responsible for the generation of all kinds of statistical reports, which is stored in pdf, rtf, html, csv formats;

(4) scheduling module (Report Scheduler): responsible for tasks such as scheduling generation and delivery of  back end reports;

(5) Web Application: is responsible for foreground applications based on J2EE architecture;

(6) Web Services Module: is responsible for providing the interface to communicate with other systems.

# 3    LNMS Detailed Design and Implementation

## 3.1    Log File Access

Network devices logs are sent to the computer where the log receiving program is in. However due to the fact that log receiving part and log analyzing part may not be on the same computer in real practice of LNMS, accessing log files from the computer where the log receiving program is in needs to be done first  in order to analyze the log. Getting log process used by LNMS is shown in Figure 2.

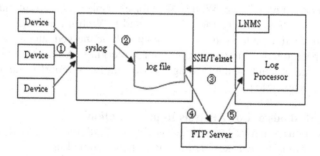

**Fig. 2.** Getting log process of LNMS

## 3.2    Log File Analysis

After log file is downloaded to local by the log file getting thread, it is processed to the analysis module and the analysis results will be stored into database [5].

### 3.2.1    Log Format

Log is the local system input, which is the direct source of all information of systems, therefore, it is essential thoroughly analyze logs of varying formats from devices. The current log formats have three kinds: Syslog, Traffic log and WELF, and the Syslog log format used here is defined by the RFC3164 format as follows:

<PRI> Mon dd hr: mm: ss hostname src = "srcIP: srcPort" dst = "dstIP: dstPort" msg = "message" note = "note" devID = "mac addr" devType = "ZW70" cat = "category "

Seen that the log format consists of 10 fields, including: (1) *PRI* indicates priority value; (2) *Mon dd hr*: mm: ss indicates the time when log is generated from device; (3) *Hostname* indicates the device name; (4) *src* indicates the source IP address and port number of the network event corresponding to the log; (5) *dst* is the destination IP address and port number; (6) *devID* indicated the last six digits of the device's MAC address; (7) *devType* indicates the device type; (8) *cat* indicates log type, which is a key field in the first analysis, with possible values such as "System Maintenance, Remote Management" "System Errors" "Access Control, TCP Reset, Packet Filter", etc. ; (9) *msg* and *note* indicate the specific content of the log, corresponding to the second and third layer analysis, which is the main body of system log analysis.

### 3.2.2    The Log Analysis Purpose and Prototype

Log which is obtained from device, contains the basic information of IP layer data stream, but what is received is a continuous string, therefore, log analysis is required to extract useful fields and achieve persistence. As the amount of logs obtained from the device is very large, a deeper analysis of the log has to be done after initial treatment but before persistence, in order to merger the logs and have preliminary statistic, which will reduce the number of visible log in the reporting engine [5].

The merging of the logs includes firstly log initial treatment of classification; then, merging the log information (cumulative of data flow number of events and other information ) which has have the same source and destination addresses in the smallest statistical unit of time. Web traffic, for example, has the minimum statistical time unit of hour, therefore for those logs related to Web traffic, logs with the same source and destination address in the same hour will be merged, and only one relevant logs will appear on the summary report after merging. After initial statistics and combination, the amount of data storage in the summary report is far less than the amount of the original log.

### 3.2.3    The Methods of Log Analysis Implementation

Log processing uses a two-layer analysis. First of all the classification of log is achieved according to *cat* key fields, and then detailed analysis is carried out according to *msg* and *note* fields. Regular Expression is a powerful tool in pattern matching and replacement which is frequently used in lexical analysis programs and the use of regular expressions greatly contributes to simplification of the writing of string operations codes.

Basically the first layer analysis separates various fields in the information of IP layer from the log (such as source IP address and port number, destination IP address and port number, time, traffic and other information) and store the results in the database; The analysis results is the second layer input, the second layer analysis calls the corresponding rules for analysis based on the specific values of each field.

The various log fields have been separated in the first layer, and the specified category of the log content is distinguished based on cat field in the second layer; then, according to the *cat* field, corresponding software are called to analyses. *Msg* and *note* fields in the result of first layer are the main input into the second layer, since *msg* and *note* fields contain more detailed log information, such as the specific method of the invasion, the login user name and other information. After the second layer of analysis, the analytical results is compared with the original data in the database , and the operation of merging is carried, when the initial statistics of the log is achieved [6].

## 3.3   LNMS Core Reporting Features

System report includes eight areas: (1) real-time monitoring of network traffic flow and a variety of real-time monitoring of network services; (2) Web site visiting records; (3) filtration of report for visiting to illegal Web sites; (4) a variety of commonly used Web services use records; (5) a variety of user-defined traffic monitoring; (6) intrusion detection analysis records; (7) error packet records; (8) equipment visiting records.

Analysis and statistic need to be carried out in varying ways taking consideration of users' requirement, such as user rankings, site ranking, site names, time granularity, the time frame and other specified conditions. Users are able to freely combine these statements to generate reports. Report generation module is the one of the core parts of the structure which is shown in Figure 3.

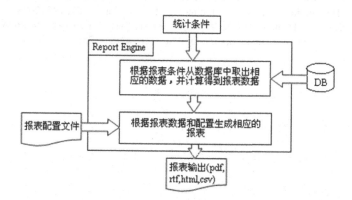

**Fig. 3.** Reports module

Report generation module consists of statistical data generation part and report generation part, where the former one is responsible for statistical data collection and generation, and the latter is responsible for generating report out based on the statistical data and configuration files.

LNMS system is structured based on J2EE, and implemented by using MVC pattern of Struts. In the design process the system configuration module, the user management module and the system's external interface with other modules have also been fully considered.

## 4    Conclusion

In this paper, the network monitoring system implementation techniques are discussed, and based on log analysis techniques a LNMS networking monitoring system has been designed and implemented, which uses Java technology, EJB encapsulated core function module with high degree of modularity and portability, to achieve a good structure. Using gateway log as data source is the main innovation in LNMS, and improving performance has been shown by experiments compared with traditional technologies of packet capture, while future research in how to support more log formats is still to be done.

## References

1. Huang, J., Li, J.: Research of Intrusion Detection Based on Firewall Log. Computer Engineering 27(9), 115–117 (2001)
2. Li, C., Wang, W., Cheng, L., et al.: Study and Implementation of Network Security Audit System Based on Firewall Log. Computer Engineering 28(6), 17–19 (2002)
3. Li, Z., Li, X.: Research and Design of Log Analysis System Based on Association Rule. Microcomputer Applications 25(3), 27–29 (2009)
4. Huang, W., Wen, C., Ou, H.: The distributing network attack indication warning system model base on analysis policy of log files. Natural Science Journal of Xiangtan University 26(4), 39–42 (2004)
5. Li, W.: The Research of Network Intrusion Forensics Systems Based on Windows Log Analysis. Heilongjiang Science and Technology Information 16, 57, 195 (2008)
6. Wen, J., Xue, Y., Duan, J., et al.: Weblog Analysis System Implemented on Association Rule. Journal of Xiamen University (Natural Science) 44(S1), 258–261 (2005)

# Study on a Kind of Mobile Phone Signals Monitoring and Shielding System

Nian-feng Li, Meng Zhang, YongJi Yang, and He Gu

College of Computer Science and Technology, Changchun University, Changchun,
Jilin Province, China
cculinianfeng@126.com

**Abstract.** A kind of composition and architecture of mobile phone signals monitoring and shielding system were introduced. The work principle of the system in this kind of architecture was provided. The key technology and implement methods in research process were analyzed. The prototype was developed and the experiment results as well as analysis were given. The results of experiment and application show that this kind of mobile phone signals monitoring and shielding system could preferably realize the monitoring and shielding function in special locations where mobile phones are not allowed to use.

**Keywords:** signal monitoring, signal shielding, wireless location.

## 1 Introduction

With the rapid development of communication technology, mobile phones have become indispensable to people's daily life and work as an important auxiliary tool. According to statistics, at the end of 2010, mobile phone number has reached 600 million or so in China. However, there are some issues can not be ignored with extensive use of mobile phones:

(1) Being a new leak channels to the security information. with the development of science and technology, positioning, tracking and surveillance to mobile phone have become important means of reconnaissance. In 2000, an report from the EU pointed out that a code-named "Echelon" global electronic eavesdropping detection network system established by the United States National Security Agency, has been stealing the intelligence and lead to at least twenty billion Euro business losses for the EU. This intelligence gathering is the main form of mobile communications content filtering monitor.

(2) becoming the new noise sources, interferes with the normal working order. Increasingly serious ring tone pollution make public places such as conference room, courts, libraries, schools and so on further environmental degradation, affecting people's normal work; In addition, many cell phone signal interaction of electromagnetic radiation has become another important source of pollution in everyday life.

(3) Being a new criminal means, threat the social harmony and stability. In some occasions, mobile phones have become tools for criminals to commit crimes, such as,

G. Lee (Ed.): Advances in Intelligent Systems, AISC 138, pp. 321–326.

some opportunistic people use mobile phones for cheating, and terrorists detonate bombs using mobile phones.

The development of cell phone signal monitoring and shielding systems can effectively solve these problems. A kind of composition and architecture of mobile phone signals monitoring and shielding system were introduced in this paper.

## 2    Composition of   Mobile Phone Signals Monitoring and Shielding System

Mobile phone signals monitoring and shielding system consists of mobile phone signal detectors, phone signal jammers, sub-station and host computer with control management software. The system workflow is: Open the cell phone signal detectors, then they will be in monitoring state, when abnormal signal detected within the deployment area (some illegal use of communications equipment), the detectors judge the signal frequency and the corresponding alarm information uploaded to the sub-station, sub-station management information and according to corresponding alarm information to start the signal jammers to block signals in deployment area, at the same time alarm information will be uploaded to the host computer, the host computer automatically displays the installation location of alarm detector and stored the alarm information in database, and remind monitor treatment to staff. Architecture of automatic monitoring and shielding system was shown in Figure 1.

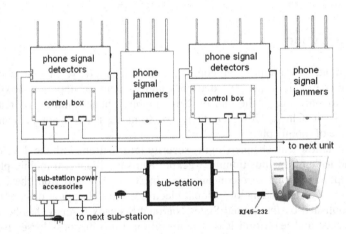

**Fig. 1.** Composition of mobile phone signals monitoring and shielding system

### 2.1    Mobile Phone Signal Detectors

Mobile phone signal detectors require accurate detection of CDMA, DCS, PHS and other mobile phone signal, when detects mobile phone signal give out alarm information to the sub-station. The working principle (shown in Figure 2) are: the wireless signals received by the antenna are filtered and amplified to give field strength meter circuit, field strength meter circuit output corresponding DC signal, the signal is amplified and sent into the microcontroller for determination and comparison, when the value is higher than the alarm value, the microcontroller give out alarm information to the sub-station.

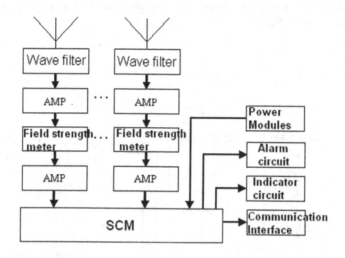

**Fig. 2.** Schematic of mobile phone signals detector

## 2.2    Phone Signal Jammers

Known by the mobile communication principle, mobile phones work in a certain frequency range and link to base station by radio waves, and use a certain baud rate and certain modulation for data transmission. In response to this communication means, phone signal shielding can be realized by scanning on certain speed from the low frequency to high frequency, then the interference will be gathered in messages received by the mobile phone. Interference signal make phone communication in this frequency range can not detected correctly the base station data packets, and thus can not establish a connection with the base station, the final performance to mobile phones were search network, no signal, no service system, and so prevent a normal mobile phone registration will eventually blocked the normal phone communication functions.

Mobile phone signal jammers need to shield signals include: a) 800MHz CDMA down; b) 900MHz GSM down; c) 1800MHz DCS down; d) 1900MHz PHS and 3G mobile phone down; e) 2100MHz 3G down. To achieve cell phone signal shielding two ways can be used: (a) active shielding methods; To interfere the normal receive mode of Mobile phones, continuous band downlink frequency signals was transmitted from jammers. In this way the actual circuit is relatively simple, and has good shielding effectiveness. But the continuous emission of interfering signals will cause a certain degree of electromagnetic pollution and waste of energy. (B) passive shielding means. Use cell phone signal detectors to receive real-time signal, when the detector receives a protective response within the signal transmitted by the phone, inform the base station to start shielding device, launch the corresponding frequency interference signals, the phone can not communicate properly. The system used in the second option.

Circuit consists of five groups of the same signal circuits, and its structure is the sawtooth wave generating circuit into a corresponding production ramp, drive VCO circuit produces a corresponding frequency signal, the frequency signal output by the amplifier AMP as an effective interference signal, the final antenna radiation to a specific spatial range (Figure 3).

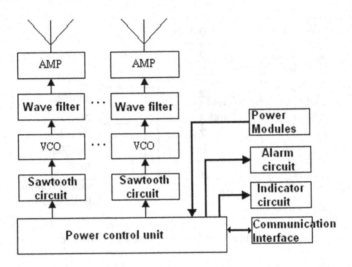

**Fig. 3.** Schematic of mobile phone signals jammer

### 2.3   The Sub-station and Control Console

(1) Sub-control management station
This section includes the ARM chip, external memory chips and liquid crystal display screen, and have functions of system information detection, monitoring state information storage, signal control, driving liquid crystal screen and displaying system status, communication with host control computer, etc.

(2) The control console
The control console worked with the mainstream PC and installed special software designed for this system, online communicate with the sub-control management station for receiving ,sending and storing the status information, and prompt the staff on duty when alarm. The software was developed with visual RAD (Rapid Application Development) tool Delphi, database using new version Microsoft's launch of the SQL Server database management system.

## 3    Key Technologies and Implementation

### 3.1   Detector Sensitivity and Immunity Problems

In this project, cell phone signal detectors require accurate detection of CDMA, DCS, PHS and others, up to 5-band cell phone signals, the frequency range from 800MHz to 2.1GHz, and not continuous, so the detector use classification methods to detect, 5-way independently electrical part to provide the detection function. A reasonable package of these five independent circuits and under the premise of ensuring a high sensitivity to improve anti-jamming capability is one of the key issues to be addressed. System uses the modified Kalman filter [1,2] and other software and hardware anti-jamming technology to solve the interference problem in systems integration.

### 3.2 Electromagnetic Compatibility of the Jammers

To play a role in shielding, jammers need to transmit interference signal, continuous emission of jamming signal would cause a certain degree of electromagnetic pollution and waste of energy [3]. Reasonable control algorithm for making jammers keep phone signal shielding within a certain time and a certain space is another key problem.

### 3.3 The Positioning of Mobile Phone Calls

Under direct wave environment using multidimensional scaling method (MDS) and non-direct wave environment using Time of Arrival (TOA) and angle of arrival (AOA) measurement enhanced location method [4,5], and with multi-sensor to achieve fast, intelligent positioning of phone calls is another key issues and innovation of project studies.

## 4 Experimental Results and Analysis

Table 1 shows the typical experimental data of the system application in a prison. Data show that: (1) phone signal around special place shielding and monitoring can be achieved by the development of this a mobile phone signals monitoring and shielding system; (2) anti-jamming capability of mobile phone signal detectors to be further improved; (3) integrated use of monitoring data for phone calls positioning algorithm could be improved.

**Table 1.** Resulting data of mobile phone signals monitoring and shielding system experiment

| Observation time | Get signal times | Effective signal times | Shielding rate | Hardware response time |
|---|---|---|---|---|
| 4 weeks | 23 times | 18 times | 100% | <1s |
| Software average response time 1.2s | | Software positioning accuracy rate 67% | | |

## 5 Conclusion

Phone signals monitoring and shielding system is a complex system of hardware and software work together, and it's difficult to develop. A kind of composition and architecture of mobile phone signals monitoring and shielding system were introduced. The work principle of the system in this kind of architecture was provided. The key technology and implement methods in research process were analyzed. The prototype was developed and the experiment results as well as analysis were given. The results of experiment and application show that this kind of mobile phone signals monitoring and shielding system could preferably realize the monitoring and shielding function in special locations where mobile phones are not allowed to use.

# References

1. Hong-yue, W.: Study on Predictive Control of Fresh Steam Temperature by Using Advanced Kalman Filters. Power Engineering 25(5), 685–687 (2005)
2. Yin-kun, W., Ming-qing, X.: A Location Algorithm of Target Based on Improved Kalman Filter. Journal of Air Force Engineering Universigty (Natural Science Edition) 3(5), 17–20 (2002)
3. Hui-xi, L., Peng, Z.: Calculation of mobile phone signal blocker's shielding range and influence to communication systems. Electronic Design Engineering 18(4), 51–54 (2010)
4. Wan-chun, L., Ping, W.: Subspace Approach in Complex Plane for Mobile Positioning with Time-of-Arrival Measurements. Journal of University of Electronic Science and Technology of China 39(3), 361–363 (2010)
5. Yuan-hua, H., Hong-sheng, L.: Distance geometry TOA-based wireless location algorithm. Computer Engineering and Applications 46(12), 112–114 (2010)

# Effects of Electron Donors on the TiO2 Photocatalytic Reduction of Heavy Metal Ions under Visible Light

Xin Zhang, Lin Song[*], Xiaolong Zeng, and Mingyu Li

Department of Environmental Engineering, Jinan University
Guangzhou, China
zhangxinhuagong@126.com, songlin113@163.com

**Abstract.** The effects on TiO2 (Degussa P25) photocatalytic reduction of Cr(VI) under visible light, using methanol, methanal and formic acid as electron donors were investigated. The results showed that the photocatalytic reduction of Cr(VI) could be encouraged by methanol, methanal and formic acid. The fastest rate of Cr(VI) photoreduction was observed in the presence of formic acid followed by methanal and methanol. Cr(VI) could hardly be reduced by TiO2 without electron donors. The conversion percent of Cr(VI) was 100% using formic acid as electron donors after 80 min. For the methanal and methanol systems, the conversion percent of Cr(VI) were 93.62% and 22.69% after 6 h, respectively.

**Keywords:** TiO2, electron donors, visible light, photocatalytic reduction, Cr (VI).

## 1  Introduction

Heavy metal ions can not be degradated and they will be in environment for a long time. On the other hand, heavy metal can make creature disease through bioaccumulation and bioenrichment. So people have been trying to find method for heavy metal ions removing. Neutralization, electrolysis, chemical oxation-reduction, extraction, adsorption, precipitation, ion exchange, membrane separation, elution, electrodialysis etc. were often used for wastewater containing heavy metal ion treatment[1]. These techniques can play a role more or less. However, the effectiveness for wastewater containing low concentration heavy metal is reduced. Moreover, most of these technologies are simple physical separation process, and pollutants did not really change into virulent or harmless substance. It is easy to cause secondary pollution[2].

$TiO_2$ is paid more attention to eliminating environmental pollutants in recent years[3-5]. It is high oxidation power, low energy, low price, easy availability, non-toxicity. Especially using in low concentration wastewater treatment, its advantages is more outstanding[6].

The efficiency of photocatalysis will be lowered by the recombination of photogenerated electron and hole. The recombination could be restrained when electron donors used as scavenger of hole were added into and the efficiency of

---

[*] Corresponding author.

G. Lee (Ed.): Advances in Intelligent Systems, AISC 138, pp. 327–333.
springerlink.com          © Springer-Verlag Berlin Heidelberg 2012

photoreduction reaction would increase. On the one hand, inorganic heavy metal ions and organic pollutants together are in ctual pollution systems usually and many organic pollutants are excellent electron donors in the waste water. Therefore, it is of great importance that the combination between the reduction of toxic heavy metal ion and the oxidation of organic pollutants oxidation through $TiO_2$ photocatalysis technology. On the other hand, the photocatalytic reaction is a group of related oxidation reduction process in essence. If metal ions used as electron receptor and organics used as electron donor are coexistence, the reduction reaction and oxidation reaction might promote mutually[7]. All the research about the effect of electron donors on the $TiO_2$ photocatalytic reduction reaction were carried out under ultraviolet light irradiation. In this paper, the effects on $TiO_2$ (Degussa P25) photocatalytic reduction of Cr (VI) under visible light irradiation, using methanol, methanal and formic acid as electron donors were investigated.

## 2    Experimental

### 2.1    Chemicals and Instruments

Degussa $TiO_2$ powders (BET area 50 $m^2g^{-1}$; anatase $TiO_2$ accounted for 80% and rutile $TiO_2$ accounted for 20%) were used as the photocatalyst. All the chemical reagents were analytical reagent and used as received. Distilled water was used throughout the experiment.

The LED (blue, wavelength range 450~475 nm, shenzhen lanbaoli photoelectric technology Co. Ltd) was used as the source of visible light. The concentration of Cr (VI) was determined by 721 spectrophotometer (shanghai precision and scientific instrument Co. Ltd). KDC-160HR high speed freeze centrifuge (Keda innovation Co. Ltd) and acidometer (HI98130, HANNA) were used through out the experiment.

### 2.2    Analysis and Characterizations

The concentration of Cr (VI) was determined by using the diphenyl carbazide colorimetric method at 540 nm (GB 7466-87, Standards of China).

### 2.3    Photocatalytic Reduction of Cr (VI)

The photocatalytic reduction of Cr (VI) was carried out in a cylindrical glass vessel, a 3w LED used as the source of visible light, which was positioned alongside the vessel. $TiO_2$ (0.04 g) was suspended into 40 mL aqueous solution of $Cr_2O_7^{2-}$ (10 mgL$^{-1}$) with the addition of electron donors (methanol, formaldehyde and formate). Prior to irradiation, the suspension was magnetically stirred in the dark for 30 min to ensure the establishment of adsorption-desorption equilibrium of electron donors on $TiO_2$ surface. After equilibration, the concentration of Cr (VI) was measured and taken as the initial concentration ($c_0$). Then the LED was opened, 2.5 mL solutions were sampled at appropriate time intervals. The samples were centrifuged immediately at 6000 rpm for 10 min and filtered through 0.45 µm filter. The filtrates were collected and analyzed. The reaction mixture was maintained in suspension by using a magnetic stirrer. Duplicate were carried out for each test, and the relative standard deviation was generally less than 5%.

## 3    Results and Discussions

### 3.1    The Photocatalytic Reduction of Cr (VI) in Methanol/TiO2 System

Cr (VI) could not be photoreduced by $TiO_2$ in the absence of methanol after 6 h under visible light irradiation. The Cr (VI) conversion was raised markedlly when methanol was added into system. As shown in fig.1, Cr (VI) conversion was raised gradually with the initial concentration of methanol increasing. When the initial concentration of methanol was 100 $gL^{-1}$, the Cr (VI) conversion could amount to be 94.71%. So the Cr (VI) conversion could be improved by increasing the concentration of methanol and it   tended to increase more

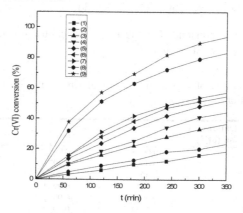

**Fig. 1.** Effects of methanol concentration on Cr(VI) conversion $Co_{(Cr(VI))}$=20 $mgL^{-1}$, $TiO_2$:1 $gL^{-1}$,initial methanol concentration:(1) 0.5 $gL^{-1}$ (2) 1 $gL^{-1}$ (3) 5 $gL^{-1}$ (4) 10 $gL^{-1}$ (5) 15 $gL^{-1}$ (6) 20 $gL^{-1}$ (7) 25$gL^{-1}$ (8) 50 $gL^{-1}$ (9) 100 $gL^{-1}$

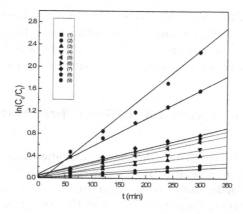

**Fig. 2.** The photocatalytic reduction of Cr(VI) kinetics curve in different initial concentration of methanol aqueous solution

$Co_{(Cr(VI))}$=20 mgL$^{-1}$, TiO$_2$: 1 gL$^{-1}$, initial methanol concentration: (1) 0.5 gL$^{-1}$ (2) 1 gL$^{-1}$ (3) 5 gL$^{-1}$ (4) 10 gL$^{-1}$ (5) 15 gL$^{-1}$ (6) 20 gL$^{-1}$ (7) 25 gL$^{-1}$ (8) 50 gL$^{-1}$ (9) 100 gL$^{-1}$ apparent when methanol concentration was lower. The trends of photocatalytic reduction of Cr (VI) were same for the different concentration curves: the initial reaction rate was faster, but as time went on, it gradually decelerated. It was most probably attributed to that Cr(VI) was reduced to Cr (III). Then Cr (III) was deposited on the surface of TiO$_2$ in the form of Cr (OH)$_3$, which covered the surface catalytic active site and resulted to reducing the activity of TiO$_2$ [6].

The kinetics curves of the photocatalytic reduction of Cr (VI) in different concentration methanol/TiO$_2$ systems could be seen in fig.2. The photocatalytic reduction of Cr (VI) conformed to the first-order kinetic equation, which can be expressed as

$$Ln \quad (C_0 / C_t) \quad = kt \tag{1}$$

k: apparent reaction rate constant; $C_0$: the initial concentration of methanol; $C_t$: the instantaneous concentration of methanol.

**Table 1.** Kinetic parameters of reduction reaction of Cr(VI) in different concentrations of methanol aqueous solution

| concentrations of methanol (gL$^{-1}$) | k | r |
|---|---|---|
| 0.5 | 0.0005 | 0.9961 |
| 1 | 0.0008 | 0.9971 |
| 5 | 0.0013 | 0.9986 |
| 10 | 0.0017 | 0.9984 |
| 15 | 0.0021 | 0.9981 |
| 20 | 0.0023 | 0.9923 |
| 25 | 0.0024 | 0.9969 |
| 50 | 0.0050 | 0.9983 |
| 100 | 0.0079 | 0.9937 |

The parameters of the first-order kinetic equation were shown in table.1. The k value was 0.0007 when the concentration of methanol was 0.5 gL$^{-1}$. It was 0.0079 when the concentration of methanol was 100 gL$^{1}$. The photocatalytic reduction of Cr (VI) could be enhanced through increasing the concentration of methanol. It was consistent with above.

## 3.2 The Photocatalytic Reduction of Cr (VI) in Different Electron Donors/TiO2 Systems

The different electron donors/TiO$_2$ exhibited marked differences on the reduction of Cr (VI). As showed in fig.3, the Cr (VI) conversion was 100% in formate/TiO$_2$ system after 80 min under visible light irradiation. For the methanal and methanol systems, the Cr (VI) conversions were 93.62% and 22.69% respectively after 6 h. It was easily found that the Cr (VI) conversion in the presence of electron donors has

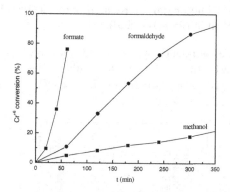

**Fig. 3.** Effects of the same concentration of methanol and methanal, formic acid on Cr(VI) conversion Concentration of methanol and methanal, formic acid: 0.5 gL$^{-1}$

the order of formate> methanal> methanol. On the one hand, methanol and formaldehyde were adhered to TiO$_2$ surface through the weak hydrogen bonding while the carboxyl of formic acid was able to form the complexes with TiO$_2$ and was adhered to TiO$_2$ surface through strong chemical adsorption [8, 9]. So formic acid was more easily absorbed on TiO$_2$ surface than methanol and formaldehyde. The photocatalytic reaction was carried out on TiO$_2$ surface, therefore only the electron donors which had already absorpsed   on surface of TiO$_2$ were effective and the more adsorbance of electron donors were, the better   photocatalytic reduction reaction was and the larger the Cr (VI) conversion was. On the other hand, the oxidability of Cr (VI) was very strong and its half-reaction of reduction as

$$Cr_2O_7^{2-} + 14H^+ + 6e = 2Cr^{3+} + 7H_2O \ (E_0 = 1.33 \ V) \tag{2}$$

Because concentration of total chromium was constant, the oxidation-reduction potential E in solution as

$$E = (E_0 + \frac{0.059 \times 14}{6} \lg[H^+]) + \frac{0.059}{6} \lg \frac{a_0 - 1/2 a_{Cr(III)}}{a^2_{Cr(III)}} \tag{3}$$

$a_{Cr(III)}$: the concentration of Cr (III), mgL-1; a0: the initial concentration of Cr (VI), mgL-1

It could be seen that the greater the concentration of H$^+$ in system was, the higher the oxidation-reduction potential of Cr (VI) was and the more easily Cr (VI) was reduced. The pH values in formate/TiO$_2$, methanal/TiO$_2$ and methanol/TiO$_2$ systems were 2.86, 5.00 and 5.54 respectively. It was easily found that the concentration of H$^+$ in formate/TiO$_2$ system was most, so Cr (VI) conversions was largest at the same condition.

### 3.3 The Mechanism of Photocatalytic Reduction in Electron Donors/TiO2 Systems Under Visible Light Irradiation

A series of control experiments were carried out. As shown in Fig. 4, the concentration of Cr (VI) was almost constant either $TiO_2$ or electron donors was presented alone under visible light irradiation. But Cr (VI) was significantly reduced when $TiO_2$ and electron donors were both presented.

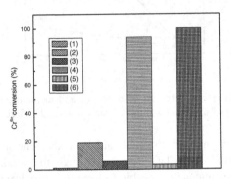

**Fig. 4.** Effects of the same concentration of methanol and methanal, formic acid on Cr (VI) conversion at different conditions

$Co_{(Cr (VI))}$=20 mgL$^{-1}$, (1) methanol: 0.5 gL$^{-1}$, without $TiO_2$, t: 6 h (2) methanol: 0.5 gL$^{-1}$, $TiO_2$: 1 gL$^{-1}$, t: 6 h (3) methanal: 0.5 gL$^{-1}$, without $TiO_2$, t: 6 h (4) methanal: 0.5 gL$^{-1}$, $TiO_2$: 1 gL$^{-1}$, t: 6 h (5) formic acid: 0.5 gL$^{-1}$, without $TiO_2$, t: 80 min (6) formic acid: 0.5 gL$^{-1}$, $TiO_2$: 1 gL$^{-1}$, t: 80 min

It infered that electron donors might be excited by visible light to produce excited state with high activity. Because a part of electron donors had been absorbed on $TiO_2$ surface after being mixing for 30 min in darkness, electron could be injected from the excited state of electron donors to the conduction of $TiO_2$ and moved to $TiO_2$ surface. Then the Cr (VI) was reduced to Cr (III) by electron. The electron donors losing electron which was similar to hole, was a cation radical with strong oxidizing. They could oxidate water and produce a series of free radicals with strong oxidizing. But when only electron donors were present in Cr (VI) solution, Cr (VI) was hardly reduced. Maybe the reason is electron and cation radical recombined soon, so electron could not arrive at Cr (VI) nearby. Because the energy band of $TiO_2$ was discontinuous and its conduction band was empty, the conditions for preservation of electron could be created when $TiO_2$ was in present in system.

## 4    Conclusion

(1) Cr (VI) could not be reduced either $TiO_2$ or electron donors was presented alone under visible light irradiation. The Cr (VI) conversion was 100% using formate as electron donors after 80min. For formaldehyde and methanol systems, the Cr (VI) conversion was 93.62% and 22.69% respectively after 6h.

(2) The rate of Cr (VI) reduction in the presence of electron donors has the order of formate> methanal> methanol.

(3) In the methanol/TiO$_2$ systems, the photocatalytic reduction of Cr (VI) conformed to the first-order kinetic equation.

**Acknowledgements.** This work was financed by Guangdong Natural Science Fund Committee (8451063201001261), Foundation for Distinguished Young Talents in higher Education of Guangdong (LYM08022), Scientific Research Cultivation and Innovation Fund, Jinan University (216113132), the Scientific and Technological Planning of Guangdong Province (No.2007A032400001), and the University-Industry Cooperation Project about Water Treatment Materials of Guangdong Province (No.cgzhzd1004).

# References

1. Li, C., Gu, G., Liu, S.: Progress in treatment of heavy metals and precious metals in wastewater by TiO2 photocatalysis. Techniques and Equipment for Environmental Pollution Control 4, 6–11 (2003)
2. Zhang, H., Tan, X., Zhao, L.: Research progress of photocatalytic reduction of heavy metal ions in waste water. Journal of Tianjin Institute of Technology 20, 28–32 (2004)
3. Fox, M.A.: Photocatalysis: Decontamination with Sunlight. Chemtech 22, 680 (1992)
4. Hoffmann, M.R., Martin, S.T., Choi, W., Bahnemann, D.W.: Environmental application of semiconductor photocatalysis. Chemical Reviews 95, 69–96 (1995),
5. doi:10.1021/cr00033a004
6. Li, S., Fu, H., Lü, G., Li, X.: Research progress of photocatalytic reduction of heavy metal ions. Photographic Science and Photochemistry 13, 325–333 (1995)
7. Fu, H., Lü, G., Li, S.: Photocatalytic Reduetion of Cr(VI) ion in the Presence of organies. Acta Physico-Chimica Sinica 13, 106–112 (1997)
8. Piairie, M.R., Evans, L.R., Stange, B.M., Martinez, S.L.: An investigation of TiO2 photocatalysis for the treatment of water contamination with metals and organic chemicals. Environment Science Technology 27, 1776–1782 (1993)
9. Li, Y., Lü, G., Li, S.: Photocatalytic hydrogen generation and decomposition of oxalic acid over platinized TiO2,. Applied Catalysis A: General 214, 179–185 (2001)
10. Peng, S., Peng, Y., Li, Y., Li, C.: Photocatalytic hydrogen generation using furfural, furfuryl alcohol and furoic acid as electron donors over Pt/TiO2. Fine Chemicals 25, 1212–1215 (2008)

# The Wireless Video Monitoring System Based on ARM / Linux

Zhiyan Sun and Li Yun

Guilin College Aerospace Industry/Department of Electronic Engineering, Guilin, China
39754097@qq.com

**Abstract.** The implementation of the three types of digital monitoring system options, and comparison of these three programs; Designed by ARM / Linux hardware platform as the core of the Remote Video Monitoring System, and through wireless network to transmit video images to the host side, in order to achieve analysis, storage and display functions, with the traditional analog surveillance systems: it significantly reduced network costs, the system greatly reduced the size of the weight, operation and maintenance easier.

**Keywords:** wireless Video, ARM, Linux.

## 1 Introduction

Video surveillance system is part of the security system, it is a comprehensive system of prevention ability. Current monitoring system has entered the digital and network era, the video image capture device from the front-end digital signal is output, and network transmission medium, based on TCP / IP protocol, using streaming media technology video web of multi- multiplexing transmission. Embedded systems using the remote video monitoring technology, in line with the characteristics of the digital network, public security, security and other industries has important practical significance.

In this paper, for low equipment costs, low operating costs and long distance applications of video surveillance systems, using ARM embedded processors to build embedded systems, Linux operating system, developed to the practical application of remote video surveillance system for low-resolution rate, low-cost, long-distance video surveillance applications [1]. System framework shown in Fig.1.

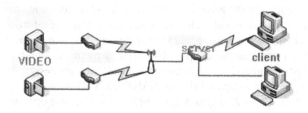

**Fig. 1.** System framework

G. Lee (Ed.): Advances in Intelligent Systems, AISC 138, pp. 335–340.

## 2    Program Monitoring Systemintroduction

The current implementation of video surveillance system network generally has three options.

### 2.1    Local Analog Video Surveillance System

Local image monitoring system mainly consists of cameras, video matrix, monitors, video recorders and other components, the use of analog video cable from the video camera connected to monitor the use of video matrix host, switch and control the use of keyboard, video tape using the time recorders; long-distance fiber-optic analog video transmission, video transmission using Optical. Traditional analog CCTV systems have many limitations, shortcomings are wired to the analog video signal transmission distance is very sensitive; wired analog video surveillance can not be networked, can only point the way to monitor the site, check evidence and cumbersome and so on.

### 2.2    PC Card-Based Remote Video Monitoring System

Proxy server is generally from a PC to act as. Running TCP / IP protocol to Internet access while connecting by a simple bus structure (RS232, RS485, etc.) with embedded systems. Part of the system only needs to communicate with the proxy code. The advantage is that you can easily solve the problem of Internet equipment, development of low degree of difficulty; drawback is that access costs are high, is not conducive to large-scale promotion. The program more suitable for large industrial equipment or more expensive Internet access needs, not suitable for low-cost device.

### 2.3    Remote Video Monitoring System Based on Embedded

Directly on the embedded processor by implementing TCP / IP protocol to achieve Internet access function does not use such a program operating system, significant savings in resources, but the processor's high performance requirements, while increasing the difficulty of development, technology is also more difficult to achieve.

Based on the above limitations of the program,   in the last article on the basis of the use of embedded ARM and Linux operating systems, operating systems, TCP / IP protocol, an improved design of ARM-based remote video monitoring system. The system is simple networking, data transmission capacity, speed, the entire system is simple. Small size, low cost, remote computer can be achieved without additional procedures, through a Web browser will be able to detect the device, and the application easier to develop and realize the complete sharing of information.

Target board features a Samsung ARM9 processor's mainstream 3C2440A. Based on the ARM920T core, clocked at up to 400 MHz, a memory management unit with (MMU), on-chip resource-rich, high cost, is an ARM9 processor for mainstream chips. Core board and standard 64MB of NAND FLASH 64MB of SDRAM, stable running Linux, WinCE, VxWorks and other embedded real time operating system [2].

System uses the 301 Series high-definition chip Vimicro USB camera, the camera uses CMOS sensor. Compared with the CCD, although the gap between the image quality somewhat, but a 500-megapixel digital camera, high five glass HD lens, the camera resolution up to 1024 × 768, video capture can basically meet the need.

Wireless communication module is a wireless network card via the USB interface to achieve. The system is using ASUS WL-167g V2 wireless card in the receiver front-end and back-end server monitoring equipment built between wireless local area networks to enable seamless point to point connection. S3C2440A the wireless card with integrated USB host interface directly connected to the work in the 2.412 ~ 2.472GHz frequency band, using OFDM, CCK, DQPSK, DBPSK modulation, etc., to comply with 802.11b / g protocol, the transmission rate of 54Mb / s, most indoor distance of 40m, the maximum distance for the outdoor 200m, to meet the requirements of video transmission.

## 3 System Hardware Design

The wireless video transmission system consists of an ARM9 processor, video capture module, network communication module, power module. The hardware components shown in Fig. 2.

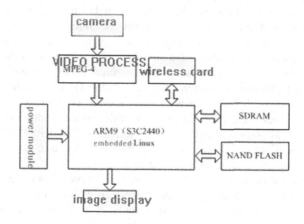

**Fig. 2.** Remote Network hardware System

## 4 Control System Soft Design

If you are experiencing this problem does not involve a paper format, the authors refer to this session can be the first website to provide a "paper format in English" requirement. If "English paper format" does not address the problem, the authors can adopt other papers of general practice,

The system's software design includes the operating system, drivers, MPEG-4 video encoding and applications. Linux provides a wealth of network and bus protocol stack, can reduce system development costs and shorten the development cycle, the system used kernel version 2.6.29 of the Linux operating system. System, the USB device driver modules, MPEG-4 video encoding and other basic modules through module cross-compiled to the target board transplantation [3].

USB device driver modules for Linux porting mainly refers to the wireless network card driver and camera-driven migration. Porting the driver to achieve the two main ways: one is directly to the driver into the kernel, the kernel boot time, directly through the initialization function for the driver to load the kernel startup is complete, while achieving a loading of the driver; two ways through the module loaded in the embedded Linux platform has been ported on the loading of the driver. The two loading methods, the former needs to change the core structure, but also in changing the driver to repeat the compilation, a larger workload. In comparison, the latter's greater flexibility. Therefore, the system's wireless network card driver to load the second approach. ASUS WL-167g V2 using the rt73 chip that has the open source Linux drivers, according to their cross-programming, to generate drive rt73.ko file. Download it to the ARM board, when the system starts to load. After successful loading, search and access in the vicinity there is the AP site, in order to achieve wireless connectivity. The use of generic camera driver linux webcam driver gspca . The driver can support most of the usb camera. After the pressure-driven cross-compile the source file to get the driver file gspca.ko driver file, the system starts immediately after loading insmod support the camera.

The way to achieve MPEG-4 encoding has two kinds-- hardware compression and software compression. Using of specialized hardware compression chip to the video signal compression MPEG-4 encoding; while software compression software compression is used to encode the video signal. The system uses software encoding and decoding video signals. Xvid is an open source MPEG-4 multimedia coder, which is based on OpenDivX written. As Xvid is the GNU GPL license for free software, which means that the software source code is publicly available. And programmers can modify the code. Its cross-compiled and downloaded to the target board, configured and optimized in the 400 MHz frequency under a resolution of 176 × 144 image of QCIF up to 2Ofps / s of the coding rate, the basic realization of real-time encoding, can achieve practical requirements.

HTTPD embedded devices using the service, the Internet or internal network can provide a Web-based graphical management interface. Use CGI (Common Gateway Interface) technology enables the browser and server interaction between [4], WebServer and CGI, in combination, is currently embedded remote monitoring of the most common and most sophisticated means. Embedded Web server is embedded systems and Internet connectivity through the Web-a key component of the CGI is the user interaction with the Web server is an important way [5]. The system sends and receives video data as follows: first, the front-end equipment is waiting for the user to connect up and the state; when the server needs data, it first entered the IP address of the front-end devices, send data requests to the front-end equipment; front-end equipment receives data request, its IP multicast address and port number to the server; the server receives, start to receive data thread. Create a user interface player, join the IP multicast group, waiting to receive data sent back to the front-end equipment to confirm the information; front-end devices receive a confirmation message, send video data to the multicast group. The operation flow shown in Figure 3. system can be connected in any network into the circumstances [6].

## 5    Experimental Results

System testing completed within the campus network environment. After camera accessed, the development board with the wireless network card connected directly to the campus network AP. PC within the campus network can view the camera via IE captured video images. Test results shown in Fig.3.

**Fig. 3.** The Actual Test Pattern

## 6    Conclusion

Embedded systems embedded Internet access to Internet technology are the development trend. This article uses the ARM embedded processors and Linux operating systems, development of practical applications can supply to remote video monitoring system in order to connect embedded systems with the Internet, have low development costs and establish a convenient and easy to use embedded image video system. Network monitoring system to support user demand, real-time monitoring, remote control, video and other centralized functions, fully meet the requirements of the user's monitor. Meet high-resolution, low-cost, long-distance monitoring applications.

## References

1. Meyer, L., Penzhorn, W.T.: Denial of service and distributed denial of service-today and tomorrow. Automatica (2004)
2. Hokezhang, Suwei: Principle and Technology of Network Processor. In: Proc. Amer. Control Conf. (2000)

3. Alspector, J., Allen, R., Jayakumar, A.: Relaxation networks for large supervised learning problems. In: Lippmann, R., Moody, J., Touretzky, D. (eds.) Neural Information Processing Systems (1991)
4. Jackel, L.D., et al.: VLSI implementations of electronic neural networks: An example in character recognition. In: Proc. IEEE Int. Conf. Syst. (1990)

# Medical Image Segmentation Based on Accelerated Dijkstra Algorithm

Dai Hong

School of Electronic and Electric Engineering,
Shanghai Second Polytechnic University Shanghai 201209, China

**Abstract.** Dijkstra Algorithm is one of the optimal path searching methods. An Accelerated Dijkstra algorithm is presented for medical image segmentation: (1) An example is used to illustrate the proposed algorithm to reduce the calculation work and increase the operating speed of the Classical Dijkstra Algorithm.(2)An Live-Wire Image Segmentation method based on this algorithm is presented to sketch the liver's contour in an abdomen image, and a morphological method is used to segment the liver image. Experimental results show that this method can run image segmentation successfully, has less interactive times than that of the manual segmentation method and run faster than the Classical Dijkstra Algorithm.

**Keywords:** Accelerated Dijkstra Algorithm, Live-Wire, medical image segmentation.

## 1    Introduction

Optimal path searching is one of the classical algorithms of graph theory aims to obtain the shortest path between two different vertexes in a given weighted graph [1]. The algorithm is an important optimal method and a focus of transportation, computer science, operational research, geographical information science(GIS), image processing areas. Image Segmentation technique is one of the most difficulties in image analysis whose goal is to extract the object of interest from background [2]. In this paper, the Dijkstra Algorithm of optimal path searching is used for medical image segmentation, and an improvement of this method is presented to accelerate its operating speed.

In Part 2, the Classical Dijkstra Algorithm is introduced and two improvements are presented to reduce calculation works and to accelerate operating speed of this method which is referred to as "Accelerated Dijkstra Algorithm", in addition, an example is used to explain the algorithm. In Part 3, The Live-Wire image segmentation method based on the Accelerated Dijkstra Algorithm is proposed to extract the liver from an abdomen image by Matlab programming, the experimental results and the performance of this algorithm are analyzed. Part 4 is conclusion.

G. Lee (Ed.): Advances in Intelligent Systems, AISC 138, pp. 341–348.
springerlink.com          © Springer-Verlag Berlin Heidelberg 2012

# 2  The Optimal Path Searching Based on the Accelerated Dijkstra Algorithm

## 2.1  The Classic Dijkstra Algorithm

Let $G=(V,E)$ be a connected weighted graph with $N$ vertexes ($N=m\times n$),where $V=\{v_1, v_2,..., v_N\}$ is the vertex set and $E=\{e_1,e_2,...,e_N\}$ is the edge set. Let $C(p,q)$ ($p=1,2,...,m$, $q=1,2,...,n$) be the weight between vertex $v_p$ and $v_q$, then the weight matrix $C=C(p,q)_{m\times n}$. The purpose of the Classic Dijkstra Algorithm is to obtain the shortest paths from the starting point to all other points of the graph[3], the steps of this method are as follows:

*Step* 1: initialization.
Let $cc$ be a $m\times n$ two-dimensional matrix which stores the shortest distances among the starting point and all other points in the graph. Let $(v_{si},v_{sj})$ is the coordinates of $v_s$, $cc(v_{si},v_{sj})=0$ and the initial shortest distances of other points in the graph are $\infty$ .Let $Q$ is the "unprocessed vertexes queue" with $Q(1)=v_s$ , $L$ is the "processed vertexes queue" with initial value is empty.

*Step* 2: calculate the shortest distances among the starting point and all other points.
Using a while loop condition: "when $Q$ is not empty", repeat the following steps:

1) Find the shortest distance of $cc(Q)$, remove the vertex $v_p$ corresponding to the shortest distance and put $v_p$ in $L$.
2) update $cc$.

$v_p$ usually has 8 neighboring vertexes $v_q(q=1\sim8)$,let $(v_{pi},v_{pj})$ and $(v_{qi},v_{qj})$ are the coordinates of $v_p$ and $v_q$, respectively. Obviously, the 8 coordinates of $v_q$ are: $(v_{pi}-1,v_{pj}-1)$,$(v_{pi}-1,v_{pj})$,$(v_{pi}-1,v_{pj}+1)$, $(v_{pi},v_{pj}-1)$, $(v_{pi},v_{pj}+1)$, $(v_{pi}+1,v_{pj})$,$(v_{pi}+1,v_{pj})$ and $(v_{pi}+1,v_{pj}+1)$. However, if $v_p$ is at the edge of the graph, then the number of neighboring vertexes <8.

Considering the difference between eight neighborhood vertexes and four neighborhood vertexes in Euclidean distance, given: $w=1$, if $v_q$ is four neighborhood of $v_p$, $w=\sqrt{2}$ if $v_q$ is eight neighborhood of $v_p$.
If:

$$\text{round}(cc(v_{pi},v_{pj})+w*C(v_{qi},v_{qj}))<cc(v_{qi},v_{qj})$$

Then:

$$cc(v_{qi},v_{qj})= \text{round}(cc(v_{pi},v_{pj})+w*C(v_{qi},v_{qj})) \tag{1}$$

Where $v_q \notin L$ and $v_q$ falls into the graph; $(v_{pi},v_{pj})$ and $(v_{qi},v_{qj})$ is the coordinate of $v_p$ and $v_q$, respectively."round" makes all paths be integers. According to (1),all temporary shortest distances from $v_q$ to $v_p$ are figured out and stored into matrix $cc$. Letting an array "dir", given: $\text{dir}(v_q)=v_p$, then $v_p$---the previous vertex of $v_q$ is recorded.
3) If $v_q \notin Q$, then put $v_q$ into $Q$.
4) Back to 1),until Q be an empty matrix.

Because only one distance is calculated at a time in the above "while" loop, N cycles must be implemented to figure out all minimum distances from all other points to the starting point.

Step3: Path backtracking.
Path backtracking is carried out from every other points to the starting point $v_s$ according to the array "dir", the shortest paths from $v_s$ to all other points are obtained. N times path backtracking are required.

## 2.2    The Accelerated Dijkstra Algorithm

Classical Dijkstra algorithm is to calculate the shortest paths from starting point to all other points, the calculating speed will decrease with the increase of N. Because only the shortest path from starting point $v_s$ to end point $v_z$ is required in interactive image segmentation, an "Accelerated Dijkstra algorithm" is presented to reduce the calculating work of the Classical Dijkstra algorithm, two improvements of the original method are as follows:

***Improvement 1:***
The "while" loop condition in "Step 2"is revised as "while the end point $v_z$ is not found in the processed Queue L, step 1),2),3) are carried out", and step 4) is revised as: "Go back to 1) until $v_z$ is found in L, then jump out the loop", so other shortest distances are not calculated. The cycle times are less than N after using the accelerated means above.

***Improvement 2:***
"Step 3" is revised as:" Path backtracking is carried out from the end point $v_z$ to the starting point $v_s$ according to "dir" ".So only one path backtracking is performed.

## 2.3    Simulation Example

An example simulated by Matlab programming is as follows: Letting a weight graph G which has N=25(5×5) vertexes, its weight matrix is C=[11 13 12 9 5;14 11 7 4 2;11 6 3 5 7;7 4 6 11 13;6 2 7 10 15], the vertexes are $v_1 \sim v_{25}$,assume that $v_1$ is located in the upper-left corner and $v_{25}$ is at the lower-right corner. Seeking the shortest path from the starting point $v_3$ to the end point $v_{15}$.

The numbers in Fig.1.(a) are the shortest distances from all points to $v_3$ by Classical Dijkstra algorithm. The cycle times of "While" and path backtracking times are 25; The numbers in Fig.1.(b) are the shortest distances from some points to $v_3$ using the Accelerated Dijkstra algorithm(Because once the end point $v_{15}$ is found in L, path searching stops, the distances of other points are initial value: infinite, which represented by "Inf"). Only the shortest path from $v_{15}$ to $v_3$ is shown in Fig.1.(b). The cycle times of "While" are 10,which is 2/5 times that of the Classical Dijkstra algorithm, path backtracking times is 1,which is 1/25 times that of the Classical Dijkstra algorithm.

Compare Fig.1.(a) to Fig.1.(b),the shortest path from $v_3$ to $v_{15}$ is the same. Obviously ,the calculation work of the Accelerated Dijkstra algorithm is less than that of the Classical Dijkstra algorithm.

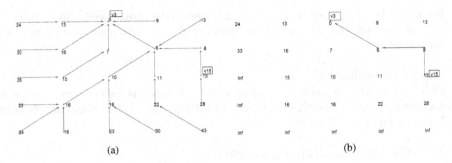

**Fig. 1.** Optimal path searching based on the Dijkstra algorithm

((a)Optimal path searching based on the Classical Dijkstra algorithm (b)Optimal path searching based on the Accelerated Dijkstra algorithm)

## 3   Image Segmentation Based on the Accelerated Dijkstra Algorithm

### 3.1   Live-Wire Image Segmetation Method Based on the Accelerated Djkstra Algorithm

Live-wire algorithm is a kind of interactive image segmentation presented by Eric N. Mortensen, et al. Image segmentation aims to extract the edge of the object, Live-wire algorithm converts edge tracking problem to the problem of optimal path searching in a weight graph using the Classic Dijkstra algorithm. The basic idea of Live-wire is: an image is viewed  as a weighted graph, first the cost function of an image is constructed, the edge of an object in the image is assigned a smaller weight (shorter distance), the non-edge is assigned a bigger weight, then two points on the edge of the object are selected manually, the Dijkstra algorithm is used to automatically generate the shortest path between them, a series of points are selected and the steps above are repeated for edge extraction of the whole target. The drawbacks of this algorithm are: image cost is heavy computation and optimal path searching is longer duration. The improvements are presented based on the "Accelerated Dijkstra Algorithm" and a cost calculation range narrowing method to enhance real-time performance of this algorithm. The steps are as follows:

*Step* 1: construct the local cost function of an image
Letting *p,q* are neighbouring pixels in an image $g_1$, the local connection cost between them (equal to the weight between two vertexes in Dijkstra algorithm) [4] is:

$$C(p,q)=\omega_G \times f_G(q)+\omega_c \times f_c(q)+\omega_D \times f_D(p,q) \tag{2}$$

Where the gradient feature function:

$$f_G(q)= 1-\frac{G(q)}{\max(G)} \tag{3}$$

In Equation *(3)*, $G(q)=\sqrt{I_x^2+I_y^2}$ ---the gradient magnitude;

$I_x$ and $I_y$ are the horizontal gradient and the vertical gradient of $g_1$, respectively:

$$I_x=g_1(x,y)-g_1(x-1,y); I_y=g_1(x,y)-g_1(x,y-1) \tag{4}$$

max($G$)---the maximum of $G$. According to $(3)$, higher gradients produce lower costs, since the gradient of the edge is highest, the total costs of target edge is minimal. $f_c(q)$ is the edge feature function at the pixel $q$:

$$f_c(q)=\begin{cases} 0 & q \text{ is an edge point detected by} \\ & \text{Canny operator} \\ 1 & q \text{ is not an edge point} \end{cases} \tag{5}$$

$f_D(p,q)$---smoothness constraint function. $\omega_G$, $\omega_c$, $\omega_D$ are the weight of $f_G(q)$, $f_c(q)$ and $f_D(p,q)$,respectively. If the connection relationships among pixels are represented by 8 neighborhood, then the shortest path is smooth enough. Letting $p$ is a 8 neighboring pixel of $q$, then the $f_D(q)$ is omitted in the cost function. Therefore given: $\omega_G=0.5$, $\omega_c=0.5$, $\omega_D=0$.

**The improvement of Live-Wire algorithm:to reduce the cost calculation range**
The original algorithm calculate the cost of the whole image, the number of the $C(p,q)$ is equal to the pixels number of $g_1$,so the calculation work of this algorithm is large. We can reduce the cost calculation range, which enough to include the starting point and the end point. Select a starting point $v_s$ and an end point $v_z$ on the target edge of $g_1$,the coordinates of them are:$(v_{si},v_{sj})$ and $(v_{zi},v_{zj})$, respectively. Then the cost calculation range of $C(p,q)$ can be reduced to the image $g_2$,letting the size of $g_2$ is $M \times M$ and the center point of $g_2$ is $v_s$ ,in order to contain $v_s$ and $v_z$ in $g_2$,the size of $M$ should at least:

$$M=2\max(|v_{si}-v_{zi}|,|v_{sj}-v_{zj}|) \tag{6}$$

and $g_2$ is located in $g_1$.
*Step* 2: Sketch the target edge.

1) The Accelerated Dijkstra algorithm is used to produce the shortest path between $v_s$ and $v_z$ automatically.

2) Take $v_z$ as the starting point, select the next end point,repeat step 1) until the last end point returns to the starting point $v_s$, thus sketches the whole target edge.

note: In this step, in order to reduce interactive operation time,a series points on the target edge can be selected previously, step 1) is repeated, and the last end point is returned to $v_s$ by programming.

*Step* 3: The morphological image segmentation of the target.

Live-Wire algorithm can only be used to sketch the target boundary. In order to extract the target from the background, a kind of "morphological image segmentation" method is presented. The basic idea of mathematical morphology is: the structure element which has some type of shape is used to measure and extract the object of corresponding shape in an image for the purpose of image analysis and recognition.

Morphology is based on set theory, its basic operations are: dilate, erode, opening and closing. A variety of useful algorithms can be combined with these basic operations, such as: skeleton extraction, boundary extraction, area filling, etc. Area filling is used for target image segmentation, the steps are as follows:

1) The binary template of target generated by area filling[5]
The expressing of area filling is:

$$I_k=(I_{k-1}\oplus B)\cap I^C \quad (k=1,2,3,\ldots) \tag{7}$$

Where $B$ is the structure element:

$$B=\begin{bmatrix} 1 & 1 & 1 \\ 1 & 1 & 1 \\ 1 & 1 & 1 \end{bmatrix} \tag{8}$$

$\oplus$ is the" dilate operation" whose function is to expand a circle of $I$. The initial value is $I_0=p_1$, $I_0$ is an arbitrary point in the target. $I^C$ is the complementary set of $I$. If $I_k=I_{k-1}$,then this algorithm will end at the $k$ step of iteration. The function of the algorithm is: as long as a point in the target is selected, the whole target region is filled with white pixels, thus the binary temple $I$ is generated.

2) Segment the target image
The original image $g_1$ is multiplied by the binary temple $I$ ,then the result image $g$ is produced:

$$g=g_1*I \tag{9}$$

### 3.2    Experiment Results Analysis

● An example of medical image segmentation
An abdomen image from Chinese Visible Human Database (resolution: 533×800) is shown in Fig.2.(a),the liver image is extracted using the presented algorithm.The following operations are based on Matlab programming:$m$ points ($m=18$)are selected in order previously, then the target edge is sketched out using Live-wire method based on the Accelerate Dijkstra algorithm(white contour),in addition, the white rectangles are the cost calculation ranges at every times of path searching;Fig.2.(b) is the target boundary, select a point in the target using the mouse; Fig.2.(c) is the binary template of the target using area filling;Fig.2.(d) is the result image.

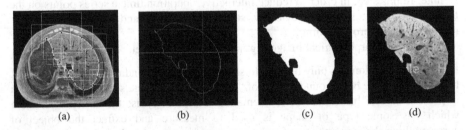

(a)                    (b)                    (c)                    (d)

**Fig. 2.** Live-wire image segmentation based on the Accelerated Dijkstra algorithm
((a)The liver edge sketched by the Accelerated Dijkstra algorithm(b)An arbitrary point is selected in the liver(c)The Binary template of liver using area filling(d)The result image)

- Performance analysis of the algorithm

The performance of the algorithm presented is compared with the "manually segmentation" method. The "manually segmentation" method is: $m$ points on the liver are manually selected and a cubic spline function is used for interpolation among all points, then the whole liver edge is sketched out. When $m=18$, The boundary can't been sketched out accurately (as shown in Fig.3.(a)) , so we have to increase the points.Fig.3.(b) shows the accurate boundary obtained   when $m=46$. The comparison of   "manually segmentation" method, Live-wire Method based on the Accelerated Dijksta Algorithm on interactive times and segmentation time are shown in Table 1.Obviously,the presented method   has less interactive times and run faster than that of the "manually segmentation" method.

Experiment environment: an Pentium IV computer whose memory is 512M; Matlab7.0 software.

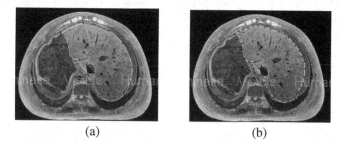

(a)                                    (b)

**Fig. 3.** Manually image segmentation ((a)The liver edge when $m=18$;(b) The liver edge when $m=46$)

**Table 1.** Performance of segmentation algorithms

| Segmetation   algorithms | Interactive times | Segmentation time (s) |
|---|---|---|
| manually | 47 | 33.7 |
| Live-wire Method based on the Accelerated Dijkstra algorithm | 18 | 12.1 |

## 4   Conclusion

This article applies optimal path searching algorithm in graph theory to image segmentation. An Accelerated Dijkstra algorithm is presented to reduce the calculation works of the Classical Dijkstra algorithm, and to accelerate its operating speed; a Live-Wire method based on the Accelerated Dijkstra algorithm is proposed to segment the liver from an abdomen image; area filling is used to extract the target image, all of the algorithms mentioned are realized by Matlab programming. The experimental result indicated that this algorithm can run image segmentation correctly, the interactive times are few and the operation time is short.

**Acknowledgements.** The Author received funding from "The key Subjects Project of Shanghai Municipal Education Commission" (grant No.: J51801).

# References

1. Long, L.J.: Application of graph theory and algorithm, 1st edn. Electronics Science and Technology University Press, Sichuan (1995)
2. Jie, T., Lian, B.S., Quan, Z.M.: Medical Image Processing and Analysis, 1st edn. Publishing House of Electronics Industry, Beijing (2003)
3. Bo, G.X., Yun, L., Bing, J.H.: An Improved Live-Wire Algorithm for Image Segmentation. System Engineering and Electronic Technology 25(8), 915–917 (2000)
4. Mortensen, E.N., Barrett, W.A.: Intelligent Scissors for Image Composition. In: Computer Graphics (SIGGRAPH 1995), Los Angeles, CA, pp. 191–198 (1995)
5. Hong, D., Feng, Y.Y.: A Fast Interactive Image Segmentation Algorithm of Talus. Computer Application and Software 26(5), 77–80 (2009)

# Study on Composition and Key Technologies of Vehicle Instrumentation Switches Fatigue Life Detection System

Nian-feng Li, Ying-hong Dong, and Ji-Ling Tang

College of Computer Science and Technology, Changchun University. Changchun,
Jilin Province, China
cculinianfeng@126.com

**Abstract.** Automotive switch is an important way for the people to interact with the cars, and their quality is directly related to vehicle quality. Vehicle instrumentation switches provided by different vendors have large differences in structure and properties. So, it is necessary to develop an intelligent switches fatigue life detection system. A kind of composition and architecture of vehicle instrumentation switches fatigue life detection system were introduced. The work principle of the system in this kind of architecture was provided. The key technology and implement methods in research process were analyzed. The prototype was developed and the experiment results as well as analysis were given. The results of application experiment show that this kind of vehicle instrumentation switches fatigue life detection system could preferably realize the fatigue life test.

**Keywords:** vehicle instrumentation switches, detection of fatigue life, parallel scheduling algorithm.

## 1 Introduction

A car is formed by thousands of parts and its quality also depends on the quality of these parts. Today's increasingly competitive in the automotive industry, improving vehicle quality performance and reducing production costs have become major tasks to automobile manufacturer. Vehicle instrumentation switch is an important way for people to interact with the car, and reflects the car's level of intelligence and humanity, and plays an important role in the automotive assembly. Vehicle instrument switch quality directly impact on the vehicle's quality, therefore, whether it is auto parts manufacturers or vehicle manufacturers, still attach great importance to the quality, especially in fatigue life. Corresponding standards have been developed in China for some car instrument switch on the quality of the specific requirements. Xinhua Changchun reported on January 28, 2010, FAW Group achieved good operating results in 2009, the annual sales of more than 1.94 million, ranking second in the domestic auto industry. The parts of nearly 200 million cars come from thousands of suppliers (such as FAW-Volkswagen have 400 external suppliers to ensure that its current supplier 2100 / day capacity). The switches from varieties suppliers have large differences in structure and performance. In the quality

G. Lee (Ed.): Advances in Intelligent Systems, AISC 138, pp. 349–354.

inspection process, it is urgent necessary to develop a new concept of versatility, intelligent switching fatigue life detection system to resolve the current vehicle instrumentation switch fatigue life testing problems.

A kind of instrument switch fatigue detection system was studied in this project and it can adapt to most vehicles at home and abroad, using ARM-based sub-control machine's processing power and high-speed communications network based on socket's ability to achieve the rapid detection and switching state parallel motion control. System provided a switching sequence programming function in order to achieve joint action on the multi-function switch test.

## 2  Architecture of Vehicle Instrumentation Switches Fatigue Life Detection System

Microelectronics and computer technology promote the extensive use of computers. The early 1980s, there have been  seen computer monitoring and control of the testing machine foreign, and formation of series products, such as the UK's Instran8000 INSTRON Series, U.S. MTS's Alpha series. Germany Shenek, Japan's Shimadzu, and Swiss w+B and other companies are also engaged in test machine research work. The intelligent automation fatigue life detection machine has greatly reduced the labor intensity of the test, and improved testing efficiency, expanded the scope of the study, and significantly improved test accuracy.

In China, the depth and breadth of fatigue test are constantly increasing, a number of research institutes, colleges have worked on design of fatigue, and have made some achievements on the fatigue test data dealing, some work has been achieved the world level. Mechanism in fatigue, fatigue failure analysis,  fatigue strength of the typical components, corrosion fatigue, basic fatigue, low cycle fatigue and surface hardening, etc. some work are beginning to do. Especially in the last ten years, scientific research units carried out a series of testing machine research word and made a lot of advanced results. For example, China Automotive Industry Corporation Chongqing Automobile Research Institute developed the steering shaft torsion fatigue test rig, Kunshan Precision Instrument Co., Ltd. China Cornell developed switch life test series machine, and Orange Co., Ltd. developed instruments- switch life test machine. At present, the switch testing machine manufacturers is nearly as many as dozens.

So, in the research level of fatigue testing machine, there is a gap between China and developed countries, but whether it is foreign or domestic, there have been more mature technologies and products, and have achieved a computerized, intelligent, digitization, automation, energy conservation and other goals. However, with the development of control technology, information technology, communication technology and other subjects, there is much room for improvement on testing machine products at home and abroad, especially in the vehicle instrumentation switches fatigue detection. Products currently available from the domestic and the foreign  exist  the following questions: (1) Poor general; performance and specifications of the main vehicle instruments are generally appropriate national standard, but the number and layout of switches on the instrument are rarely limited, in recent years, embedded systems technology enables automotive electronic instruments (such as radio, etc.) functions develop very rapidly, species change is rapid, the existing testing machine can not meet the test requirements of newer electronic products. (2) Lack of capacity in parallel and testing speed difficult to be increased; existing testing machine is mostly in single machine mode, even if some testing systems provide multi-position detection capability, operation also is in the serial mode. According to national   switch   standards GB16915.1-2003, GB16915.2-2003, etc., assume that on each position switches need to be detected 1000000 times, by the existing testing machine at least 800 hours are needed, because the existing testing machine can only performance   10 to 20 times per minute. Test efficiency is very low. (3) The switch state detection is inadequate. Query through the Internet, the testing machine mostly use open-loop control mode at present and only perform switch operation without switch state detection, and the testing personnel must be on duty.   (4) Lack of functional test for multi-switch-joint movement.

State regulations limit the reliability and durability of vehicle instrumentation switches and these indicators can be simulated by indoor switches life test methods to evaluate. The study found that different types of combination switch varied in the shape and size, but the work action can be summarized as three kinds, rotating, flip and press. Based on this, a kind of automotive instrument switch fatigue-testing system to be able to adapt to the present majority of domestic and foreign wound be studied on, which consists of the host control unit, ARM-based sub-control unit, implementing agencies (such as cylinders, motors, fixtures, etc.) and field sensors four major sections. ARM-based sub-control unit is responsible for controls the switching, collection of the trial data and communication with the host. The host unit is responsible for overall completing the test row and the data processing. And ultimately provides an analysis of product performance and quality discrimination. System architecture is shown in Figure 1.

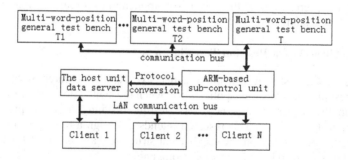

**Fig. 1.** System architecture

# 3   Key Technologies and Implementation

## 3.1   Parallel Scheduling Algorithm for Multi-work-Position and Multi-switch

The host computer unit worked with the mainstream PC, installed special software designed for this system, online communicate with the ARM-based sub-control unit for receiving, sending, intelligent processing and storing the switch status information, schedule three-position switch parallel action and prompt the staff on duty when alarm. The software was developed with visual RAD (Rapid Application Development) tool Delphi and new version Microsoft's launch of the SQL Server database management system was used as database.

Research goal is to develop capable of rotating, flipping and pressing three switch operation detection and three-work-position test platform(each work-position can simultaneously perform 10 switches test). To improve the detection rate of the specimen, the detection must be parallel among the work-position and every switch if the switches function without the constraints of special requirements. So, the parallel scheduling algorithm directly affects the speed of the test detection system. Of course, the speed of vehicle instrument switches mechanical action and the communication speed between the host computer and the ARM-based sub-control unit will restrict the overall speed of the detection system. But in the case of these objective factors, looking for better parallel scheduling algorithm is the main way to improve test speed.

### 3.2    The ARM-Based Sub-control Unit

Computer becomes the core of the new century and information technology reflects the mainstream of technological development and trends. Continue to broaden the field of computing disciplines and cross-disciplinary integration with other disciplines, foster some new frontier points. Embedded technology is a new field for the development of computer science. Embedded technology has been widely used in scientific research, engineering, military technology and other aspects, and become the main force in the development of the post-PC era IT field.

Consider the system information processing capacity requirements of state inspection, enforcement action and the needs of communication speed, project intends to use ARM-based sub-control unit. Interface circuit design, operating system migration, the preparation of the driver and application program design are the key issues need to be addressed.

### 3.3    The Three-Work-Position Test Platform

Versatility of the test platform plays a decisive role in research work. Currently available test platform has low applicability, when the switch position of the specimen changed in the distribution, a new type testing machine must be provided, and production costs increased. This research intends to design a three-position universal test platform, and    study on the mechanical structure, save test cost for the vehicle assembly business or related accessories companies.

## 4    Experimental Results and Analysis

Table 1 shows the typical experimental data of the system application in a factory. Data show that: (1) Developed vehicle instrument switches fatigue detection system can complete the rotating, flipping and pressing three test operation and the detection are parallel among the work-position. (2) Switch travel distance is not less than 20mm. (3) Switch closure time and holding time are adjustable. (4) System is suitable for a variety of appearance products; (5) The multi-work-position and multi-switch parallel scheduling algorithm and mechanical structure need to be improved.

**Table 1.** Resulting data of system experiment

| Observation time | Specimens | Action Types | Testing times | Hardware response time |
|---|---|---|---|---|
| 4 weeks | 20 | 3 kinds | $1{\sim}10^6$ adjustable | <1s |
| Parallel scheduling algorithms efficiency | | | Action sequences degree programmable | |
| >90% | | | 100% | |

## 5    Conclusion

Vehicle instrument switches fatigue detection system is a complex system of hardware and software work together, and it's difficult to develop. A kind of composition and architecture of switches fatigue detection system were introduced in this paper. Combined with this kind of architecture, work principle of the main components was provided. The key technology and implement methods that must be solved in the research process were analyzed. The results of experiment and application show that this kind of vehicle instrumentation switches fatigue life detection system could preferably realize the fatigue life test. System architecture, key technology and implement methods provided by this article have certain significance for the development of other similar systems.

## References

1. Park, J.-H., Chun, Y.-B., Kim, Y.-J., et al.: A Study On The Fatigue Behaviour Of Electroplated Nico Thin Film For Probe Tip Applications. Materialwissenschaft und Werkstofftechnik 40(3), 187–191 (2009)
2. Ghidini, T., Dalle Donne, C.: Fatigue life predictions using fracture mechanics methods. Engineering Fracture Mechanics 76(1), 134–148 (2009)
3. Pinkaew, T., Senjuntichai, T.: Fatigue Damage Evaluation Of Railway Truss Bridges From Field Strain Measurement. Advances in Structural Engineering 12(1), 53–69 (2009)
4. Yusof, F., Withers, P.J.: Real-time Acquisition Of Fatigue Crack Images For Monitoring Crack-tip Stress Intensity Variations Within Fatigue Cycles. The Journal of Strain Analysis for Engineering Design 44(2), 149–158 (2009)
5. Chen, Q., Xu, L., Salo, A., et al.: Reliability Study of Flexible Display Module by Experiments. In: 2008 International Conference on Electronic Packaging Technology & High Density Packaging, vol. 2, pp. 1086–1091 (2008)
6. Kenedi, P.P., de Souza, L.F.G., Cordeiro, C.A.: Conception of a Fatigue Test Device for High-Strength Steel Wires. In: SAE Brasil 2007 Congress and Exhibit, pp. 23061–23067 (November 2007)
7. Park, J.-H., Myung, M.S., Kim, Y.-J.: Specimen Size Effect On Fatigue Properties Of Surface-Micromachined AI-3%Ti Thin Films. In: International Conference on Integration and Commercialization of Micro and Nanosystems, pp. 661–665 (2007)
8. Shimamoto, A., Hwang, D.-Y., Nemoto, T.: Development of Biaxial Servo Controlled Fatigue Testing System. In: 1st International Conference on Advanced Nondestructive Evaluation, Part I, pp. 57–62 (2006)

9. Ferraris, E., Fassi, I., De Masi, B., et al.: Polysilicon fatigue test-bed monitoring based on the 2nd harmonic of the device current measurement. In: MEMS, NANO and Smart Systems 2005, pp. 55–60 (2005)
10. Boussalis, H., Liu, C., Rad, K., et al.: Integrated Embedded Architectures and Parallel Algorithms for a Decentralized Control System. In: The 20th IEEE International Symposium on Intelligent Control (ISIC 2005), vol. 2, pp. 1567–1572 (2005)

# Application of Neural Network in the Velocity Loop of a Gyro-Stabilized Platform

Shiqiang Ma and Yuan Ding

Department of Mechanical Engineering, Changchun University, Changchun, Jilin, China
msqwxmy@163.com

**Abstract.** Based on the self-learning property of neural network, a controller of neural network is attentively put forward for the velocity loop of a stabilized platform in this paper. The experimental results are given in both the controller of neural network and the traditional controller designed with traditional frequency domain methodology. Our experiment contrast results show that such neural network control structure is very effective for improving the low velocity property of stabilized platform and it is valuable for practical engineering.

**Keywords:** neural network, self adaptation, stabilized platform, velocity loop.

## 1   Introduction

In photoelectric airborne scout and measuring devices, it always needs to point the optical axis of a photoelectric sensor to the target, so as to complete the target acquisition, tracking and measuring. A stabilized platform is the major unit for ensuring the optical axis stabilization. Sensors such as visible TV camera system, infrared camera system and laser rangefinder, etc. are generally installed on a stabilized platform, in other words, integral stabilized platform method is adopted. Basically, electro-mechanical multi-frame structures are used for these platforms, which include two- axis platform structure form and three-axis platform structure form, etc. The control system in the platform is one of the critical factors for ensuring the platform performance. For controlling one axial in this platform, double closed loop control structure is generally adopted, which structure diagram is as shown in Figure 1.

In this diagram, velocity gyro, velocity correction steps, power amplifier, electrical motor and platform load, etc. form the speed loop. Tracker, position correction steps and speed loop composite the position loop. In practical control system design, it is found that magnitude margin and phase margin as required by the system have been considered in the design procedures and the system may be ensured to be relatively stabilized in a certain range by using of traditional frequency domain design methodology. However, such traditional frequency domain correction methodology cannot ensure the system to be always provided with optimal control performance because the controlled stations are impossible to provide accurate measurement and further due to the features of controlled stations change along with the changes of

G. Lee (Ed.): Advances in Intelligent Systems, AISC 138, pp. 355–361.

external environment and conditions. Especially when the carrier is in different postures, it may lead a change of orthocenter of the platform and the friction torque between systems of axes will also be changed, and further will cause a difference between the features that are based on in the correction loop design and the features of controlled stations. Therefore, it is expected that the control parameters may change self-adaptively subject to a required index so as to ensure the system to have an optimal control performance from beginning to end. Aiming at abovementioned problem, this paper tentatively shows a method to realize the self adaption adjustment of control parameters in low speed by using of self-teaching features of neural network and on the basis of traditional correction methodology. The practical application result shows that this method may effectively improve the low speed feature of stabilized platform's speed loop, overcome the influence to the system performance caused by nonlinear disturbance such as friction and wire-winding, etc. and lower down the requirements to the mathematical model's precision of controlled station. It may has a extensive practical value.

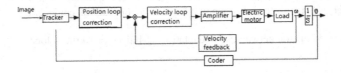

**Fig. 1.** Stabilized platform control system diagram

## 2    The General Design Method of Digital Controller

In a digital control system, the controller design method that is widely used in engineering is the continuous and systematical discretization design method. When sampling frequency of the system is high (normally as 5-10 times than the cutoff frequency of open loop of the system), the discretization system is regarded approximately as a continuous system, by using of design methodology such as traditional frequency method, the transfer function of correction steps is designed. Then by using of Impulse invariance design method, Zero poles matching method or bilinear variance design method, etc. continuous correction steps transfer functions is discretized to get corresponding digital control algorithm. Bilinear variance method is a very common discretization method in engineering.

If the transfer function D(s) of continuous system correction loop is given as:

$$D(s) = \frac{U(s)}{E(s)} = \frac{B_0 + B_1 s + B_2 s^2 + \cdots + B_m s^m}{A_0 + A_1 s + A_2 s^2 + \cdots + A_n s^n}. \tag{1}$$

By using of bilinear variance method, make impulse transfer function D(z) of discretized D(s) as:

$$D(z) = D(s)\Big|_{s=\frac{2}{T}\frac{1-z^{-1}}{1+z^{-1}}} = \frac{u(z)}{e(z)} = \frac{b_0 + b_1 z^{-1} + b_2 z^{-2} + \cdots + b_m z^{-m}}{1 + a_1 z^{-1} + a_2 z^{-2} + \cdots + a_n z^{-n}}. \tag{2}$$

then there is a constant corresponding relationship between the coefficients in D(s) and the coefficients in D(z). If is given, can be uniquely determined after the system sampling frequency is designated. Therefore, the transfer function D(s) of correction steps is got through continuous system design method, and impulse transfer function D(z) can be determined, further,

D(s) and D(z) will have same order.

Transform z inversely, then we get the recurrence control algorithm for digital realization after sorting:

$$u(k) = \sum_{i=0}^{m} b_i e(k-i) - \sum_{j=1}^{n} a_j u(k-j) \ (k = 0, 1, 2, \cdots). \tag{3}$$

In this formula, "k" represents the present sampling time, (k-i) and (k-j) represents the past sampling time which is apart from present sampling time by i or j sampling periods.

Based on the traditional frequency domain design method, each control parameter in the formula (1) is a constant, and cannot ensure the system to perform optimally when controlled station parameters are variable or under condition that there are various disturbances, in other words, the system does not have self adaption adjustment ability. If each control parameter in the formula (1) can be adjusted self-adaptively with the change of controlled station parameters or with that of disturbance so that some performance index may approach to the most optimal value all the time, then the system control performance will surely be improved in great extent. Neural network has ability that may approach to any nonlinear function, therefore, it is an effective method for controlling parameters self-adjustment. In order to meet the requirements of concurrent control system, a neural network is applied in the automatic adjustment of system control parameters.

## 3   Neural Network Realize Self-adjustment of Control Parameters

On the basis of self-adaption linear (Adaline) neural network theory, formula (1) may constitute the neural network structure as shown in Figure 2.

In formula (1), e(k), e(k-1), ... , e(k-m) and u(k-1), u(k-2), ... , u(k-n) as input variables of neural network, and these variables may all be got by measurement or calculation in the system; b0, b1, ..., bm and a1, ..., an as metrics of the neural network. When the system is running, the neural network adjusts metrics automatically on the basis of LMS (Least mean-square) learning algorithm and enables the system performance indexes to reach or approach the optimal for all the time.

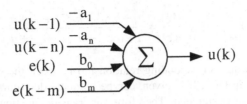

**Fig. 2.** Network structure of Adaline neural network

Using the neural network in Figure 2 to set up control structure diagram of stabilized platform velocity loop is as shown in Figure 3.

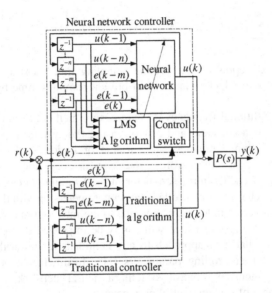

**Fig. 3.** Control structure diagram of stabilized platform velocity loop

In the Figure, r(k) as the velocity loop input variable, y(k) as the velocity loop output variable, and the difference the two is the error "e(k)". In order to overcome the shortcoming that the neural network controller converges very slow when obvious error "e(k)" comes up, switch between two types of control models, traditional controller control and Adaline neural network controller control, is adopted subject to the e(k) value. That is, in case of obvious error, traditional control structure is adopted, while in case of little error, Adaline neural network control structure is adopted. In this way, it may get a better control performance.

Adaline neural network uses LMS learning algorithm to train network, and the function of control system performance index as formula (4), the metric vector's adjustment value as formula (5), the metric vector of present sampling time(k) as formula (6):

$$E_P(k) = \frac{1}{2}[r(k) - y(k)]^2 \qquad (4)$$

$$\Delta W = -\frac{\eta}{\|X\|^2}\frac{\partial E_p}{\partial W} = \frac{\eta}{\|X\|^2}[r - y]\frac{\partial y}{\partial u}X^T . \qquad (5)$$

$$W(k) = W(k-1) + \Delta W(k-1) . \qquad (6)$$

In above formula, X is as the input variable vector $[x_1, x_2, ..., x_{n+m+1}]^T$ of neural network, W is as the metric vector $[w_1, w_2, ..., w_{n+m+1}]^T$. In the PWM control system, the control variable "u" is approximately linear with output variable "y"; and the $\partial y/\partial u$ value may be determined by experiment(s). $\eta$, as the learning rate, normally is a positive number, which may influence the convergence rate/velocity and effectiveness of the algorithm. Its value can be adjusted by means of self adaption. The flow of self adaption adjustment method is as shown in Figure 4. Our purpose is to shorten the learning time as possible as could in precondition of convergence. The basis guiding ideology for adjustment is to increase learning rate $\eta$ under condition of learning convergence so as to shorten the learning time, while decrease $\eta$ till converge when $\eta$ is too big and causes the performance index a failure of convergence. In the Figure, increasing coefficient is as $\alpha > 1$, and decreasing coefficient as $0 < \beta < 1$.

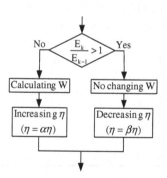

**Fig. 4.** Flow chart of Adaline adjustment method

The initialized status of neural network metric (control parameter) may have great influence on the performance of control system. An incorrect selection may slow down the network convergence velocity when two models are switched over and then destroy the system performance. In engineering practice, the mathematic model of controlled station can be roughly determined by experiments, then based on this rough mathematic model, the system correction step may be designed in traditional correction method. Discretization in double linear transformation method can get impulse transfer functional expression, this way, we may get control parameter, and take them as the initial value of network metrics. Such initial values selected in this way may ensure a very smooth and steady switchover when switching between two types of models.

## 4    Experiments as Examples

In a gyro stabilized platform, apply abovementioned control method in the platform velocity loop control system. In the experiment, the electromechanical time constant is measured as $T_m = 0.042s$, and the electromagnetic time constant is so small that it may be ignored. Firstly, traditional frequency correction method is adopted. For several times of experiments and comparison, the correction steps structure and parameters are as shown in Formula (2). The system has a preferable control performance.

Discretize the Formula (2) in double linear transformation method, a neural network with 5 input variables can be constitute, in which, initial metrics of input variables respectively as $-a_1 = 1.998$, $-a_2 = -0.9978$, $b_0 = 0.000954$, $b_1 = -0.001859$ and $b_2 = 0.000906$. It is measured in the experiments that the learning rate initial value is as 0.0045, $\eta$ increasing coefficient $\alpha = 1.05$ and $\eta$ decreasing coefficient $\beta = 0.75$, two types of control models switching point: when error $e(k) > 2\,°/s$, traditional correction steps are adopted, otherwise, neural network correction steps are adopted. System step function response is shown as Figure 5. From which, presumably, the switch between two types of control models are relatively smooth and steady.

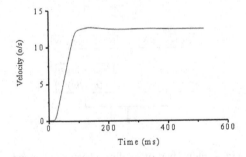

**Fig. 5.** Step Response of Velocity Loop

When the system is running in low speed (  ), the response based on traditional correction steps are shown as in Figure 6(a), and the response based on neural network correction steps are as shown in Figure 6 (b). It is thus clear that a neural network may improve remarkably the low speed stationarity of a system.

# 5    Conclusion

The feature of stabilized platform's velocity loop may have great influence on the control performance of whole platform. To adopt self adaptation linear neural network control structure may effectively improve the low velocity performance of velocity loop when the accurate mathematical model of system is not given and various disturbances involve. The determination of neural network metric initial value and the selection of learning rate initial value may have major influence on the control effect. Firstly, to adopt traditional correction method and to use the control parameters that we have got as the initial value of neural network metric may enable the system to have a faster response speed. It is an effective selection method for metric initial value and has a practical value in engineering. Meanwhile, to adjust the learning rate by means of self adaption may ensure the learning time to be shortened effectively in precondition of algorithm convergence.

# References

1. Haykin, S.: Neural Networks: A Comprehensive Foundation, 2nd edn. Tsinghua University Press (2001)
2. Xu, L.: Neural Network Control. Harbin Institute of Technology Press (1999)
3. Li, S.: Theory and Application of Intellectual Fuzzy Control. Harbin Institute of Technology Press (1990)

# The Design for the Assembly Press of Side Ring for Trucks

Guangguo Zhang[1], Lei Zhang[2], and Yuanzhi Tang[1]

[1] Dept. of Mechanical Engineering , Hubei Automotive Industries Institute, Shiyan, Hubei Province 442002 P.R. China
[2] Technical Center of Dongfeng Motor Corporation, Wuhan, Hubei Province 430058 P.R. China
{qyjxxzgg,annefly1987110}@sina.com, zhzhangang@126.com

**Abstract.** In order to meet the demands of automation,solve the current problems that the assembly press of side ring for trucks with poor efficiency, unreliable,high cost and so on,we design a new,simple and high reliability assembly press of side ring which is dedicatly,based on the analysis of the relevant technical parameters and the requirements of the site conditions.Test and field trial shows that the pressure is stable and reliable,clean,the assembly quality improved obviously,it also can reduce the labor intensity and production cost, improve the production efficiency.

**Keywords:** Side ring, Assembly press, Design.

## 1    Introduction

With the development of society, with high efficiency and high quality automatic assembly line can adapt the request of modernization. Block circle is the wheels of light is an important part of it prevent tues from rim, also stop sediment into internal damage tube. Wheel assembly used for inner tire of the press is wheel assembly, at the same time such machines are usually applied on the production line. In the groove conveyor line will be rim, inner tues, layering to being connected location, lifting device jacked up workpiece, press the upper portion of the pressure head work and finish the conjunction of workpiece loading. In order to increase the production efficiency design a reliable; Good stability; Easy maintenance of the press is very significant.

According to the special equipment factory original dongfeng automobile assembly pressed into the press circle to ensure reliability block rim, push rod easy card to die, fault more difficult to repair and maintenance, time is long, low efficiency of problems, the author through the in-depth research and analysis, and the research, design, a new model of wheel assembly press.To the production, achieve the purpose of improving productivity, in order to obtain more economic benefit.

## 2    Technical Requirements and Key Technology Index

**Technical requirements.** Precision stability, convenient operation, advanced performance, safety and reliability, long service life.

G. Lee (Ed.): Advances in Intelligent Systems, AISC 138, pp. 363–368.
springerlink.com          © Springer-Verlag Berlin Heidelberg 2012

Non-standard parts structure design simple easy to processing.

Casting through strict artificial limitation and vibration aging, ensure that the pressure strength and stability of the installed

The use of advanced super audio quenching process, makes the installation service life is longer.

Cylinder floor for 34 mm thick, the pressure head high from the ground for 1270 mm.

To cope with the assembly line, the bottom of the shaft from the cylinder first shaft shoulders for 740 mm.

After the car shall not damage processing from rim.

Production speed for each minute after being connected a block circle.

Environment temperature-10 °C ~ + 40 °C.

**The key technology and technical indexes.** This subject is the key to solve problem is oriented stem verticality, cylinder pressure control and spring of preloaded. Guide bar is lead cylinder at the vertical movement, ensure that the pressure head and block circle of vertical degree. Cylinder in athletic process pressure changes, when no stem cavity volume increases pressure decrease. When the piston moves to the top, if not convenient increase pressure can lead to press does not successfully completed work being connected. Spring of preloaded is to ensure that the steel talons can fall to block circle of flange, workers through the naked eye can easy to implement. So to ensure that can block circle is convenient fast pressure into a car from rim. Should strive to do: to ensure that the stiffness of the lever is greater than the stiffness of the block circle; Spring or critical in place, can make the steel talons can fall in the block circle of flange; Guide to stem bottom of vertical degree can't too low, with specific and fixed the center of the board to ensure alignment lever can pressure in the block on the edge circle; To limit the rim of the five degrees of freedom, the Z axis rotation does not limit not influence processing; Pressure installed simple structure, appearance beautiful, repair, maintenance and easy to adjust; Pressure installed stable work, work continuously 6 ~ 8 hours may not appear any fault; Pressure installed control manual control.

## 3   Pressure Installed Structure Design

**General structure and working principle.** Through the analysis of the existing technical parameters and requirements, to meet the effective and reliable will block pressed into the circle the requirements of the rim. According to the structure of block circle and felloe, block the inner diameter of the circle of the outer diameter than rim is small, so the only put apart to block circle circle will block pressed into the rim, at the same time, round and pressed into the block to rim, this will be in vertical and horizontal directions to put pressure on the block circle. According to the structural characteristics of the wheels and in order to facilitate the operation, and the pressure was installed design of structure. The localization way side a short pin, the face positioning limit of rim X, Y, Z axis movement, sell a limited the X, Y axis rotation. In a horizontal direction to force, from up to down using hydraulic drive a disc will block circle open, of which the disk uniform with six can suppressing the steel talons.

In vertical direction to force, from down to up using pressure raised wheel, the end of the six lever block adopts circle, push rod and the steel talons stagger installation.

To guarantee the disc and rise of piston rod vertical degree and the center of the disk and the center of the piston rod straightness, applied to the design of the disc down and wheel up the same orientation stem guide. Also in design a debugging great guarantee the center of the disk and the piston rod center vertical degree. Pressure installed working principle Fig.1. 1-Guide stem,2- Check ring,3- Truck Rim,4- Guide stem,5- Cylinder,6- Clip specific,7- Steel Talons,8-Compression bar.

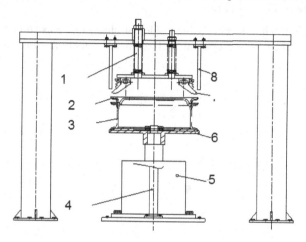

**Fig. 1.** Pressure installed working principle diagram

Pressure by three major components installed main: 1-frame roof structure, 2-cylinder structure, 3-clip specific guidance rod structure. Structure as shown in Fig .2.

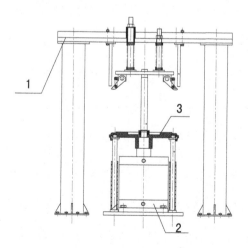

**Fig. 2.** Pressure installed structure

**Clip specific design.** The specific into fluctuation two parts, through the welding joined up. Hole center line and the requirements of the vertical bottom is to ensure that rim can vertical rising steadily. Through the aperture phi 90 and with the cooperation of the piston rod to guarantee the rim is picked up the process of the transverse shaking, so its processing precision demand is high. And rim have important surface contact, the selection of surface roughness is 0.8. This fixture is suitable only for the same type of wheel, to ensure the rim of the accurate positioning and the diameter of the convex platform processing the demand is higher. Clip specific structure as shown in Fig.3.

**Guide stem design.** Guide stem design guide bar chart shown as shown in Fig. 4. Its role is to combine the core shaft guarantee fixed board can climb straight up, so it's the

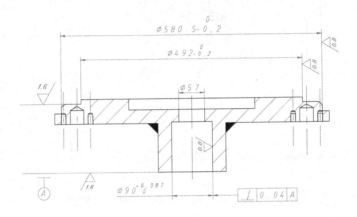

**Fig. 3.** Clip concrete structure

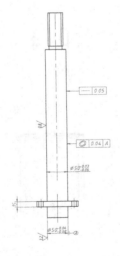

**Fig. 4.** Guide bar chart

straightness demand is high. If the orientation of the stem and scratch on cylindrical center were in the same line, after installation will tilt took less than guidance. Installation guide stem upside and orientation set of contact, the shaft and the hole with higher accuracy.

**The speed of the cylinder control and adjust.** In athletic process, control and adjust the speed of the cylinder is very important. When workers will block circle with stable after car from rim, start lifting cylinder. Cylinder rose steadily to open up the pressure head, open up the pressure head is automatically will open to lever block circle circle and play a role in each (block circle center and rim center in the same vertical). Cylinder continue to rise pressure head is successfully pressed into the circle will block rim, at this time, there will be a empty clay mugs slipped sound. Workers can start after manual directional control valve cylinder drop, prevent the piston and cylinder collided at the bottom have buffer the design. The throttle valve type design one-way two-way throttling control loop. Its principle of work the following Fig. 5.

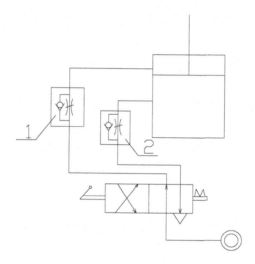

**Fig. 5.** Wo-way speed loop

When the manual directional control valves, from compressed air to air, the reversing valves and throttle valve into the cylinder of have no stem cavity, drive the pistons to move up. By adjusting the throttle valve 2 opening, can control the different exhaust velocity, and also can control the piston velocity. Similarly, cylinder down through this way also can adjust the throttle control the speed of 1 opening.

## 4   Summary

The structural design of the air cylinder to provide pressure, the pressure was stable and reliable, clean no pollution. The wheel to block the successful development of the press circle installed, solve the special equipment factory original dongfeng automobile assembly pressed into the press circle to ensure reliability block rim, push rod easy card to die, fault more difficult to repair and maintenance, time is long, low efficiency. Realize the automation of the assembly lines, which reduces the homework strength, improve the production efficiency, greatly reduce the maintenance and maintenance cost, for the benefit of the enterprise to obtain higher.

## References

1. Cheng, D.: Handbook of mechanical design. Chemical Industry Press, Beijing (2004)
2. FAG. The Design of Rolling Bearing Mountings. Publ. No. WL 00 200/5 EA
3. Gan, L.: Geometric tolerance and quantity. Higher Education Press, Beijing (2003)
4. Liu, H.: Material Mechanics, 4th edn. Higher Education Press, Beijing (2005)

# Web Service Composition Based on AXML

Junfu Zhao and Shengmei Ma

Engineering Training Center, Inner, Mongolia Technology University, Baotou,
Inner Mongolia, China
junfu_zhao@yahoo.com.cn

**Abstract.** With the development of the Web services application, It is necessary
to resolve sharing and interaction in massive application services. The dynamic
composition of Web services provides new solution to solve bottleneck brought
by information sharing and application corporation. Through Active XML
(AXML, for short) is introduced in the paper, Expand Web services, Dynamic
composition Web services architecture is proposed based on AXML. The
architecture supports Dynamic services composition building in time,
deployment and delivering in dynamic environment is implemented

**Keywords:** Active XML, Web services, Web services, Composition.

## 1 Introduction

With the development of Internet technology, Web services is a new distributed
computing model, realize the true meaning of the platform independent and language
independent. The Web service is defined as using the standard XML technology and
other services to interact with the software module and self described application.
More and more enterprises will be their own business as Web services, enterprise
service speed, service scope, the service quality raised taller requirement. In the face
of more and more Web services, how to rapidly deploy and discovery services,
according to dynamic Web services composition [1],and published as a completion of
specific tasks of the new service, in order to improve the service immediately, reduce
the development cost of service has become the key problem needed to be solved.

In the Web services architecture between each module, the message is passed to
the XML format. However, during the process of data exchange some data needs to
be updated in real time, so that the existing XML cannot be timely response; in this
context, INRIA ( the French Institute of Automation ) put forward a kind of
distributed data management language — Active XML [2], which is characterized in
the XML embedded in Web service call to achieve access to some constant the
updated data, can by calling the Web service on XML document some data in
real-time to give accurate description.

The Web service combination has been in the related fields of study to gain
attention, related technology [3] has appeared such as: WISE, e-flow. However, most

G. Lee (Ed.): Advances in Intelligent Systems, AISC 138, pp. 369–375.
springerlink.com     © Springer-Verlag Berlin Heidelberg 2012

of them need to deal with the underlying programming details, designers need to consider the interaction, message mapping between, serving the calling method. Based on the Active XML dynamic Web service composition is the use of AXML dynamic characteristics on Web services to extend the relevant services, the concept and definition of unity, the establishment of service connection between the flexible group, achieve and improve portfolio quality. In this paper, the Web Services Description Language ( WSDL ) based on the introduction of AXML, Web services tailored to the specific environment of dynamic combination.

## 2    Web Service Composition Analysis

In a heterogeneous system running on different platforms Web services may be provided by different providers, by a different programming language, in order to meet the request of service according to the specific application background and the need for reasonable Web service combination. The Web service combination potentially shorten development time, reduce the workload of the development of new applications. Service composition can be divided into static and dynamic combination. Static combination is at design time or at the loading, which some little change in the operating system is effective, but the lack of customization flexibility. Dynamic portfolio is in the run-time combination, it can bring more flexibility. Dynamic composition of Web services need to solve the following problem [1]: 1)how can the basic Web services related data are updated in real time, accurate description of the Web service; 2)how to achieve the basic service coordination, dynamic interaction. 3)how to ensure the orderly execution of Web services composition.

## 3    The Main Technology of Web Sevices Composition

In view of the above problems this paper first introduced the AXML [7] (dynamic XML ), this distributed data management language, on the Web service related data updated in real time. Secondly, through the given Web service composability conditions on Web service combination judgment, guarantee the basic service coordination, dynamic interaction. Then, put in a data structure to solve the problem of combining the Web service order execution.

### A.. Web service related data updated in real time

WSDL is a Web service description language XML, WSDL by defining a set of grammar to describe the Web service. However, WSDL only from the syntax of Web services for expression, the paper introduces AXML to the WSDL extension, using the dynamic property of AXML [2], the Web service related data updated in real time. The following is a Active XML to the WSDL extension, and then realize the Web service update simple example:

```
<service name =" weather forecast information consultation"/>
<purpose function=" weather forecast">
.....
<binding name ="SOAP"/>
<message name ="query">
<parameter name= "date" type="string" .... />
<parameter name= "region" type="string" ..../>
<parameter name= "temperature" type="float" unit= "degree" />
</message>
<operation name= "receiveSpecialquery" mode="request-response" />
<input name= "query" />
.......
<operation/>
<flow source=" ..\\..\newspaper.axml " target="replyqueryresult " >
```

**Fig. 1.** Extended WSDL document

In Figure 1, extended WSDL document, the top element < Service >, including service is the name of the weather forecast information consultation attribute. In < purpose > indicates that the service function of weather forecast, < binding > element tag specified in the binding type is of type SOAP, in the < message > tag specifies the message name, along with a list of news 3 parameters and the parameters of each name, type and other information. To provide a query to the one day, in a certain area of the temperature parameter information as input information, use the < flow > element tag description service operation process control, first in the weather service WSDL by specified in Figure 2 the message.axml document URL address to perform the AXML document, thus the dynamic access to a regional temperature, then from < replyqueryresult > to provide advisory services results.

```
<?xml version="1.0" encoding="ISO-8859-1">
  <newspaer xmlns="http://lemonde.fr"
            xmlns:rss="http://purl.org/rss"
            xmlns:axml="http://purl.org/net/axml">
  <title>china daily</title>
  <date>2006-04-16</date>
  <weather>
  <axml:call service="forecast@weather.com">
        <city>Beijing</city>
  </axml:call>
  </weather>
  </newspaper>
```

**Fig. 2.** AXML document

Figure 2 is a message.axml called the AXML document by call service, calls the forecast@weather.com to obtain the temperature in Beijing in April 16, 2006.

*B .The Web service can be integrated judgment*

Access to the Web service real time update, the service composability are judged, determines that a service is combined, to realize dynamic Web services composition. Presented here is a level 4 can be combined decision [4], ensure correctness.

*1)mode composability*

In order to make the two services can be combined and interactive operation, in the client / server ends of the operation must be consistent with the model, according to the existing call form defines the following 4 call operation mode:

*Notification:* is unidirectional, initiated the request message but do not accept any response message.

*Receive:* operation receives an input message and deal with it, but does not produce any output message.

*Invoke-response:* operation generates an output message, then a corresponding input message.

*Request-response:* request response type, operating to receive a message, process it and produces a corresponding response output.

*2)binding composability*

In ensuring the operation mode can be combined, also must ensure that due to the different services may support different binding protocols ( such as S0AP, HTTP ), these services are able to understand each other communication protocols, so as to ensure the normal interaction.

*3)operation can be combined*

Make sure that the interactive operation between the relevant objective and classification. If two operations for different functions, they will not make a mapping between.

*4)message mapping consistency*

A message contains a plurality of parameters, each parameter has a corresponding data type. The Web service message interaction needs to be checked is passed to the data type of the parameter with the recipient required parameters corresponding to the type of data type of parameters, ensure that the transmission was included in the received data type.

# 4   AXML Based on Dynamic Web Service Compostion Syetem Structre and Realization

## 4.1   AXML Based on Dynamic Web Service Composition Architecture

Dynamic Web service composition Architecture [6] as shown in Figure 3, is divided into 4 layers. The data storage layer, contains different information library. Every information corresponding to a it can be access service components. In the combined service layer, service component is about every single data source wrapper. This wrapper is the role of extraction and packaged query results into an AXML document, the document is returned to the upper call. It provides a unified access interface to use WSDL language to the relevant categories of service components described as a kind of service, through a combination of decision rules are combined to generate a new service. Inference layer to application layer service calls came in service logic for processing, at the same time resolution request is reasonable, and the rules of the service request of standardization and constraint set. After the logic of reasoning process, the user's request is reconstructed for information to the request source.

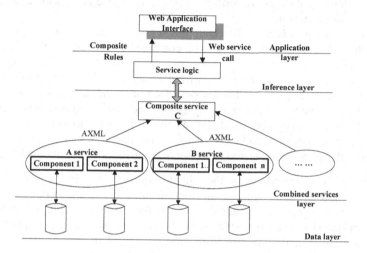

**Fig. 3.** Web service composition architecture

## 4.2    Dynamic Web Service Composition Process and Its Combination

Dynamic Web service composition synthesis module [5] is a dynamic Web service combination structure of the core, due to limited space in this paper only to the synthesis module in detail, on the other part does not give a detailed introduction. Dynamic Web service combination structure diagram as shown in Figure 4, the synthesis module mainly includes 4 modules:

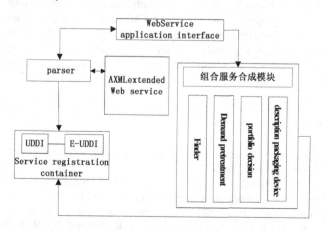

**Fig. 4.** Web service composition structure diagram

*1) needs preconditioning:* accept user input query condition, and carries on the concept of standardization.

*2) service finder :* the use of service attribute description information search to meet the requirements of multiple services, and analytical services component of the WSDL document, the service will bind type, message type, parameter information, operation information extraction storage for the next service component combinations can be used for determining.

*3) Service composability judgement:* the service composability judging device through the 3.2 section mentioned can be combined to find rules of service can be combined to judge.

*4) composite service description wrapper:* combined service generates new description. The description file has a series of < Link > elements, each with a < Link > element combined service in an operation and the corresponding component connected services operation. And the statement in order to perform an activity between the two operations to be completed by the interaction. The newly generated description file has to be registered to the service registration container in UDDI.

Synthesis module according to the client request, find the registered service component, and they are spliced together, form a combined service. Then, the combined services will be registered in UDDI for the user to call. When the Web application interface receives the service request, the service request to the corresponding dynamic update of Web service concept of unified standardization, standardization of service request as input conditions into next level service finder, the service finder will list all eligible service description, and call the service can be combined judging module of these services portfolio decision, will once again screened service description is provided to a combined service description of packing module, by a wrapper for each service description are combined, forming the composition service description, presented to the private UDDI registration center registration.

**Acknowledgment.** In the Web service development application trends, in order to be able to according to business needs, fast, flexible integration of various existing Web services, this paper proposes based on the AXML dynamic Web service composition, given the combination structure of the system, while the Web service composition execution order, synthesis module, combination of realization is introduced in detail in this paper. For dynamic composite service validation and security issues, and further research and exploration.

# References

1. Yue, K., Wang, X.: Web service core support technology. Research Summary of Computer Software 3(15), 428–442 (2004)
2. Abiteboul, S., Benjelloun, O., Milo, T., Manolescu, I., Weber, R.: Active XML:Peer to Peer Data and Web Services Integration(demo). In: Proc. of VLDB (2002)
3. eFlow. In: Proceedings of the CASISE Conference, Stockholm, pp. 13–31 (2000)
4. Medjahed, B., Bouuettaya, A., Elmagarmid, A.K.: Composing Web Services on the Semantic Web. The VLDB Journal 12(4), 331–351 (2003)

5. Benatallah, B., Dumas, M.: Definition and Execution of Composite Web Service: The SELF-SERV Project. Bulletin of the IEEE Computer Society Committee on Data Engineering, The VLDB Journal (2003)
6. Thakkar, S., Knoblock, C.A., Ambite, J.L., et al.: Dynamically Composing Web Services from On-line Sources (2002), http://www.aaai.org
7. Jin, R., Shi, H., Gao, Y., Zhao, J.: Based on the Web Services AXML in workflow process modeling. Micro Computer Information 2(3), 253–255 (2006)

# Design and Application of Remote Intelligent Monitoring System Based on CDMA

Xu Liang

JiLin Institute of Architecture Engineering, ChangChun City, China

**Abstract.** This paper designed a CDMA-based wireless communication technology for remote intelligent monitoring system. Using the CDMA network's SMS services, packet data services and Intermet network TCP / IP protocol, to achieve the wireless transmission of data, and to complete real-time monitoring of distribution systems, lighting systems and electrical equipment in the buildings.

**Keywords:** CDMA, remote monitoring system, DGTS-800.

## 1 Introduction

CDMA is the abbreviation of the Code Division Multiple Access. It is a new wireless communications technologies, which is developed based on a branch of digital technology - spread-spectrum communication technology, Compared with GSM, CDMA has the advantages of the system capacity, the base station coverage area, high-quality voice, soft switching technology, low rates of dropped calls, small transmit power, good secrecy, high data transfer rate and facilitate the transition to third generation mobile communication, etc[1].

This paper describes a remote data transmission system, based on CDMA technology and Atmega128 microcontroller ,the system can receive and send data through the form of short message, or by TCP / IP protocol[2], the use of Internet network and a PC for data transfer.

## 2 Remote Intelligent Monitoring System

Remote intelligent control system consists of CDMA wireless data transmission terminal [2] and the control center of two parts, can achieve data acquisition and remote wireless transmission function. CDMA wireless data transmission terminal is the hub of CDMA remote intelligent monitoring system, its performance reflects the level of the whole system. The system structure is shown below:

### 2.1 CDMA Wireless Data Transmission Terminal

CDMA wireless data transmission terminal can complete remote data transmission through short message or a CDMA network in the form of packet data services by using the data monitoring system composed of single-chip. It's core control device is the ATmega128 MCU, which main peripheral circuit are:

(1) data acquisition interface circuit used to implement data collection measurement points.

(2) relay control circuit mainly used to implement volume control switch.

(3) real-time clock circuit that's core devices are SD2403 mainly used to provide an accurate time signals to the system.

(4) CDMA MODEM, which core components are DTGS-800 mainly used to implement and monitor the wireless data transmission of the center.

(5) anti-crash external timer reset circuit used to prevent a runaway process leading to the phenomenon of death.

(6) LCD display circuit, which core components are RT12864M for real-time display of data in the terminal.

System solutions diagram in Figure 2.

It has two major tasks, one is responsible for collecting the Measurement point data and processing, packaging and store the results according to protocol, the other is to

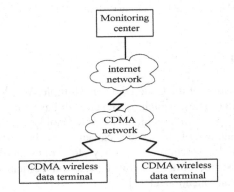

**Fig. 1.** System structure

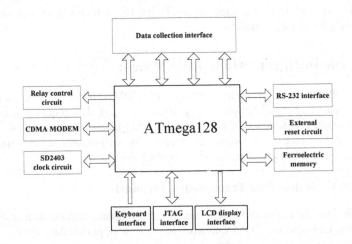

**Fig. 2.** System solutions diagram

complete the communications with the PC of the system control center, and to sent the measurement point data to the PC in the monitoring center.

## 2.2   Monitoring Center

Control center formed by a single PC and CDMA MODEM requires a PC connected to the Internet in real time. It has a VPDN (Virtual Private Network) access, to respond to the remote subsystem VPDN connection requests, and to complete the data transmission between wireless data terminal and control center real-time by IP address. It also can call the data of CDMA wireless data terminal, and processing of, display, store and print the data, and to Issue a variety of control instructions to the wireless data terminal for setting communication and control parameters.

## 2.3   CDMA MODEM

CDMA MODEM is the basis for wireless data transmission, which uses the DTGS-800 wireless data module as the core, and has a SIM card socket, supports AT command serial interface, power connector, antenna and other components. Interface block diagram shown in Figure 3 [3].

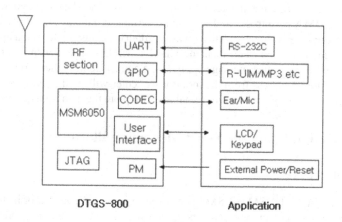

**Fig. 3.** Interface block diagram of DGTS — 800

### 2.3.1   Features
DTGS-800 is the latest compact AnyDATA wireless data module, and work in the cellular band. The DTGS subsystem - 800 includes a CDMA processor (MSM6050), integrated codec, microphone amplifier and an RS-232 serial interface, which supports forward-link data communication and has a transmission rate up to 153kbps.

### 2.3.2   AT Command Associated with the Short Message [4]
AT + SMSP = <param.>; set the encoding format to send SMS
AT + SMSR; read and delete the last one unread message

AT + UGSM?; 0 not use the command set; 1 uses it.

AT + CMGS =; send a short message.

### 2.3.3    AT Commands Related to Packet Data Services

AT + CAD?; Query network services status

AT + CRM = 130; set the serial communication protocol

AT + DIP = "xxxx"; set the destination IP address, CRM = 130 valid

AT + DPORT = "x"; set the destination port, CRM = 130 valid

### 2.4    System Works

First, the CDMA wireless data transmission terminal processing of and stores the measurement data Periodically, and prepares to transfer data at any time.

System for data transmission is divided into two ways, one is directly sent its own IP address from the monitoring center to CDMA wireless data terminal with the form of short message, when the CDMA wireless data transmission terminal receives the IP address, it will take the initiative to apply for a connection. When the connection is established, CDMA wireless data transmission terminal can store the data packet sent to the CDMA wireless network, and then to the control center with TCP / IP transport protocol through the Internet. On the contrary, the monitoring center can also be transmitted via the Internet packet data, then transmit the data through the CDMA network to CDMA terminals, and can send various commands to control the data collection terminal. The other is CDMA wireless data transmission terminal applys to the monitoring center to send data, then the CDMA wireless data transmission from the terminal in the form of short message apply to the control center, after monitoring center to allow, then the IP address from the monitoring center will send a short message to the CDMA wireless data transmission terminal and complete the data transmission.

## 3    The Design Using the Short Messaging Platform

### 3.1    To Send Data

By sending the AT + UGSM = 1 SMS command to use the second set of instruction set. Unicode encoding the data to be sent, and then using AT + CMGS command to send.

By reading the return value in the input buffer, using InStr () function to determine whether the last two characters is "OK" or not, by which to determine whether messages sent successfully. If not successfully, passed the Sleep () function to wait for 0.1 seconds before the caller re-issued. Specific process shown in Figure 4.

### 3.2    Receive Data

Monitoring Center PC through the serial port on VB6.0 connect with CDMA MODEM, using MSComm control to transmit and receive data through the serial port. By InStr ()function to monitor the receive buffer if the received string "CMT:" or not, to determine whether there is new short message. If there is any new message, to read the message through MSComm1.Output = "AT + SMSR" and be saved automatically as needed. Specific workflow shown in Figure 5.

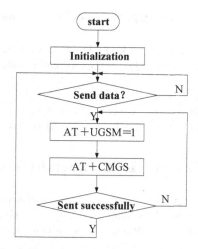

**Fig. 4.** Short message data process flow chart

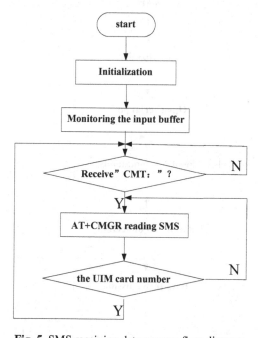

**Fig. 5.** SMS receiving data process flow diagram

Receive SMS main procedures are as follows:

ReceiveDate = MSComm1.Input 'read the data in input buffer
m = InStr (ReceiveDate, "10001") 'query The location of the string 10001 first occurrence
If m = 0 Then 'if no 10001, Handling in accordance with cell phone number
k = Len (ReceiveDate)  'the string length returned

j = InStr (ReceiveDate, "+ CMGR:")  'query the position of first occurrence for string + CMGR:

SSMS = Mid (ReceiveDate, j + 48, k - j - 48 - 6) 'extract the message content

tel = Mid (ReceiveDate, j + 16, 11)   'extract the call number

Else

k = Len (ReceiveDate)  'the string length returned

tel = Mid (ReceiveDate, m, 5)  'extract the call number

SSMS = Mid (ReceiveDate, m + 26, k - m - 26 - 6) 'remove the message content

## 4    The Design of Using Packet Data Service Transceiver Platform

The main task is to achieve point to point network connections by using of packet data services for data transmission. The system uses Anydate module DTGS-800 embedded TCP / IP protocol can establish a PPP connection with the monitoring center by sending

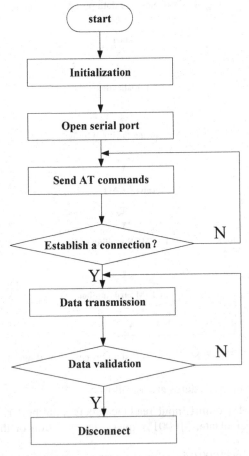

**Fig. 6.** Packet Data Service to send and receive data process flow diagram

AT commands, so that complete point to point network connections, you can remote control the wireless terminal and control center data communication. Specific process shown in Figure 6.

First, DTGS-800 using AT commands to establish network connection needs a destination IP address, so before the application to establish a PPP connection, you must be monitoring center PC's IP address in the form of short message to notify the remote control terminal, so that other terminals can be connected with it.After the success of the network connected, remote data terminal is assigned a temporary IP address by the network, which can obtain an IP address by AT + CBIP? Order, and transmit it to the monitoring center PC, so control center and remote data terminal can transmit data.

## 5    Conclusion

This design achieved real-time two-way remote communication by using Atmega128 microcomputer control system and CDMA MODEM, which can communicate with the monitoring center in the form of short message by CDMA  network .You can also use TCP / IP protocol to send and receive data, in order to establish the monitoring center and microcomputer control system for remote wireless communication function, microcontroller can establish communication through AT command and control system CDMA module, to achieve the purpose of system functions.

This system is a good direction for the future of international and domestic wireless monitoring system, which has a great development potential. In today's increasingly high degree of automation of the information age, CDMA remote intelligent monitoring system will play an irreplaceable role In many areas, such as a large field of well control, large-scale coal mine of the control and hydrological monitoring of the control [5].

## References

1. Qu, C.: Navigation and Positioning System Based on GPS & CDMA. In: 2007 Second IEEE Conference on Industrial Electronics and Applications (2007)
2. Zhou, G.-S.: Design of Real-time wireless data transmission based on CDMA module. Control and Automation Publication Group 25(5-2), 5–6 (2009)
3. DTGS-800 Reference Manual Application Information (April 8, 2004)
4. Any DATE CDMA MODEM AT Manual. Ver3.1 (March 26, 2004)
5. Tang, X.-Y.: Design on Remote Supervisory Control System Based on ARM. Communications Technology 9(40), 39–41 (2007)

# Design and Application of LED Beam Display Based on LED Number Display Interface

Hong Deng[1], Xiuhui Chang[1], and Yanling Zhang[2]

[1] College of Information Engineering, Hebei United University, Tangshan, Hebei, China
{ahong53,huitai55}@sina.com
[2] Institute of Electrical Engineering, Graduate Student, Yanshan University
Qinhuangdao, Hebei, China
yanlingcome@126.com

**Abstract.** This paper introduces a design actualizing LED beam display. The design directly used LED number display interface on single-chip micro-computer emulation experiment system (Dais-958B+ teaching instrument). The handy hardware design is to obtain font and bit interface line through LED number display socket, and connect the line to LED beam display. The paper presents the overall idea, schematic diagram of hardware circuit and source program of software design to realize the beam display function.

**Keywords:** LED number display, LED beam display, display interface.

## 1    Introduction

LED bar graph is more intuitive and more realistic than LED digital display in displaying temperature, liquid and other applications.

The design innovation is the use of LED digital display characters with bit interface on Dais-958B+ teaching instruments (single-chip microcomputer simulation system) platform. To achieve multi-point LED bar graph directly. The hardware design method is simple, the characters and bit interface cable from the LED digital display outlet connected to the LED bar graph display can be realized beam display. be strictly followed.

## 2    Design Concept

### 2.1    Structural Analyses of LED Digital Display and LED Bar Graph Display

8-segment LED digital display consists of 7 segments (a-g) and the decimal point (h), it is divided into two different types, common cathode and common anode. "Common cathode" means the sections of 8-segment LED cathode are connected together as a common electrode; "common anode" means is the sections of 8-segment LED are connected together as anode common pole [1].

Structure and the schematic diagram of common cathode 8-segment LED digital display are shown in Figure 1 [2].

G. Lee (Ed.): Advances in Intelligent Systems, AISC 138, pp. 385–390.
springerlink.com          © Springer-Verlag Berlin Heidelberg 2012

Through the analyzing of Figure 1 shows, rank a-h segments of the Common cathode 8-segment LED digital display from low to high, is the 8 sections of the common cathode LED bar graph display.

Structure and the schematic diagram of common cathode 8-segment LED bar graph display are shown in Figure 2.

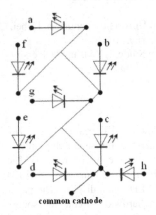

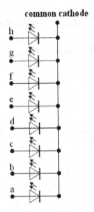

**Fig. 1.** Structure and the schematic diagram of Common cathode 8-segment LED digital display

**Fig. 2.** Structure and the schematic diagram of common cathode 8-segment LED bar graph display

## 2.2   Realization of the Design of LED Bar Graph Display

We got the design of LED bar graph display from the above analysis: 6 bits 8-segment LED digital display 6 × 8 = 48 segment LED, from the lower to the higher

**Table 1.** The Code Table of Bright LED Bar Graph Displays

| h | g | f | e | d | c | b | a | Code (H) | LED |
|---|---|---|---|---|---|---|---|---|---|
| 1 | 1 | 1 | 1 | 1 | 1 | 1 | 1 | FF | OFF |
| 1 | 1 | 1 | 1 | 1 | 1 | 1 | 0 | FE | L1 ON |
| 1 | 1 | 1 | 1 | 1 | 1 | 0 | 0 | FC | L1—L2 ON |
| 1 | 1 | 1 | 1 | 1 | 0 | 0 | 0 | F8 | L1—L3 ON |
| 1 | 1 | 1 | 1 | 0 | 0 | 0 | 0 | F0 | L1—L4 ON |
| 1 | 1 | 1 | 0 | 0 | 0 | 0 | 0 | E0 | L1—L5 ON |
| 1 | 1 | 0 | 0 | 0 | 0 | 0 | 0 | C0 | L1—L6 ON |
| 1 | 0 | 0 | 0 | 0 | 0 | 0 | 0 | 80 | L1—L7 ON |
| 0 | 0 | 0 | 0 | 0 | 0 | 0 | 0 | 00 | L1—L8 ON |

segment (a, b, ... h) and from lower to higher bits (D0, D1, ... D5) are brighten followed ,until all 48 segments be brighten , just match with the 6 × 8 = 48 points LED bar graph display from lower to higher (L1, L2, ... L48) were brighten, until all the 48 points LED be brighten. The code table and set of code tables of bright LED bar graph displays are shown in Table 1, Table 2.

**Table 2.** Code table of LED Bar Graph Display Group

| group | D7 | D6 | D5 | D4 | D3 | D2 | D1 | D0 |
|-------|-----|-----|-----|-----|-----|-----|-----|-----|
| Code  | XX  | XX  | 20H | 10H | 08  | 04  | 02  | 01  |

## 2.3    The Hardware Design of LED Bar Graph Display

Dais-958B$^+$ single-chip microcomputer simulation system device has six 8-segment common cathode LED digital display, the design in this case used 6 × 8 = 48 segments to constitute a common cathode LED light beam display. The hardware circuit diagrams of LED bar graph display are shown in Figure 3 [3].

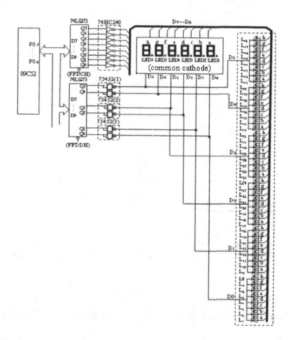

**Fig. 3.** The hardware circuit diagram of LED bar graph display

As Figure 3 shows, AT89C52 connect to the display interface IC via bus, which one group of 74LS273 and 74HC240 are shaped latch and drive, another group of 74LS273 and 75452 × 3 (3 pieces of 75452), are word-bit latch and drivers respectively [4].

Latch is required, as the data bus and address bus of 89C52 microcontroller are time-multiplexing, and the use of dynamic scanning display. Data bus connected to the shape port and word-bit port, which is 74LS273 eight D-latch input port, shaped port address is FFDCH, word-bit port address is FFDDH. 74LS273 eight D-latch of shaped port connect D7-D0 ports to each LED digital display of the h, g, f, e, d, c, b, a sections correspondingly via 74HC240 bus driver (RP) [5]. 74LS273 eight D-latch of Word-bit port connect the D5-D0 ports to each common cathode of digital display LED6-LED1 correspondingly through 75452 inverting driver [6]. LED bar graph display consists of the 48-point LED, and the connection to common cathode of 6 bits 8 segments LED digital display is: segment corresponds with the point, and bit corresponds with the group.

## 3    Application Design

### 3.1    Hardware Design

Hardware circuit diagram of LED bar graph display using in A/D conversion is shown in Figure 4, chip-select address Y0 of ADC0809 in the platform of Dais-958B$^+$ chip microcomputer simulation system is FFE0H.

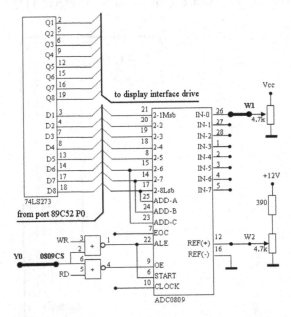

**Fig. 4.** Hardware circuit diagram of LED bar graph display using in A/D conversion

### 3.2    Software Design

Function: potentiometer W1 on the experimental apparatus provide analog input voltage. Write program to convert analog into digital. Analog input voltage began to rise from 0V, and $6 \times 8 = 48$ points LED bar graph display were lighted from low to high (L1, L2, ... L48), and vice versa.

Used single-chip control ADC0809 0 channel to achieve A / D conversion, the MCS-51 assembly language source code is as follows.

```
            ORG    0010H
            AJMP   MAIN
; ======== main ========
MAIN:   MOV A,#00H
            MOV DPTR,#0FFE0H
            MOVX @DPTR,A            ;Sampling of ADC0809-
            MOV  R7,#80H             ;-0 channel
WAINT: DJNZ  R7,WAINT
            MOVX A,@DPTR           ;remove the samples
            NOP
            NOP
            MOV  B,#08H
            DIV  AB
            MOV  R4,A
            MOV  R5,B               ;calculate the results
            LCALL DIS               ;call display subroutine
            SJMP  MAIN              ;return
;======== display ========
DIS:     MOV   R1,#30H            ;R1 points to 30H unit
            CJNE  R4,#00H,DIS1
LOOP1:  MOV  A,R5
            MOV   @R1,A
LOOP0:  INC   R1
            MOV   @R1,#00H
            CJNE  R1,#36H,LOOP0
            AJMP  DIR
DIS1:    MOV   @R1,#08H
            INC   R1
            DJNZ  R4,DIS1
            CJNE  R1,#36H,LOOP1
            AJMP  DIR
DIR:     MOV  R0,#30H
            MOV  R3,#01H
            MOV  A,R3
LD0:     MOV  DPTR,#0FFDDH
            MOVX @DPTR,A
            MOV  DPTR,#TAB
            MOV  A,@R0
DIR0:    MOVC A,@A+DPTR
DIR1:    MOV  DPTR,#0FFDCH
            MOVX @DPTR,A
            ACALL DELAY
            INC   R0
            MOV  A,R3
```

```
        JB    ACC.5,LD1
        RL    A
        MOV   R3,A
        AJMP  LD0
LD1:    RET
; ======== delay ========
DELAY: MOV   R6,#4
D1:     MOV   R7,#150
        DJNZ  R7,$
        DJNZ  R6,D1
        RET
; ====== display code ======
TAB: DB 0FFH,0FEH,0FCH,0F8H,0F0H
    DB 0E0H,0C0H,080H,00H
; =====================
        END
```

## 4    Conclusion

The hardware design is simple, beam display can be realized by connecting characters and bits interface cable from the LED digital display outlet to the LED bar graph display. Software program can be extended to the LED bar graph display of multi-channel A/D conversion after being modified.

## References

1. Deng, H., Chang, X.: General 8 Chapter LED Number Display Device. Chinese Patent of Invention, 201010269524.2 (September 02, 2010)
2. Deng, H., Li, Z.: A Design of Large Eight Burst LED Display of Compatibly Common Cathode and Positive. Electrical Measurement and Instrumentation 35(386), 41–42, 37 (1998)
3. Deng, H., Zeng, Y., Wang, J.: Tutorial on Single-chip Microcomputer Experiment and Applicable Design, 2nd edn., pp. 136–138. Metallurgical Industry Press, Beijing (2010)
4. Texas Instruments. The bipolar digital integrated circuits data book for design engineers, PART1, TTL & Interface Circuits; PART 2, TTL & Bipolar Memory, pp.15–48. USA Texas Instruments Inc. (1982)
5. Deng, H.: The projection type number display set. China Invention Patent, ZL200410012485.2 (April 05, 2006)
6. Lin, C., Chen, L., Yuan, L., et al.: Fundamentals of Single-chip Microcomputer and Applications, p. 191. China Machine Press, Beijing (2009)

# Author Index